Weakly Interacting Molecular Pairs: Unconventional Absorbers of Radiation in the Atmosphere

NATO Science Series

A Series presenting the results of scientific meetings supported under the NATO Science Programme.

The Series is published by IOS Press, Amsterdam, and Kluwer Academic Publishers in conjunction with the NATO Scientific Affairs Division

Sub-Series

I.	**Life and Behavioural Sciences**	IOS Press
II.	**Mathematics, Physics and Chemistry**	Kluwer Academic Publishers
III.	**Computer and Systems Science**	IOS Press
IV.	**Earth and Environmental Sciences**	Kluwer Academic Publishers
V.	**Science and Technology Policy**	IOS Press

The NATO Science Series continues the series of books published formerly as the NATO ASI Series.

The NATO Science Programme offers support for collaboration in civil science between scientists of countries of the Euro-Atlantic Partnership Council. The types of scientific meeting generally supported are "Advanced Study Institutes" and "Advanced Research Workshops", although other types of meeting are supported from time to time. The NATO Science Series collects together the results of these meetings. The meetings are co-organized bij scientists from NATO countries and scientists from NATO's Partner countries – countries of the CIS and Central and Eastern Europe.

Advanced Study Institutes are high-level tutorial courses offering in-depth study of latest advances in a field.
Advanced Research Workshops are expert meetings aimed at critical assessment of a field, and identification of directions for future action.

As a consequence of the restructuring of the NATO Science Programme in 1999, the NATO Science Series has been re-organised and there are currently five sub-series as noted above. Please consult the following web sites for information on previous volumes published in the Series, as well as details of earlier sub-series.

http://www.nato.int/science
http://www.wkap.nl
http://www.iospress.nl
http://www.wtv-books.de/nato-pco.htm

Series IV: Earth and Environmental Sciences – Vol. 27

Weakly Interacting Molecular Pairs: Unconventional Absorbers of Radiation in the Atmosphere

edited by

Claude Camy-Peyret

Université Pierre et Marie Curie & CNRS,
Paris, France

and

Andrei A. Vigasin

Wave Research Center, General Physics Institute,
Russian Academy of Sciences, Moscow, Russia

Kluwer Academic Publishers

Dordrecht / Boston / London

Published in cooperation with NATO Scientific Affairs Division

Proceedings of the NATO Advanced Research Workshop on
Weakly Interacting Molecular Pairs: Unconventional Absorbers of Radiation in the
Atmosphere
Fontevraud, France
29 April–3 May, 2002

A C.I.P. Catalogue record for this book is available from the Library of Congress.

ISBN 1-4020-1595-X (HB)
ISBN 1-4020-1596-8 (PB)

Published by Kluwer Academic Publishers,
P.O. Box 17, 3300 AA Dordrecht, The Netherlands.

Sold and distributed in North, Central and South America
by Kluwer Academic Publishers,
101 Philip Drive, Norwell, MA 02061, U.S.A.

In all other countries, sold and distributed
by Kluwer Academic Publishers,
P.O. Box 322, 3300 AH Dordrecht, The Netherlands.

Printed on acid-free paper

Printed in the Netherlands.

Contents

Part II   LABORATORY STUDIES

Part III   ATMOSPHERIC APPLICATIONS

Preface

The Advanced Research Workshop entitled "Weakly Interacting Molecular Pairs: Unconventional Absorbers of Radiation in the Atmosphere" was held in Abbaye de Fontevraud, France, from April 29 to May 3, 2002. The meeting involved 40 researchers from 14 countries. The goal of this meeting was to address a problem that the scientific community is aware of for many years. Up now, however, the solution for this problem is far from satisfactory. Pair effects are called unconventional in the title of this meeting. In specific spectral domains and/or geophysical conditions they are recognized to play a dominant role in the absorption/emission properties of the atmosphere. Water vapor continuum absorption is among the most prominent examples. Permanently improving accuracy of both laboratory studies and field observations requires better knowledge of the spectroscopic features attributable to molecular pairs which may form at equilibrium. The Workshop was targeted both to clarify the pending questions and, as far as feasible, to trace the path to possible answers since the underlying phenomena are yet incompletely understood and since a reliable theory is often not available. On the other hand, the lack of precise laboratory data on bimolecular absorption is often precluding the construction of reliable theoretical models. Ideally, the knowledge accumulated in the course of laboratory studies should correlate with the practical demands from those who are carrying out atmospheric field measurements and space observations.

The Workshop' co-directors have been pursuing several goals. First, we are hoping that the interested reader might be curious to discover the state-of-the-art methods, theories and techniques used for the study of weakly interacting molecular pairs — an interdisciplinary domain of science. Our second goal is to point out at serious deficiencies in the understanding of bimolecular phenomena occurring in the atmosphere in order to stimulate new laboratory and theoretical investigations. Third, our motivation is to open new ways in which laboratory measurements and theoretical developments could be more efficiently used for remote sensing of the atmosphere. In other words, our ultimate goal consists of promoting opportunities for bridging the gap between laboratory experiments, sophisticated theories and field observations in the interests of atmospheric science and applications.

The papers in the present volume are tentatively grouped in three categories, namely, Theory, Laboratory Studies, and Atmospheric Applications. The proportion between the number of pages among these categories does not match the weight of the corresponding topics as they were presented in the course of our ARW meeting. It seems indicative, however, of the actual distribution of efforts among the various research groups. Atmospheric applications follow the already existing laboratory data and theoretical developments with a significant delay. One of the goals of the present volume was to reduce the gap between purely academic studies and their potential applications. We leave to the readers of this volume to judge in what extent we have succeeded in doing so.

The theory of bimolecular absorption is marking progress in various directions. Leaving the low temperature dimeric spectra aside, the Workshop concentrated on the theory of collision-induced and other absorption continua. The papers by Gustafsson and Frommhold and by Brown and Tipping point out the need of thorough calculations of higher-order terms in the expansion of the induced dipole moment function in order to evaluate the absorption intensity in a reliable way. A rigorous definition for various types of pair states comprising true bound, resonance, and free pairs is required to establish the role played by pair states in continua and collision-induced absorption of radiation as has been demonstrated in the papers by Vigasin and by Baranov *et al.* According to Tonkov and Filippov the account of line-mixing effect may play a crucial role as far as absorption continua are concerned. In the present volume the use of new computational schemes is discussed which allow for prediction of the spectra of highly excited or unstable molecular pairs. These include *ab initio* calculations (Bussery-Honvault *et al.*) along with variational anharmonic vibrational calculations (Pavlyuchko *et al.*) and three-dimensional trajectory simulations (Ivanov). The problem of evaluating the concentration of dimers in gases at equilibrium is addressed in the papers by Slanina *et al.* and Vigasin. The uncertainty in the abundance of dimers in the atmosphere is one of the major limitation for the dimer hypothesis.

The papers by Baranov *et al.*, McKellar, Sneep and Ubachs, Aquilanti *et al.*, Shvetsov *et al.*, Carleer *et al.* and others demonstrate that significant breakthroughs have been achieved recently in laboratory studies of the mechanisms and spectroscopic manifestations of intermolecular interactions. The use of high resolution infrared absorption techniques is quite frequent although visible, UV and submillimeter ranges draw much attention as well. Two main directions of the recent laboratory studies are the use of traditional long-path absorption Fourier transform spectroscopy and the use of Cavity Ring Down Spectroscopy (CRDS). The advantage of the latter consists in the possibility to combine high sensitivity with small volume of the sample under study in an absorption cell. Promising perspectives are offered by the use of the resonator techniques in the submillimeter wave region. The measurements of absorption coefficients in the frequency range from 45 to 200 GHz are available nowadays with a sensitivity of about $4 \times 10^{-9}\,\mathrm{cm}^{-1}$ (as reported by Shvetsov *et al.*). Similar sensitivity (about $4 \times 10^{-8}\,\mathrm{cm}^{-1}$) characterizes the CRDS method (see e. g. the paper by Sneep and Ubachs) as applied to the study of oxygen CIA in the visible and the water vapor continuum in the infrared ranges. The high sensitivity and resolution appropriate to FTIR experiments in long-path absorption cells allow for detection of distinct sharp features in structureless collision-induced spectral profiles as observed at lower spectral resolution. In the papers by Baranov *et al.* fine structure patterns have been detected recently in dipole forbidden infrared absorption spectra of pure and mixed O_2 and CO_2 in the fundamentals and lower overtones. Interestingly, no analogous fine structure was observed in the $1.27\,\mu$m induced band of pure oxygen. The nature of these fine structure patterns in collision-induced absorption in gases is a subject of controversies in the recent literature. These features can be thought of to be either due to formation of dimers or to the line-mixing of dipole forbidden transitions. New in-depth experimental measurements and theoretical modeling of the underlying phenomenon are needed to clarify this problem. The nature of the water continuum absorption is raising much controversies too. The discussion showed that despite great efforts undertaken recently for in depth studies of the water vapor continuum in controlled laboratory conditions, this phenomenon is still a matter of significant incertitude and controversies in so far as both its qualitative nature and its quantitative details are concerned.

In the up-to-date version of Clough's model a weak "bump" is appearing on a smooth continuum background. Some researchers are asking themselves, could it be the dimeric signature? Preliminary results are reported by McKellar on the search of the water dimer features in the spectra taken in a room temperature long path absorption cell. The expectations behind this experiment were based on the knowledge of the location of the water dimer lines derived earlier in molecular beam experiments. The search in room temperature water vapor failed to locate any of the dimeric features in the range where they were expected. An extensive search for weak water vapor absorptions in the infrared and in the visible ranges was undertaken by Carleer and associates. The cell used has a 50 m base length. A phenomenon unexplained yet of the water vapor nucleation in the cell was reported which hinders accurate spectroscopic measurements of the line intensities. The size distribution of tiny water droplets in a cell was found to be remarkably sharp with mean diameter of about 0.8 μm. Outside the laboratory, the water vapor continuum plays an important role in atmospheric and space retrievals as shown by Maurellis *et al.* The measurements of the so-called "oxygen dimer" and/or collision-induced oxygen absorption in the atmosphere were presented by several groups, namely by Pfeilsticker, Camy-Peyret, and associates. The results of these observations show that significant progress have been made in many aspects of field measurements. Most of them are due to the use of high resolution spectroscopic instruments installed on flying platforms such as research aircrafts, stratospheric balloons, and satellites. Spectacular examples of the high quality atmospheric spectra retrieved with this type of instruments are given in the corresponding papers.

It is a great pleasure to the Editors of the present volume to thank the NATO Scientific Affairs Division for the support which made the ARW meeting in Fontevraud possible. In addition the encouragement and advice from the concerned NATO authorities, the support of CNRS department Sciences Physiques et Mathématiques (SPM) and Délegation Régionale Paris–B were essential at various stages of this project. The invaluable contribution from all the participants which render this meeting both fruitful, instructive, and internationally friendly at the same time is greatly acknowledged. On a more personal level we are pleased to extend our thanks to Nicole Gasgnier, Sebastien Payan, and Daniel Curie for their permanent technical assistance during all stages of this ARW project. Our special thanks go to Sergei Lokshtanov for his invaluable help in producing the camera-ready script.

C. CAMY-PEYRET

A. VIGASIN

Contributing Authors

Vincenzo Aquilanti is Professor of Chemistry at the University of Perugia. He is author of nearly 300 papers on various aspects of atomic and molecular physics and of theoretical chemistry.

Béatrice Bussery-Honvault is Chargé de Recherche at the Centre National de la Recherche Scientifique (CNRS) and is presently attached to the Laboratoire de Physique des Atomes, Molécules et Surfaces (PALMS) of the University of Rennes (France). She earned her Ph. D. in Atomic and Molecular Physics and her Habilitation from the University of Lyon in 1994 and 1997, respectively. She has contributed to the theoretical study of the electronic and rotational-vibrational structure of the O_2–O_2 dimer involved in the visible spectroscopy conducted by Lyon's and Grenoble's teams.

Claude Camy-Peyret is presently director of the Laboratory for Molecular Physics and Applications (LPMA), a joint institute of University Pierre et Marie Curie (UPMC) and Centre National de la Recherche Scientifique (CNRS). He earned his Ph. D. at UPMC in 1975 and has been active since that time in the field of high resolution vibration-rotation spectroscopy of H_2O, O_3, NO_2 and many other asymmetric rotors. He is now involved in the use of infrared spectroscopy and radiative transfer models for measurements of the tropospheric and stratospheric composition from the ground, from balloons, and from satellite.

David Cappelletti is presently researcher at the Dipartimento di Ingegneria Civile et Ambientale of University of Perugia and is associated researcher also with Istituto Nazionale Fisica della Materia (INFM). He earned his Ph. D. in 1993 at University of Perugia and is active in the field of dynamics and structure of weakly bound molecular complexes.

Maxim N. Eremenko is currently a post-graduate student in co-supervision between Tomsk State University of Control Systems, Russia, and Université Pierre et Marie Curie, Paris, France.

Gerald T. Fraser is leader of the Optical Thermometry and Spectral Methods Group of the Optical Technology Division of the National Institute of Standards and Technology. He earned his Ph. D. degree in Physical Chemistry from Harvard University in 1985. His research interests include applications of molecular spectroscopy to intermolecular and intramolecular forces, plasma diagnostics, and atmospheric sciences.

Lothar Frommhold is a professor of physics at the University of Texas. He received his Dr. rer. nat. degree in experimental physics (1959) and his Dr. habil. degree in applied physics (1964) at the University of Hamburg. He was awarded with various research and visiting professorships in Europe, the U.S.A., and Columbia (S.A.). He is an Elected Fellow of the American Physical Society, the author of the monograph "Collision-induced Absorption in Gases" (Cambridge, 1994), and co-editor of conference proceedings. He also compiled the first bibliography of collision-induced light scattering.

Sergey V. Ivanov is currently Senior Research Scientist in the Department of Advanced Laser Technologies of Institute on Laser and Information Technologies, Russian Academy of Sciences (ILIT RAS) in Troitsk, Moscow Region. He earned his Ph. D. degree from Moscow State University, Physical Department, in 1985 and has been active since that time in the field of infrared spectroscopy of small atmospheric molecules. His present research interests include numerical simulation of intermolecular interactions, non-LTE effects in spectroscopy, spectroscopic methods of gas detection.

Maria B. Kiseleva is presently an assistant at the Department of Physics of StPetersburg State University, Laboratory of Molecular Spectroscopy. She earned her Ph. D. in 1999 at the Department of Physics of StPetersburg State University. Since 1995 she is doing research in the field of electronic absorption spectroscopy, in particular in electronic spectra induced by intermolecular interactions.

Ahilleas Maurellis completed his Ph. D. at the University of Kansas in space plasma physics and has since worked on the development of new spectral sampling methods for satellite remote sensing of trace gases at the FOM-Institute for Atomic and Molecular Physics (AMOLF) and the Space Research Organization of the Netherlands. He currently participates as a retrieval and instrument simulation specialist in a variety of international remote-sensing collaborations while maintaining a special focus on the radiative and spectral properties of water vapor and the role of this molecule in climate change.

A. R. W. McKellar is a Principal Research Officer at the Steacie Institute for Molecular Sciences of the National Research Council of Canada, in Ottawa. He received his Ph. D. in Physics from the University of Toronto in 1970. He is an experimentalist whose research has mostly involved infrared spectroscopy of unusual molecules, such as molecular ions, free radicals, and weakly-bound complexes of astrophysical, atmospheric, and chemical interest.

Anatoly I. Pavlyuchko is currently a professor of physical chemistry at the Volgograd Technical University. He completed his Ph. D. in optics at the Moscow Pedagogical University in 1982. He earned his Doctoral degree in physical chemistry from the Chemical Faculty, Moscow State Unversity in 1993. In 1998 he co-authored with L. A. Gribov a book entitled "Variational Methods for Solving Anharmonic Problems in the Theory of Vibrational Spectra of Molecules" published by 'Nauka' in Moscow.

Klaus Pfeilsticker is presently Privatdozent with the Institute of Environmental Physics at the University of Heidelberg, Heidelberg Germany. He earned his Ph. D. and his habilitation at the University of Heidelberg in 1986 and 1998, respectively. He has been active since then in the field of UV/vis spectroscopy of atmospheric trace gases. Presently he is involved in the space-borne (GOME and SCIAMACHY), airborne (the LPMA/DOAS stratospheric balloon) and ground-based spectroscopic studies that aim to understand better the photochemistry and radiative properties of the atmosphere.

Alexander A. Shvetsov is a research scientist at the Institute of Applied Physics, Russian Academy of Sciences, Nizhnii Novgorod. He graduated from Gorky State University in 1971. From 1971 till 1977 he worked for the Nizhny Novgorod Research Radiophysics Institute. Since 1977 till now he is with the Institute of Applied Physics, Russian Academy of Sciences. His main fields of activity cover microwave spectroscopy including microwave remote sensing of the atmosphere and the underlying surface, development of instruments and techniques for passive remote sounding in the millimeter and submillimeter ranges.

Kevin Smith is currently a project scientist at the Space Science and Technology Department of the Rutherford Appleton Laboratory in Oxfordshire, UK. He completed his Ph. D. in experimental molecular spectroscopy at the University of Strathclyde in 1997 and since then he has been working on high-resolution, Fourier transform infrared, and visible spectroscopy of atmospheric molecules in the laboratory and field. His current interests include satellite remote sensing of gases and spectroscopy for atmospheric radiative transfer and climate applications.

George Tabisz is presently Professor in the Department of Physics and Astronomy in the University of Manitoba in Winnipeg, Canada. He received his Ph. D. from the University of Toronto in 1968. His interests lie in the study of molecular interactions and dynamics through optical techniques, chiefly laser light scattering and far infrared absorption, and in the theory of spectral line shapes. In 1995 he co-edited (with M. N. Neuman) a book in the NATO ASI series entitled "Collision- and Interaction-Induced Spectroscopy," published by Kluwer.

Richard H. Tipping is presently a professor in the Department of Physics and Astronomy at the University of Alabama. He earned his Ph. D. from the Pennsylvania State University in 1969. His current research interests include far-wing spectral line shapes and collision-induced absorption, with application to broadband absorptions in planetary atmospheres.

Mikhail V. Tonkov is currently heading the Sector of Molecular Spectroscopy of Atmospheric Gases at the Institute of Physics, StPetersburg State University. He earned his Ph. D. and Doct. Sci. degrees from StPetersburg State University, Physical Department, in 1968 and 1985, respectively. He is active in the field of molecular spectroscopy and the spectral evidences of intermolecular interactions in the gas phase.

Wim Ubachs is professor of Physics at the Laser Centre Vrije Universiteit Amsterdam. His research interests are in molecular spectroscopy, non-linear optics and extreme ultraviolet laser based radiation sources. In recent years his research focus has also been directed towards problems of atmospheric physics.

Andrei A. Vigasin is currently a Leading Research Scientist in the Wave Research Center, General Physics Institute, Russian Academy of Sciences in Moscow. He earned his Ph. D. (1978) and Doct. Sci. (1995) degrees from Moscow State University, Physical and Chemical Departments, respectively. His main scientific interests are presently in the domain of spectroscopy of weakly interacting molecules. In 1998 he co-edited (with Z. Slanina) a book entitled "Molecular Complexes in Earth's, Planetary, Cometary, and Interstellar Atmospheres" published by The World Scientific.

Wim J. van der Zande is currently program head of the quantum dynamics program of atomic and molecular systems in the FOM Institute for Atomic and Molecular Physics. He will take up a professorship at the Katholic University of Nijmegen in the Department of Molecular and Laser Physics from the beginning of 2003. He earned his Ph. D. in 1988 at the University of Amsterdam and has been active in molecular spectroscopy and molecular dynamics. He entered the field of atmospheric sciences in 1996.

I

THEORY

SPECTRA OF TWO- AND THREE-BODY VAN DER WAALS COMPLEXES

M. Gustafsson, L. Frommhold
Physics Department, University of Texas,
Austin, Texas 78712-1081, U.S.A.

Abstract   Spectra of all interacting pairs, triples, ... of molecules that are in excess of the simple sum of the spectra of the *non-interacting*, individual molecules are called interaction-induced spectra. These occur in absorption, emission and light scattering (Raman) spectra. If reliable intermolecular potential and induced dipole or polarizability surfaces are known, such spectra can be computed. Close agreement of computed spectra and laboratory measurement is then observed. When those surfaces are not available from the fundamental theory, reasonable empirical induced dipole surfaces can be obtained from laboratory measurements so that reliable temperature interpolation (and even extrapolation) of such spectra for the applications is often possible. We review the existing cases of such spectra.

Keywords: supramolecular absorption / supramolecular emission / supramolecular Raman processes / spectroscopy of collisional pairs / collision-induced dipoles / collision-induced polarizability invariants / interactions of binary molecular complexes with light / interactions of ternary complexes with light

1. Introductory remarks

The spectra of dense gases and gas mixtures differ from the spectra of the same gases observed at low densities. As one increases the gas densities from near zero, first the familiar "allowed" rotovibrational and electronic bands may appear in the appropriate rotovibrational and electronic frequency bands, with intensities that increase linearly with density. At intermediate densities, in general new absorption bands appear which (initially) increase in intensity as density square, cube, ... or in any case nonlinearly. These bands are due to van der Waals complexes of two or more molecules which may be "free," i.e., a transient collisional complex, or "bound," i.e., a weakly bound van der Waals

3

C. Camy-Peyret and A.A. Vigasin (eds.),
Weakly Interacting Molecular Pairs: Unconventional Absorbers of Radiation in the Atmosphere, 3–22.
© 2003 *Kluwer Academic Publishers. Printed in the Netherlands.*

molecule. Such absorption bands are found in all molecular gases, even in gases that contain only infrared-inactive molecules. Rotovibrational supramolecular absorption bands of common, dense gases, such as oxygen, nitrogen, ... and their mixtures were discovered in 1949 by Welsh and associates [1, 2]; early observations of absorption bands in compressed air and oxygen [3] are now understood to be of a supramolecular nature [1, 4–6].

Many absorption bands by binary complexes of molecules have been thoroughly investigated both theoretically and experimentally. A brief summary will be given below; detailed information may be found elsewhere [1, 2, 7–11]; see also existing bibliographies [12–14]. For ternary spectra some guiding ideas will be reviewed below. A few semi-quantitative data exist at present towards an understanding of the ternary spectra [15, 16]. Experimental evidence abounds of important N-body spectral contributions, with $N > 2$, even at densities that are much lower than condensed matter densities.

Supramolecular spectra have also been called CIA spectra, or pressure-induced spectra. Besides the supramolecular, absorption bands just mentioned, a supramolecular, Raman process exists which results from polarizability increments induced by molecular interactions. This process is called collision-induced light scattering (CILS) [2, 7–10]; for a bibliography see Ref. [14].

2. Supramolecular properties

A complex of two or more molecules may be considered another molecule, with new properties and specific spectra that go beyond the sum of spectra of the (non-interacting, or well separated) individual molecules that make up the complex. We will call such a complex a supramolecule, even if it exists for short times only, e. g., $\approx 10^{-12}$ s in the case of a collisional complex. The term "molecule" will be reserved for the individual parts of the supramolecule as usual.

For example, supramolecules usually possess an electric dipole moment, even if its individual constituent molecules are nonpolar. These dipoles arise from mechanisms that are universal and well-known from the studies of intermolecular forces. Four different, universal mechanisms are known to generate supramolecular dipoles.

- *Dispersion force-induced dipoles.* Intermolecular dispersion forces displace the electronic charge clouds slightly during interactions, with the result of generating an electric dipole moment of the supramolecule, Fig. 1 (left, ii).

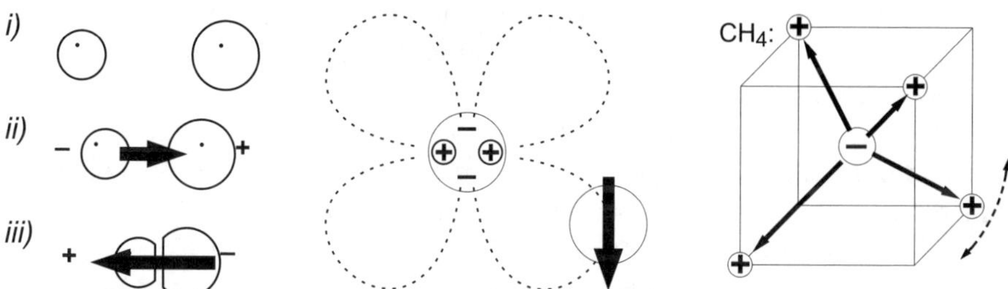

Figure 1.   Interaction-induced dipoles. Left: i) at large separations, a collisional pair such as He–Ar will be nonpolar; ii) at intermediate separations, dispersion forces induce a dipole moment in any dissimilar pair; iii) at near range, exchange forces induce a dipole of opposite polarity. Center: the electric field of the permanent electric quadrupole moment of molecules such as N_2 polarizes an interacting partner. Right: collisions with another particle may cause frame distortion of molecules such as CH_4 induce a dipole moment momentarily.

- *Exchange force-induced dipoles.*   Intermolecular exchange forces cause a slight displacement of the electronic charge clouds relative to their nuclei, with the result of generating a net dipole moment in the supramolecule, Fig. 1 (left, iii). The polarity of that induced dipole is the opposite of the dispersion force-induced dipole, because dispersion forces are attractive and caused by an enhancement of electronic charge in the space between the molecules. Exchange forces, on the other hand, are repulsive, and cause a depletion of electronic charge in that region.

- *Multipole-induced dipoles.* Molecules are surrounded by an electric field which may be described by a classical multipole expansion. Interacting atoms or molecules will be polarized in the fields of the permanent multipoles, Fig. 1 (center).

- *Collisional frame distortion-induced dipoles.* Molecules of a high degree of symmetry often have strong internal dipoles which are arranged so that the net dipole moment is zero in the unperturbed frame. As an example, we mention CH_4 which possesses four very strong internal dipole moments along the H^+C^- branches. These add up to zero, owing to the tetrahedral symmetry of the unperturbed molecule. A collision may temporarily displace a proton and thus generate a transient, sizable supramolecular dipole moment, Fig. 1 (to the right).

Note that some or all of the mechanisms mentioned will induce electric dipole moments whenever molecules or dissimilar atoms interact — regardless of what kind of particles, or how many particles actually interact at any given moment, and whether particles are bound (van der Waals molecules) or free (in collisional interaction). Supramolecular dipoles, like intermolecular forces, are omnipresent.

We note that supramolecular dipole strengths are typically fairly weak. Even at close range they amount to only 10^{-3} Debye or so. They fall off rapidly with increasing separation. In other words, supramolecular dipoles are roughly three orders of magnitude weaker than the permanent dipole moments of common molecules, such as H_2O or CO.

Supramolecular dipoles cause absorption in certain regions of the electromagnetic spectrum. These dipoles are also responsible for the second, third, ... virial coefficients describing dielectric properties, the Clausius—Mossotti and Lorenz—Lorentz equations [17, 18].

Binary dipole surfaces: *ab initio* calculations. For supramolecules with a limited number of electrons, e.g., H_2–H_2, H_2–He, etc., very accurate quantum chemical calculations of the dipole surfaces exist [19, 20]; for a summary of work prior to 1993 see also [9]. Such *ab initio* results show clearly the various dipole components: exchange force-induced dipoles show an exponential decrease with increasing separation R; dispersion force-induced dipoles fall off as R^{-7}, and multipole-induced dipoles have their characteristic long-range behavior R^{-N}, with $N = 3$ for dipolar induction, $N = 4$ for quadrupolar induction, etc. Quantum chemical studies of the kind provide some guidance for the development of empirical dipole models for bigger systems that cannot be treated by the demanding quantum chemical methods.

Classical multipole approximation. Dipole surfaces of binary multi-electron supramolecules, such as the ones found in the Earth's atmosphere (N_2–N_2, N_2–O_2, etc.) may often be represented by the (semi-)classical multipole-induced dipole approximation [17, 21, 22], which neglects (or else models empirically) the quantum effects of the exchange forces. Generally applicable expressions have been given by Poll and Tipping (see, for example, pp. 191 ff. of Ref. [9]).

If one or both of the interacting molecules are highly polarizable, the multipole-induced dipole components will typically be much stronger than the overlap and exchange contributions so that the latter may be neglected [22], or perhaps be represented by small and relatively inexact, empirical corrections. In that case, the induced dipole components may be computed from the knowledge of molecular multipole strengths [23] and polarizabilities, using classical electrodynamics [24, 25]. Elaborate tensor calculations by Hunt and collaborators also account for nonuniformity of the local electric field, the gradient of the field, the dispersion dipole, and hyperpolarizabilities [22, 26–30].

We note that the electronic supramolecular spectra arise similarly by multipolar induction; in that case, the electric multipoles are due to electronic configurations [31].

Dipoles of ternary systems. Early attempts to model ternary dipoles in terms of pairwise-additive dipole components were not very successful. In more recent times, experimental as well as theoretical evidence has emerged that substantial irreducible ternary dipole components exist, at least for some systems [15, 32–39].

Pairwise-additive ternary dipole components were reported for all systems for which reliable pair dipoles were known [35]. The comparison of the spectral moments based on these ternary dipoles with measurements of the ternary spectral moments of the absorption spectra of a number of gases and mixtures demonstrated rather consistently a significant shortfall of the calculated spectral moments [35]. This was taken as an indication that irreducible ternary dipole components exist in most — if not all — systems considered and certainly in unbound systems consisting of three H_2 molecules. The significance of the irreducible dipole contribution increased rapidly with increasing temperature, suggesting that the ternary dipole components arise from close encounters, i. e., from triple collisions with more or less overlapping electron clouds.

The interaction-induced dipole of three interacting molecules consists of the vector sum of the pairwise additive dipole components, plus an irreducible ternary dipole surface [15, 35, 38, 40–43]. More work is necessary for a better understanding of the latter, but several studies suggest that the principal irreducible dipole component is of the exchange quadrupole-induced dipole (EQID) type [15, 33–35, 38, 44–46]. During a binary collision at near range exchange forces displace the electronic molecular clouds relative to the nuclei to the far sides of the binary complex, thus generating momentarily a strong quadrupole moment. The third molecule is polarized in the electric field of the exchange force-induced quadrupole [15], Fig. 2 (at left).

EQID is a quantal mechanism. Previously, it had been suggested that the irreducible dipole could be the classical permanent quadrupole-induced dipole-induced dipole which, however, was found to be not nearly as strong as EQID in three interacting H_2 molecules [15], Fig. 2 (to the right).

Pair polarizability increments. Molecules are polarized in an external electromagnetic field. The field-induced dipole oscillates with the frequency of the incident radiation and it may also be modulated with the rotovibrational molecular frequencies. These dipoles give rise to Rayleigh and Raman scattering of light.

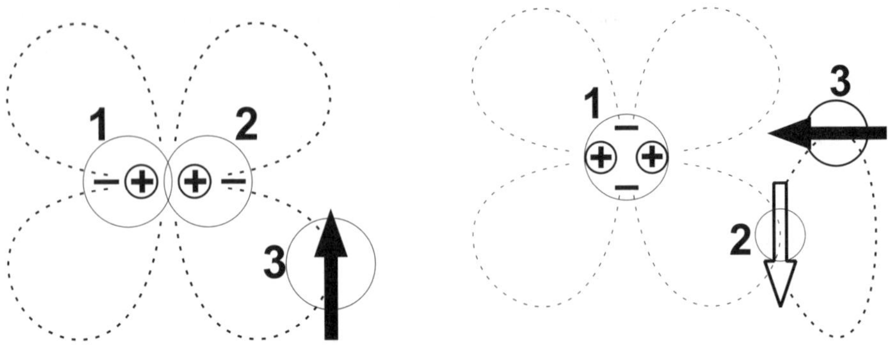

Figure 2.   Irreducible ternary induced dipoles. At left: EQID, the exchange force-induced quadrupole (1+2) — induced dipole (in the molecule 3). At right: QIDID, the dipole (2) — induced dipole (3); the former (2) being induced by the field of the permanent quadrupole moment of a molecule (1) such as N_2, O_2.

The polarizability invariants, trace and anisotropy, of the supra-molecule differ from the simple sums of these quantities of the (non-interacting) individual molecules, for two reasons. One reason is the local fields of each molecule differ from the external radiation field, due to the presence of the field-induced dipoles of molecules nearby. This effect of local field distortion has lead to the classical dipole-induced dipole (DID) expression of the supramolecular polarizability invariants [47, 48]. The other reason is the slight rearrangements of the molecular electronic clouds due to the intermolecular exchange and dispersion forces. Quantum chemical calculations of the pair polarizability increments exist for a few binary systems, e. g., He–He [49–52]. The DID model is reasonably successful for the more highly polarizable molecular gases. Details may be found in various review articles [21, 22, 51, 53–59] and a bibliography [14]. We note that the study of the Kerr and dielectric second virial coefficients also provide valuable information concerning interaction-induced polarizability invariants [60, 61].

3. Supramolecular spectra

Almost immediately after the discovery of supramolecular absorption by H. L. Welsh and associates [62] it was realized that interaction-induced absorption of electromagnetic radiation would be important in just about any cool and dense environment. Herzberg pointed out the first direct evidence of the existence of H_2 molecules in Uranus and Neptune's atmospheres, the supramolecular $S_3(0)$ line of H_2 [63,64] (which is dipole-forbidden in the non-interacting H_2 molecule). Most notably, Herzberg also could argue that sizeable helium concentrations exist in these atmospheres. Nonpolar species such as H_2 and He are difficult to detect in such remote, cool environments; supramolecular signatures helped to reveal such contributions. Since then the understanding of the interaction-induced absorption spectra in the atmospheres of the planets and their big moons, which are composed of molecules such as H_2, He, CH_4, Ar, N_2, CO_2, was systematically expanded. Relevant supramolecular absorption data have been collected in data bases for applications in the planetary sciences [65–67]; see also references cited in Table 1. In more recent years, it was realized that interaction-induced absorption is of major significance in the atmospheres of late stars [68], white dwarfs [69–73], low-mass stars and brown dwarfs [74, 75], the hypothetical "population III" stars [76, 77], and low-metallicity cool stars [78]. There can be no doubt that the Earth's atmosphere is similarly affected by interaction-induced spectral signatures in all spectral bands.

Characteristic features. Supramolecular absorption and Raman spectra are due to interaction-induced dipole moments and polarizability invariants, respectively, of two or more interacting molecules. At fixed frequency ω and temperature T, in gases a virial expansion of the absorption coefficient, $\alpha(\omega; T)$, is often possible in terms of powers of the gas density n. For unmixed gases one may write [79, 80, 9]

$$\alpha(\omega; T) = \alpha^{(1)}(\omega; T)\, n + \alpha^{(2)}(\omega; T)\, n^2 + \alpha^{(3)}(\omega; T)\, n^3 + \cdots . \quad (1)$$

The leading coefficient $\alpha^{(1)}$ of the expansion represents the monomer absorption spectra, that is the "dipole-allowed" optical transitions of (non-interacting) individual molecules. For infrared inactive gases this term vanishes. The remaining expansion coefficients $\alpha^{(2)}$, $\alpha^{(3)}, \ldots$ represent the two-, three-, etc. body supramolecular contributions, which are often separable in actual measurements. In mixtures of gases a, $b, \ldots$ one can write down similar expressions, with binary, $\alpha^{(aa)}$, $\alpha^{(ab)}$, $\alpha^{(bb)}$, and ternary contributions, $\alpha^{(aaa)}$, $\alpha^{(aab)}$, $\alpha^{(abb)}, \ldots$ besides the "allowed" terms $\alpha^{(a)}$, $\alpha^{(b)}$.

Other characteristics of supramolecular spectra are:

- Whereas in a few cases collision-induced rotational lines are spectroscopically resolvable, more often individual rotational lines overlap substantially, owing to the short duration Δt of collisions (Heisenberg's uncertainty relation $\Delta\omega \Delta t \geq \frac{1}{2}$). The "lifetime" Δt of a collisional complex amounts to 10^{-12} s or so which in most cases renders rotational bands quasi-continuous;

- Rotovibrational bands quite generally appear at high densities, even in infrared inactive gases. Moreover, bands at sums and differences of rotovibrational frequencies of the individual molecules are also observed. These are due to simultaneous transitions in interacting molecules;

- Rotovibrational bands of van der Waals dimer, rotovibrational bands may be detected near all rotovibrational transition frequencies of the individual molecules. However, these are typically very weak and strongly pressure broadened. At any instant the number of collisional pairs is usually much greater than the number of bound dimers so that the integrated dimer intensities are much weaker than the collision-induced contributions. The latter appear as a strong background of the bound dimer signatures;

- Supramolecular dipoles are typically quite weak, roughly two to three orders of magnitude weaker than the dipoles of common polar molecules. Accordingly, line intensities of supramolecular spectra are typically insignificant at low densities, but they may be quite striking as gas densities are increased.

Frequency bands. Supramolecular dipoles typically vary in time in a way that reflects the relative motion within the supramolecule, and the rotovibrational degrees of freedom of the individual molecules. Since the supramolecular spectra are the Fourier transforms of the dipole autocorrelation function, typically we have a translational band in the microwave and far infrared regions of the electromagnetic spectrum, owing to the translational relative motion of individual molecules. Interaction-induced dipoles are small when the collisional partners are widely separated. As two molecules approach each other, the dipole strengths generally increase and eventually fall off again as the molecules recede. Purely translational spectra in the microwave and far infrared region exist for pairs of dissimilar atoms. Molecular translational spectra typically more or less coincide with the induced purely rotational band at microwave frequencies and in the far infrared.

Interaction-induced dipoles are also modulated by the rotovibrational motions of the constituent molecules, which gives rise to various rotovibrational bands of the interacting molecules — bands that may be forbidden in the non-interacting molecules. Moreover, double (and triple, ...) rotovibrational or electronic transitions generally occur in supramolecules which generate new spectral absorption bands at sums and differences of the bands of molecules that make up the supramolecule by simultaneous transitions in two (three, ...) molecules in the infrared region.

Intercollisional dips. Isotropic induced dipole components may cause quite striking absorption dips, typically at zero frequency shifts (e. g., Q-branches). These are largely due to correlations of interaction-induced dipoles in subsequent collisions that lead to destructive interference [81–85]. Intercollisional absorption dips may be considered many-body effects. Intercollisional dips are "sharp" features when the time t between collisions is much greater than the mean duration Δt of a collision (Heisenberg's uncertainty relation). We note that virial expansions of the line shapes, Eq. (1), are not valid at such frequencies so that two- and three-body, etc., contributions can not be separated experimentally at frequencies where the dips appear. At small enough frequency shifts, the absorption dips are true many-body effects. The shapes of the absorption dips have been modeled by inverted Lorentzian profiles [81, 82].

***Ab initio* calculations of interaction-induced spectra.** Calculations of spectral profiles and intensities of supramolecular spectra are often necessary if laboratory measurements do not exist at temperatures as needed for the applications, or when laboratory measurements had to be taken at densities much higher than atmospheric densities. If reliable intermolecular potential and induced dipole functions are available, the spectra of interacting pairs of molecules may be computed from a quantum formalism. For most of the existing calculations, the isotropic interaction approximation has been used [9]. If the anisotropy of the interaction potential may be neglected, the computational procedures are much simplified: complete line shape calculations can then be done in fractions of one hour on personal computers. Such calculations involve as a first step the determination of the rotovibrational energies of the bound dimer. Reliable free to free, free to bound, bound to free and bound to bound transition dipole matrix elements can then readily be computed and complete spectra be composed for comparison with measurements and for various applications in planetary and Earth science [9].

Accounting for the anisotropy of the intermolecular interactions complicates the computational procedures substantially. Some such calculations exist [19, 20] based on the close coupled (CC) scheme. Such calculations are very computation intensive, but may be marginally feasible for the massive molecular systems one is concerned with in studies of the Earth's atmosphere. It is therefore of interest to compare the results with simplified quantum calculations that can be done in minutes even on small, personal computers. Close coupled computations, on the other hand, may take months of computer time.

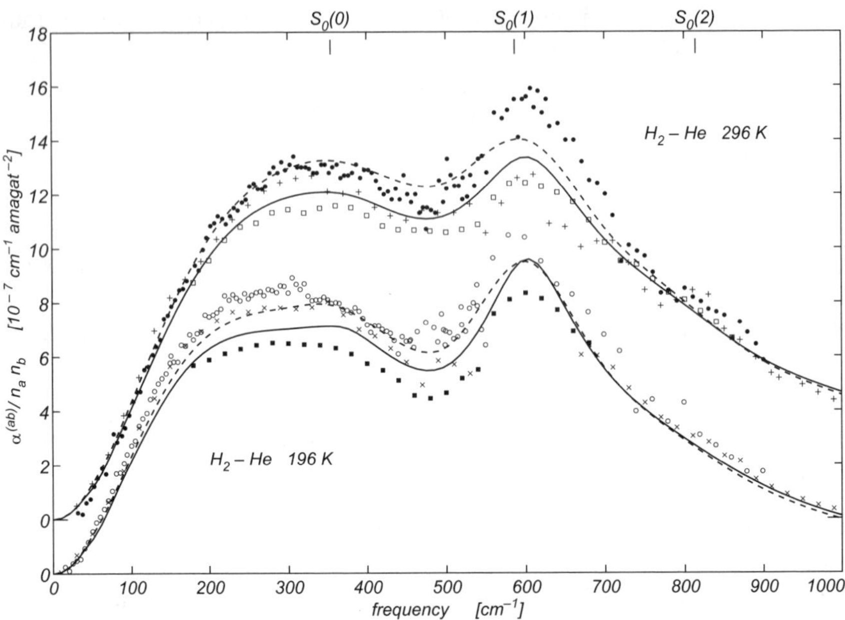

Figure 3.   The binary absorption coefficient $\alpha^{(ab)}$, normalized by the H_2 and He densities n_a, n_b, as function of frequency in the rototranslational band of H_2, at the temperature of 298 K (upper trace), and 196 K (lower trace, shifted). Solid and dashed curves represent calculations with and without accounting for the anisotropy of the H_2–He interaction potential, respectively. The dipole surface and line profile calculations are from Ref. [19]; measurements ● and ○ from Ref. [86]; +, □, ×, and ■ from Ref. [87].

The results obtained with the isotropic interaction approximation (IIA) are dashed in Fig. 3. At the frequencies shown, for the most part the IIA results are above the CC results by up to 20%. In the far wings (not shown), at frequencies from 1000 to 2400 cm^{-1}, the CC results for hydrogen are higher than IIA by a factor of two [88]. We note that the interaction anisotropies of systems such as N_2–N_2, H_2–O_2, etc., are much greater than that of H_2–He. In such cases, much greater spectroscopic effects of the anisotropy may therefore be expected than are evident in the figure. The most reliable potential and dipole surfaces presently available have been employed. In other words, these results are considered state of the art computations, with an accuracy believed to be in the ±6% range on an absolute scale. For comparison, the available measurements are also shown. The differences with the most recent measurement amount to typically less than 10%, the combined uncertainties of theory and measurement.

We note that measurements of supramolecular absorption spectra are quite demanding because of the smallness of interaction-induced dipoles. In laboratory measurements one may be tempted to use high gas densities, to make up for that deficiency. One must be careful, however, to avoid ternary and higher-order contributions that may not be present to such an extent in atmospheric research, where gas densities are generally lower.

We note that the measurement of the so-called enhancement spectra of gas mixtures is prone to somewhat greater uncertainty than measurements in pure gases. This is so because in gas mixtures, e. g., of helium and hydrogen, one must subtract the H_2–H_2 contributions from the sum of absorption due to the H_2–H_2 and H_2–He pairs. The spectra are fairly similar and subtraction of comparable intensities renders the end result more uncertain. This may be an explanation for the striking differences between the measurements seen in Fig. 3. Calculations such as the ones shown in the figure may help to resolve such uncertainties and make possible cautious extrapolations of laboratory measurements to other temperatures and the far wings of the spectra, for applications in atmospheric physics.

Superimposed van der Waals dimer bands — often unresolved owing to strong pressure broadening of these weakly bound molecules — appear near the rototranslational transition frequencies [89–92]. We note that the H_2–He supramolecule is one of the very few van der Waals complexes that do not form bound dimers, owing to the shallowness of the specific well and the low supramolecular weight.

Electronic spectra. Electronic interaction-induced spectra occur at visible and ultraviolet frequencies; long familiar ones are those of O_2–X pairs, with X representing another O_2 or an N_2 molecule [3, 6]; see also [93].

Approximate calculations of supramolecular spectra. Quantum line shape calculations based on the multipolar induction model are also known [9]. Figure 4 shows as an example the rototranslational band of the supramolecular absorption spectrum of nitrogen [94]. Small empirical corrections of the quantal effects have been applied.

Interaction-induced Raman spectra are also known to exist; see a bibliography for details [14]. The Raman process redistributes radiative energy, especially at frequencies in the visible and ultraviolet. Raman intensities are proportional to frequency to the fourth power.

Ternary spectral components. Virial expansions of supramolecular absorption spectra have long been known. Ternary (and sometimes even higher-order) contributions have been isolated in a number

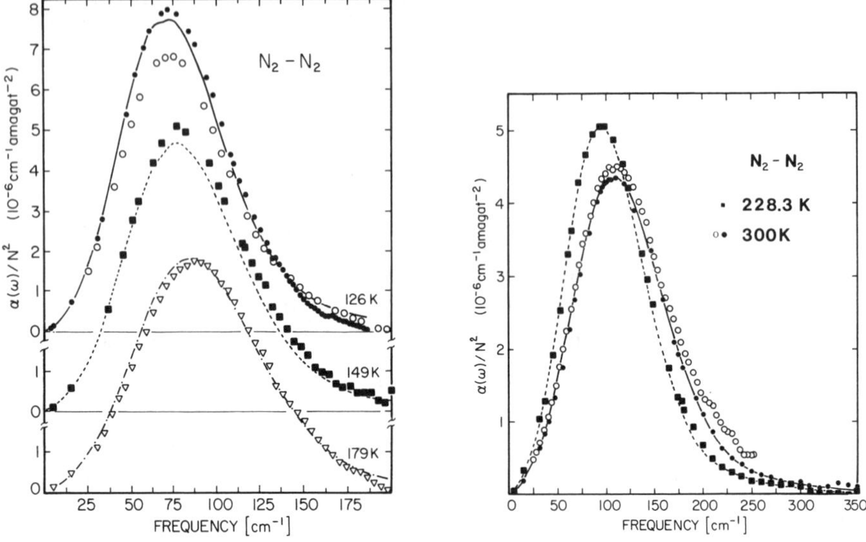

Figure 4. The rototranslational absorption band of nitrogen at various temperatures. Solid lines: quantum calculations [94]; various measurements: dots, circles, etc.

of cases. A theoretical study [33, 44, 45] of the intercollisional dip [81] of rare gas mixtures pointed out that without the assumption of an irreducible ternary dipole component in rare gas triples the observed spectra could not be reproduced theoretically. Moreover, in 1995 Reddy and associates pointed out a spectral feature in the interaction-induced second overtone band of H_2 [95] that must be considered a simultaneous vibrational transition of the type $Q_1(J_a) + Q_1(J_b) + Q_1(J_c)$. In other words, three H_2 molecules (a, b, c) in collisional interaction, in the presence of a single photon, underwent a simultaneous vibrational transition, with quantum numbers v_a, v_b, v_c changing from 0 to 1 simultaneously. (The rotational quantum numbers J_a, J_b, J_c, were either 0 or 1.) Simultaneous triple transitions with absorption of one photon have long been predicted and are believed to be due to irreducible dipole components.

Remarkably, the three independent observations (3rd virial coefficient, intercollisional dip, triple transition) which suggest the existence of irreducible dipole components have been explained *by a single, semi-quantitative dipole model* [15, 39, 46]. Contrary to the earliest suggestions of the nature of ternary irreducible dipole components, it appears now that the exchange quadrupole-induced dipole (EQID) component

Table 1. Available computer models of binary dipoles and absorption spectra, based on quantum line shapes calculations. The letters e and a in the dipole column stand for *empirical* and *ab initio*, respectively; the RT, F, O in the next column stand for rototranslational, fundamental and overtone bands, respectively, of the molecular complex mentioned first in column 1.

pair	dipole	band	frequencies, cm^{-1}	Ref.
CH_4–CH_4	e [96]	RT	0–750	[97]
CH_4–He	e [98]	RT	0–500	[98]
CO_2–CO_2	e	RT	0–250	[99]
H_2–CH_4	e [100, 101]	RT	0–1000	[100]
H_2–H_2	a [102]	RT	0–2000	[66, 103]
	a [104, 105]	F	3500–5500	[66, 103]
	a [103]	O	7500–9500	[66]
H_2–He	a [19, 106]	RT	0–2000	[66, 107, 108]
	a [19, 109]	F	3500–5500	[66, 110–112]
	a [19]	O	7500–9500	[66, 110]
H_2–N_2	e [113, 114]	RT	0–1000	[114]
N_2–N_2	e [94]	RT	0–300	[94]

mentioned in Section 2 above is a principal contributor to the irreducible ternary dipole [33, 15] — at least in the case of compressed hydrogen.

Modeling of supramolecular spectra. The individual $Q_v(J)$, $S_v(J)$, etc., "lines" and double transition profiles of the interaction-induced spectra differ very much from the lorentzian or gaussian profiles often used to model ordinary spectral lines and bands. Induced lines are very broad and distinctly asymmetric, especially in the vibrational bands [9, 115]. Moreover, interaction-induced lines usually possess dimer features near the line centers. Nevertheless, reasonably simple analytical profiles exist which approximate the interaction-induced profiles remarkably well, often over peak-to-wing intensity ratios of three or four orders. The so-called BC and EBC profiles [9, 115–117] are easy to compute and represent virtually all measured and calculated interaction-induced spectral lines closely, especially when some care is taken to also model the dimer bands separately [9, 115].

These BC and EBC profiles have been used to generate accurate analytical representations of laboratory measurements of interaction-induced spectra. Such representations of measured spectra are useful, for example, for frequency and temperature interpolations, and even for cautious extrapolations, for applications in the atmospheric sciences.

Quantum calculations of such spectra have also thus been approximated so that the results can be reproduced accurately in split seconds on personal desktop computers, for similar purposes. Numerous references of such spectral models are given in Table 1; see also [118].

Acknowledgments

The support of the Robert A. Welch Foundation, grant F-1346, is gratefully acknowledged.

References

[1] Welsh, H. L. (1972) Pressure induced absorption spectra of hydrogen. In: A. D. Buckingham and D. A. Ramsay, editors. *MTP Int. Review of Science— Physical Chemistry, Series one, vol. III: Spectroscopy,* Chapter 3. Butterworths, London.

[2] van Kranendonk, J., editor. (1980) *Intermolecular Spectroscopy and Dynamical Properties of Dense Systems — Proceedings of the Int. School of Physics "Enrico Fermi," Course LXXV.* North-Holland Publ. Company, Amsterdam.

[3] Janssen, J. (1885) Analyse spectrale des éléments de l'atmosphère terrestre, Compt. Rend. Acad. Sci. Paris, 101, 649–651; (1886) Sur les spectres d'absorption de l'oxygène, Compt. Rend. Acad. Sci. Paris, 102, 1352–1353; (1888) Sur les spectres de l'oxygène, Compt. Rend. Acad. Sci. Paris, 106, 1118–1119; (1889) Sur l'origine tellurique des raies de l'oxygène dans le spectre solaire, Compt. Rend. Acad. Sci. Paris, 108, 1035–1037.

[4] Tabisz, G. C. (1971) Pressure dependence of the electronic transition $^1\Sigma_g^+ \leftarrow {}^3\Sigma_g^-$ in the absorption spectrum of compressed oxygen, Chem. Phys. Letters, 9, 581–582.

[5] Tabisz, G. C., Allin, E. J., and Welsh, H. L. (1969) Interpretation of the visible and near infrared absorption spectra of compressed oxygen as collision induced electronic transitions, Can. J. Phys., 47, 2859–2871.

[6] McKellar, A. R. W., Rich, N. H., and Welsh, H. L. (1972) Collision-induced vibrational and electronic spectra of gaseous oxygen, Can. J. Phys., 50, 1–9.

[7] Special issue: Collision-Induced Phenomena, (1981) Can. J. Phys., 59, 1403–1407.

[8] Birnbaum, G., editor. (1985) *Phenomena Induced by Intermolecular Interactions.* Plenum Press, New York.

[9] Frommhold, L. (1993) *Collision-induced Absorption in Gases.* Cambridge University Press, Cambridge.

[10] Tabisz, G. C., and Neuman, M. N., editors. (1995) *Collision- and Interaction-Induced Spectroscopy.* NATO ASI series B, vol. 452. Kluwer, Dordrecht.

[11] Vigasin, A. A. and Slanina, Z., editors. (1998) *Molecular Complexes in Earth's, Planetary, Cometary, and Interstellar Atmospheres.* World Sci., Singapore.

[12] Rich, N. H. and McKellar, A. R. W. (1976) A bibliography on collision induced absorption, Can. J. Phys., 54, 486.

[13] Hunt, J. L. and Poll, J. D. (1986) A second bibliography on collision induced absorption, Molec. Phys., 59, 163–164; first update: (1990) Guelph-Waterloo Program for Graduate Work in Physics, Dept. of Physics, University of Guelph, Guelph, Canada.

[14] Borysow, A. and Frommhold, L. (1989) Collision induced light scattering — a bibliography, Adv. Chem. Phys., 75, 439–499; Ulivi, L. and Frommhold, L. (1993) A second bibliography on collision-induced light scattering.

[15] Moraldi, M. and Frommhold, L. (1996) Dipole moments induced in three interacting molecules, J. Molec. Liquids, 70, 143–158.

[16] Birnbaum, G. and Guillot, B. Cancellation effects in collision-induced phenomena, In: Ref. [10].

[17] Buckingham, A. D. (1967) Permanent and induced molecular moments and long-range interactions. In: I. Prigogine and S. A. Rice, editors. *Adv. Chem. Phys., Vol. 12*, pp. 107–142. Wiley Interscience, New York.

[18] Buckingham, A. D. (1980) The effects of collisions on molecular properties, Pure and Appl. Chem., 52, 2253–2260.

[19] Gustafsson, M., Frommhold, L., and Meyer, W. (2000) Infrared absorption spectra by H_2–He complexes: The effect of the anisotropy of the interaction potential, J. Chem. Phys., 113, 3641–3650.

[20] Gustafsson, M. and Frommhold, L. (2001) The HD–He complex: Interaction-induced dipole surface and infrared absorption spectra, J. Chem. Phys., 115, 5427–5432.

[21] Hunt, K. L. C. *Ab initio* and approximate calculations of collision induced polarizabilities. In: Ref. [8].

[22] Hunt, K. L. C. Classical multipole models: Comparison with *ab initio* and experimental results. In: Ref. [8]

[23] Stogryn, D. E. and Stogryn, A. P. (1966) Molec. Phys., 11, 371–393.

[24] Tipping, R. H. and Poll, J. D. In: Ref. [9], pp. 191 ff.

[25] Tipping, R. H. and Poll, J. D. (1985) Multipole moments of hydrogen and its isotopes. In: K. N. Rao, editor. *Molecular Spectroscopy: Modern Research*, Vol. 3, Chapter 7. Academic Press, New York.

[26] Bohr, J. E. and Hunt, K. L. C. (1987) Dipoles induced by van der Waals interactions during collisions of atoms with heteroatoms or centrosymmetric linear molecules, J. Chem. Phys., 86, 5441–5448.

[27] Bohr, J. E. and Hunt, K. L. C. (1987) Dipoles induced by long-range interactions between centrosymmetric linear molecules: theory and numerical results for H_2–H_2, H_2–N_2, and N_2–N_2, J. Chem. Phys., 87, 3821–3837.

[28] Li, X. and Hunt, K. L. C. (1994) Transient collision-induced dipoles in pairs of centrosymmetric, linear molecules at long range: Results from spherical-tensor analysis, J. Chem. Phys., 100, 9276–9278.

[29] Hunt, K. L. C. and Li, X. Dielectric properties of dense fluids. In: Ref. [10].

[30] Li, X., Champagne, M. H., and Hunt, K. L. C. (1998) Long-range, collision-induced dipoles of T_d–$D_{\infty h}$ molecule pairs: theory and numerical results for CH_4 or CF_4 interacting with H_2, N_2, CO_2, or CS_2, J. Chem. Phys., 109, 8416–8425.

[31] Julienne, P. S. Collision-induced radiative transitions at optical frequencies. In: Ref. [8].

[32] Mountain, R. D. and Birnbaum, G. (1987) Molecular dynamics study of inter-collisional interference in collision induced absorption in compressed fluids, J. Chem. Soc. Faraday Trans. 2, 83, 1791–1800.

[33] Guillot, B., Mountain, R. D., and Birnbaum, G. (1989) Triplet dipoles in the absorption spectra of dense rare gas mixtures: I. Short range interactions, J. Chem. Phys., 90, 650–662.

[34] Guillot, B. (1989) Triplet dipoles in the absorption spectra of dense rare gas fluids: II. Long range interactions, J. Chem. Phys., 91, 3456–3462.

[35] Moraldi, M. and Frommhold, L. (1989) Three-body components of collision in-duced absorption, Phys. Review A, 40, 6260–6274.

[36] Moraldi, M. and Frommhold, L. (1992) Second and third virial coefficients of collision-induced absorption and light scattering. In: J. J. C. Teixeira-Dias, editor. *Molecular Liquids: New Perspectives in Physics and Chemistry*, C: Math. and Phys. Sci., Kluwer Academic Publ., Amsterdam, p. 423.

[37] Moraldi, M. and Frommhold, L. (1995) Triple transition $Q_1(j_1) + Q_1(j_2) + Q_1(j_3)$ near 12,466 cm^{-1} in compressed hydrogen, Phys. Rev. Letters, 74, 363–366.

[38] Moraldi, M. and Frommhold, L. Irreducible three-body dipole moments in hy-drogen. In: Ref. [10].

[39] Moraldi, M. and Frommhold, L. (1995) Irreducible dipole components of three interacting H_2 molecules and the triple Q_1 transition near 12,466 cm^{-1}, J. Chem. Phys., 103, 2377–2383.

[40] van Kranendonk, J. (1957) Theory of induced infrared absorption, Physica, 23, 825–837.

[41] van Kranendonk, J. (1959) Induced infrared absorption in gases. Calculation of the ternary absorption coefficients of symmetrical diatomic molecules, Physica, 25, 337–342.

[42] Li, X. and Hunt, K. L. C. (1996) Nonadditive, three-body dipoles and forces on nuclei, J. Chem. Phys., 105, 4076–4093.

[43] Li, X. and Hunt, K. L. C. (1997) Nonadditive three-body dipoles of inert gas trimers and $H_2\cdots H_2\cdots H_2$: Long-range effects in far infrared absorption and triple vibrational transitions, J. Chem. Phys., 107, 4133–4153.

[44] Guillot, B. (1987) Theoretical investigation of the dip in the far infrared ab-sorption spectrum of dense rare gas mixtures, J. Chem. Phys., 87, 1952–1961.

[45] Guillot, B., Mountain, R. D., and Birnbaum, G. (1988) Theoretical study of the 3-body absorption spectrum in pure rare-gas fluids, Molec. Phys., 64, 747–757.

[46] Moraldi, M. and Frommhold, L. (1994) Three-body induced dipole moments and infrared absorption: The H_2 fundamental band, Phys. Review A, 49, 4508–4519.

[47] Michels, A., de Boer J., and Bijl, A. (1937) Remarks concerning molecular in-teractions and their influence on the polarizability, Physica, 4, 981–994.

[48] Kielich, S. (1960) A molecular theory of light scattering in gases and liquids, Acta Phys. Polonica, 19, 149.

[49] Dacre, P. D. and Frommhold, L. (1982) Rare gas diatom polarizabilities, J. Chem. Phys., 76, 3447–3460.

[50] Moszynski, R., Heijmen, T. G. A., Wormer, P. E. S., and van der Avoird, A. (1996) *Ab initio* collision-induced polarizability, polarized and depolarized Raman spectra of the helium diatom, J. Chem. Phys., 104, 6997–7007.

[51] Moszynski, R., Heijmen, T. G. A., Wormer, P. E. S., and van der Avoird, A. (1997) Theoretical modeling of spectra and collisional processes of weakly interacting complexes, Adv. Quantum Chem., 28, 119–140.

[52] Korona, T., Moszynski, R., and Jeziorski, B. (1997) Convergence of symmetry-adapted perturbation theory for the interaction between helium atoms and between a hydrogen molecule and a helium atom, Adv. Quantum Chem., 28, 171–188.

[53] Buckingham, A. D. (1972) *Theory of Dielectric Polarization*, Vol. 2 of *Series one*, Butterworths, London, p. 241.

[54] Sutter, H. (1972) Dielectric polarization in gases. In: M. Davies, editor. *A Specialist Periodic Report—Dielectric and Related Molecular Processes Vol. 1*, Chapter 3, Chem. Soc., London, p. 65.

[55] Buckingham, A. D. (1976) Gaseous molecules in electric and magnetic fields, Ber. Bunsen Ges. Phys. Chem., 80, 183–187.

[56] Buckingham, A. D. (1987) General Introduction—Faraday Symposium 22 on Interaction Induced Spectra in Dense Fluids and Disordered Solids. J. Chem. Soc. Faraday Trans. 2, 83 (10), 1743.

[57] Hunt, K. L. C., Liang, Y. Q., and Sethuraman, S. (1988) Transient, collision-induced changes in polarizability for atoms interacting with linear, centrosymmetric molecules at long range, J. Chem. Phys., 89, 7126–7138.

[58] Hunt, K. L. C. (1990) Infrared and vibrational Raman band intensities: a new physical interpretation based on nonlocal polarizability densities. In: L. Frommhold and J. W. Keto, editors. *Spectral Line Shapes*, Vol. 6, American Institute of Physics, New York, p. 513.

[59] Hunt, K. L. C. and Li, X. Collision-induced dipoles and polarizabilities for S state atoms or diatomic molecules. In: Ref. [10].

[60] Buckingham, A. D. (1967) Permanent and induced molecular moments and long-range intermolecular forces, Adv. Chem. Phys., 12, 107–142.

[61] Bose, T. K. A comparative study of the dielectric, refractive and Kerr virial coefficients. In: Ref. [8].

[62] Crawford, M. F., Welsh, H. L., and Locke, J. L. (1949) Infrared absorption of oxygen and nitrogen induced by intermolecular forces, Phys. Review, 75, 1607.

[63] Herzberg, G. (1952) The atmospheres of the planets, J. Roy. Astron. Soc. Can., 45, 100.

[64] Herzberg, G. (1952) Spectroscopic evidence of molecular hydrogen in the atmospheres of Uranus and Neptune, Astrophys. J., 115, 337–340.

[65] Borysow, A. and Frommhold, L. (1991) Pressure-induced absorption in the infrared: a data base for the modeling of planetary atmospheres, J. Geophys. Res., 96, 17501–17506.

[66] Birnbaum, G., Borysow, A., and Orton, G. S. (1996) Collision-induced absorption of H_2-H_2 and H_2-He in the rotational and fundamental bands for planetary applications, Icarus, 123, 4–22.

[67] Trafton, L. M. (1998) *Planetary Atmospheres: The role of collision-induced absorption*, Chapter 6 in: A. A. Vigasin and Z. Slanina, editors. *Molecular Complexes in Earth's, Planetary, Cometary, and Interstellar Atmospheres*, World Scientific, Singapore, pp. 177–193.

[68] Linsky, J. L. (1969) On the pressure-induced opacity of molecular hydrogen in late-type stars, Astrophys. J., 156, 989–1005.

[69] Shipman, H. L. (1977) Masses, radii and model atmospheres for cool white-dwarf stars, Astrophys. J., 213, 138–144.

[70] Liebert, J. (1980) White dwarf stars, Ann. Rev. Astron. Astrophys., 18, 363.

[71] Liebert, J., Lebofsky, M. J., and Rieke, G. H. (1981) Infrared photometry and the atmospheric composition of cool white dwarfs: The lowest luminosity candidates, Astrophys. J., 246, L73–L76.

[72] Mould, J. and Liebert, J. (1978) Infrared photometry and the atmospheric composition of cool white dwarfs, Astrophys. J., 266, L29–L33.

[73] Jørgensen, U. G., Hammer, D., Borysow, A., and Falkesgaard, J. (2000) The atmospheres of cool, helium-rich white dwarfs, Astronomy and Astrophysics, 361, 283–292.

[74] Burrows, A., Hubbard, W. B., and Lunine, J. I. (1989) Brown dwarfs, Astrophys. J., 345, 439.

[75] Burrows, A. Marley, M., Hubbard, W. B., Lunine, J. I., Guillot, T., Saumon, D. Freedman, R., Sudarsky, D., and Sharp, C. (1997) A nongray theory of extrasolar giant planets and brown dwarfs, Astrophys. J., 491, 856–875.

[76] Palla, F. (1985) Low-temperature Rosseland mean opacities for zero-metal gas mixtures. In: G. H. F. Diercksen, W. F. Huebner, and P. W. Langhoff, editors. *Molecular Astrophysics—State of the Art and Future Directions*, D. Reidel Pub. Cy., Dordrecht, p. 687.

[77] Stahler, S. W., Palla, F., and Salpeter, E. E. (1986) Primordial stellar evolution: the protostar phase, Astrophys. J., 302, 590–605.

[78] Borysow, A., Jørgensen, U. G., and Zheng, Ch. (1997) Model atmospheres of cool, low-metallicity stars: The importance of collision-induced absorption, Astronomy and Astrophysics, 324, 185–195.

[79] Moraldi, M., Celli, M., and Barocchi, F. (1989) Theory of virial expansion of correlation functions and spectra: Application to interaction induced spectroscopy, Phys. Review A, 40, 1116–1126.

[80] Moraldi, M. (1990) *Spectral Line Shapes, Vol. 6*, Chapter "Virial expansion of correlation functions for collision induced spectroscopies", AIP Conference Proceedings 216. American Institute of Physics, New York, pp. 438.

[81] van Kranendonk, J. (1968) Intercollisional interference effects in pressure induced spectra, Can. J. Phys., 46, 1173–1179.

[82] Lewis, J. C. Intercollisional interference—Theory and experiment. In: Ref. [8].

[83] Lewis, J. C. Intercollisional interference effects. In: Ref. [2].

[84] Lewis, J. C. and van Kranendonk, J. (1970) Intercollisional interference effects in collision-induced light scattering, Phys. Rev. Letters, 24, 802–804.

[85] Lewis, J. C. and Herman, R. M. (2001) Low density intercollisional interference. In: J. Seidel, editor. *Spectral Line Shapes*, Am. Inst. Phys., vol. 11., pp. 457.

[86] Birnbaum, G. (1978) Far-infrared absorption in H_2 and H_2–He mixtures, J. Quant. Spectroscopy and Rad. Transfer, 19, 51–62.

[87] Birnbaum, G., Bachet, G., and Frommhold, L. (1987) Experimental and theoretical study of the far-infrared spectrum of H_2–He mixtures, Phys. Review, A 36, 3729–3735.

[88] Gustafsson, M. and Frommhold, L., work in preparation.

[89] McKellar, A. R. W. (1990) Infrared spectra of van der Waals molecules. In: L. Frommhold and J. W. Keto, editors. *Spectral Line Shapes Vol. 6*, AIP Conference Proceedings 216, Am. Inst. Phys., New York, pp. 369.

[90] McKellar, A. R. W. and Schäfer, J. (1991) Far-infrared spectra of hydrogen dimers $(H_2)_2$ and $(D_2)_2$, J. Chem. Phys., 95, 3081–3091.

[91] McKellar, A. R. W. Infrared studies of van der Waals complexes. In: Ref. [10].

[92] McKellar, A. R. W. (2003) Infrared spectra of weakly-bound complexes and collision-induced effects involving atmospheric molecules, this volume, 223–233.

[93] Julienne, P. S. Collision-induced radiative transitions at optical frequencies. In: Ref. [8].

[94] Borysow, A. and Frommhold, L. (1986) Collision-induced rototranslational absorption spectra of N_2–N_2 pairs, Astrophys. J., 311, 1043–1057.

[95] Reddy, S. P., Xiang, Fan, and Varghese, G. (1995) Observation of the triple transition $Q_1(J_1) + Q_1(J_2) + Q_1(J_3)$ in molecular hydrogen in its second overtone region, Phys. Rev. Letters, 74, 367–370.

[96] Borysow, A. and Frommhold, L. (1987) Collision induced rototranslational absorption spectra of binary methane complexes $(CH_4$–$CH_4)$, J. Molec. Spectrosc., 123, 293–309.

[97] Borysow, A. and Frommhold, L. (1987) Collision induced rototranslational absorption spectra of CH_4–CH_4 pairs, Astrophys. J., 318, 940–943.

[98] Taylor, R. H., Borysow, A. and Frommhold, L. (1988) Concerning the rototranslational absorption spectra of He–CH_4 pairs, J. Molec. Spectrosc., 129, 45–48.

[99] Gruszka, M. and Borysow, A. (1997) Rototranslational collision-induced absorption of CO_2 for the atmosphere of Venus at frequencies from 0 to 250 cm^{-1} at temperatures from 200 to 800 K, Icarus, 129, 172–177.

[100] Borysow, A. and Frommhold, L. (1986) Theoretical collision induced rototranslational absorption spectra: H_2–CH_4 pairs, Astrophys. J., 304, 849–865.

[101] Borysow, A., Frommhold, L., and Dore, P. (1986) Rototranslational far infrared absorption spectra of H_2–CH_4 pairs, J. Chem. Phys., 85, 4750–4752.

[102] Meyer, W., Frommhold, L., and Birnbaum, G. (1989) Rototranslational absorption spectra of H_2–H_2 pairs in the far infrared, Phys. Review A, 39, 2434–2448.

[103] Borysow, A. (1991) Modeling of collision-induced infrared absorption spectra of H_2–H_2 pairs in the fundamental band at temperatures from 20 to 300 K, Icarus, 92, 273–279. *Erratum:* (1993) Icarus, 106, 614.

[104] Meyer, W., Borysow, A., and Frommhold, L. (1989) Absorption spectra of H_2–H_2 pairs in the fundamental band, Phys. Review A, 40, 6931–6949.

[105] Meyer, W., Borysow, A., and Frommhold, L. (1993) Collision-induced first overtone band of gaseous hydrogen from first principles, Phys. Review A, 47, 4065–4077.

[106] Meyer, W. and Frommhold, L. (1986) Collision induced rototranslational spectra of H_2–He from an accurate *ab initio* potential surface, Phys. Review A, 34, 2771–2779.

[107] Borysow, J., Frommhold, L., and Birnbaum, G. (1988) Collision induced rototranslational absorption spectra of H_2–He pairs at temperatures from 40 to 3000 K, Astrophys. J., 326, 509–515.

[108] Borysow, A. and Frommhold, L. (1989) Collision-induced infrared spectra of H_2–He pairs at temperatures from 18 to 7,000 K: Overtone and "hot" bands, Astrophys. J., 341, 549–555.

[109] Frommhold, L. and Meyer, W. (1987) Collision induced rotovibrational spectra of H_2–He pairs from first principles, Phys. Review A, 35, 632–638. *Erratum:* (1990) Phys. Rev., A, 41, 534.

[110] Borysow, A., Frommhold, L., and Moraldi, M. (1989) Collision induced infrared spectra of H_2–He pairs involving $0 \leftrightarrow 1$ vibrational transitions and temperatures from 18 to 7,000 K, Astrophys. J., 336, 495–503.

[111] Borysow, A. (1992) New model of collision-induced infrared absorption spectra of H_2–He pairs in the 2–2.5 μm range at temperatures from 20 to 300 K: an update, Icarus, 96, 196.

[112] Borysow, A., Frommhold, L., and Meyer, W. (1990) Collision induced absorption by H_2–He pairs in the fundamental band: rotational states dependence, Phys. Review A, 41, 264–270.

[113] Dore, P., Borysow, A., and Frommhold, L. (1986) Rototranslational far infrared absorption spectra of H_2–N_2 pairs, J. Chem. Phys., 84, 5211–5213.

[114] Borysow, A. and Frommhold, L. (1986) Theoretical collision induced rototranslational absorption spectra for modeling Titan's atmosphere: H_2–N_2 pairs, Astrophys. J., 303, 495–506.

[115] Moraldi, M., Borysow, A., and Frommhold, L. (1988) Rotovibrational collision induced absorption by nonpolar gases and mixtures (H_2–He pairs): About the symmetry of line profiles, Phys. Review A, 38, 1839–1847.

[116] Birnbaum, G. and Cohen, E. R. (1976) Theory of line shapes in pressure induced absorption, Can. J. Phys., 54, 593–602.

[117] Borysow, J., Trafton, L., Frommhold, L., and Birnbaum, G. (1985) Modelling of pressure-induced far infrared absorption spectra: Molecular hydrogen pairs, Astrophys. J., 296, 644–654.

[118] Brown, A. and Tipping, R. H. (2003) Collision-induced absorption in dipolar molecule–homonuclear diatomic pairs, this volume, 93–99.

BIMOLECULAR ABSORPTION
IN ATMOSPHERIC GASES

A. A. Vigasin

Wave Research Center,

General Physics Institute, Russian Academy of Sciences,

Vavilova 38, Moscow, 119991, Russia

Abstract   Intermolecular interactions are known to affect the absorptivity of the
atmosphere in various aspects. Pair effects give rise to distortions
in the line profiles, make dipole forbidden bands to appear in the spectra, modify the bandshapes at elevated density etc. Three categories
of molecular pairs can be distinguished in a gas which are true bound,
metastable, and free pairs. Although somewhat overlapping, the spectroscopic manifestations of these entities are essentially different. Their
partial contributions are subject to strong variations as a function of
temperature and intermolecular potential energy. The present paper
is targeted at reviewing bimolecular absorption phenomena taking infrared spectra of N_2, O_2, CO_2, and H_2O as an example.

Keywords: weakly bound molecules / dimers / resonances / collision-induced absorption / line-mixing / bandshapes / pressurized gases / water vapor
continuum

1. Introduction

Extensive efforts, both experimental and theoretical, undertaken
during the past decades resulted in significant progress in the understanding of the nature of intermolecular interactions. The main properties of dozens of weakly bound (van der Waals) dimers have been
characterized both spectroscopically and quantum-chemically in quite
extensive details. Dissociation energies for van der Waals dimers are
known to vary within several orders of magnitude viz. from 10 cal/mole
to 1.5 kcal/mole. This implies a substantial population of a large number of rovibrational states of these dimers at near room temperature.
Moreover, pair states having an energy in excess of the dissociation
threshold are also significantly populated. Any macroscopic property,

23

C. Camy-Peyret and A.A. Vigasin (eds.),
Weakly Interacting Molecular Pairs: Unconventional Absorbers of Radiation in the Atmosphere, 23–48.
© 2003 *Kluwer Academic Publishers. Printed in the Netherlands.*

e. g. absorption of radiation by weakly interacting pairs at typical atmospheric conditions, is therefore the result of a statistical average over the states of a broad ensemble of thermally excited intermolecular pairs. Rigorous definitions are then needed for the various types of pair states in order to elaborate reliable spectroscopic models describing bimolecular absorption in the atmosphere. Three distinct types of pair states are known to exist in a gas at equilibrium (see e. g. [1, 2]). These are true bound or dimeric states, metastable or quasibound states, and free pair states. Intuitively, true bound states are easy to identify with the dimers conventionally probed in molecular beams and free jet expansions. These techniques allow for very efficient cooling of the internal degrees of freedom in a gas. In a cold gas, however, only ground states play a role. In contrast, in the atmosphere many intermolecular vibrational and rotational states of dimers are readily populated. This implies that thermally averaged parameters for the dimers at ambient temperature can deviate significantly from those characterizing their ground states. In what follows we suggest a formal definition for true dimeric states. Metastable pair states can be identified with collisional resonance states. To avoid confusion one has to keep in mind that free pair states should not be identified with continuum states of isolated monomers in a gas.

The deviation of a macroscopic property F in a gas from ideality can be represented in terms of a virial expansion

$$F = A\,\rho + B\,\rho^2 + C\,\rho^3 + \ldots \qquad (1)$$

Here A, B, and C are the first, second, and third virials, respectively, ρ stands for a gas density in $\mathrm{kg\,m^{-3}}$. Assuming for example that for one mole of a gas F stands for a gas pressure i. e. $F(\rho) = p$, the first virial is then $A = RT$, where R is the gas constant, T is temperature. Equation (1) reduces thus to a virial equation of state. Truncation the expansion (1) to the first virial gives rise to the equation of state for a perfect gas in which intermolecular interactions are supposed to be of infinitesimal duration. It can be shown (see e. g. [3]) that the term B is determined entirely by the pair intermolecular potential characterizing the interaction energy of any selected pair of monomers as a function of monomer–monomer separation and orientations, if the latter are applicable. The third term C in (1) is mainly due to triple interactions between monomers, though a minor contribution from pairs of monomers is also present. In what follows we shall restrict ourselves to molecular pairs only, that is no terms higher than that proportional to the density squared will be considered below. In terms of absorption coefficient $\kappa(\nu)$ we can write a virial expansion analogous to (1) in the form:

$$\kappa(\nu) = \alpha_1(\nu)\,\rho + \alpha_2(\nu)\,\rho^2 + \ldots \tag{2}$$

Here the first term relates to absorption by isolated monomers and the second term is due to absorption by molecular pairs. Equations (1) and (2) differ in that the first term in expansion (1) never vanishes if F stands for a thermodynamic property; it can disappear, however, from (2) if specific selection rules state that transitions are forbidden in isolated monomers. The second terms both in (1) and in (2) can be subdivided into contributions from bound, quasibound and free pair states (see e.g. [1]). The free term relates to pairs of monomers, the interaction of which is not binding. The relevant absorption varies as the density squared in contrast to the first term in (2) which relates to independent monomers and scales linearly with density.[1] Hence, free pair states give rise to a correction to the absorption by individual monomers, which is associated with the purely repulsive pair interaction between unbound monomers.

Discrimination between the various types of pair states requires consideration of the phase space accounting for both spatial positions and impulses of the interacting monomers. When such a discrimination is made, any observable has to be averaged over a specific domain in the phase space. This is of special value when spectral shapes are concerned since these latter are sensitive to the lifetime of a selected excited state. Typical lifetimes vary by many orders of magnitude from true dimers to free pairs, that explains why spectroscopic models appropriate to various types of pair states are likely to be significantly different. It is the statistical average of the spectroscopic profiles over the totality of the pair states which can only produce simulated spectra in agreement to the observed ones.

To quantify the number density of true bound molecular pairs the law of mass-action is applicable. The conventional rigid rotor–harmonic oscillator approximation should be avoided since it fails to describe properly the thermally excited states of loosely bound pairs which are often highly populated.

In the forthcoming paragraphs we shall concentrate on the formalism of statistical mechanics describing the subdivision of the various pair states in the phase space. The problem of resonance states is

---

[1]For the sake of simplicity we restrict ourselves to consideration of pairs formed from similar monomers only. In case of dissimilar monomers the second term in (2) has to include also the product of the partial densities of the different species present in the mixture.

addressed as well as their expected spectral manifestations. Modifications affecting absorption bandshapes in pressurized gases are discussed taking infrared allowed bands of CO_2 and H_2O together with collision-induced absorption (CIA) spectra in N_2, O_2 and CO_2 as examples. The problem of the water vapor continuum absorption in the atmosphere is briefly covered.

2. Statistical mechanics of interacting molecules

Before going into the details of statistical mechanics it is worth to point out the difference between the "statistical" and "spectroscopical" definitions of short-lived pair states. Whatever be the criterion adopted in statistical physics definition for decaying states it may not fit the spectroscopic reality since no unique definition can be suggested for such states in spectroscopy. Formal statistical cutoff for quasistable states demands their lifetimes to spontaneous decay be limited. Spectroscopical methods all have their intrinsic characteristic time, viz. the typical period of the electromagnetic wave used as a spectroscopic probe. This time varies many orders of magnitude from e.g. 10^{-15} s in the ultraviolet up to 10^{-7} s in the radiofrequency ranges. Consequently, the use of various spectroscopic methods to probe molecular pairs may result in the identification of various fractions of short-lived states either with stable or with quasistable species. In what follows we ignore this difference and only the "statistical" definition of bimolecular states is adopted below.

2.1. Definition of bimolecular states

Rigorous partitioning in the phase space of structureless monomers originates from the statistical physics theory outlined by Stogryn, Hirshfelder and Hill [1,3]. Similar ideas were adopted later in an attempt to extend this theory to polyatomics [2,4,5] though accounting for the anisotropy of interactions makes the situation drastically more complicated. The main complication arises due to the opening of new channels to form metastable states. Indeed, the scattering of structureless particles may result in the formation of shape resonances only. Figure 1 shows that discrete quantum states can form above the dissociation limit when the effective intermolecular potential accounting for centrifugal energy is used:

$$U_{\text{eff}} = U(R) + \frac{h}{8\pi^2 c\mu} \frac{L(L+1)}{R^2}. \tag{3}$$

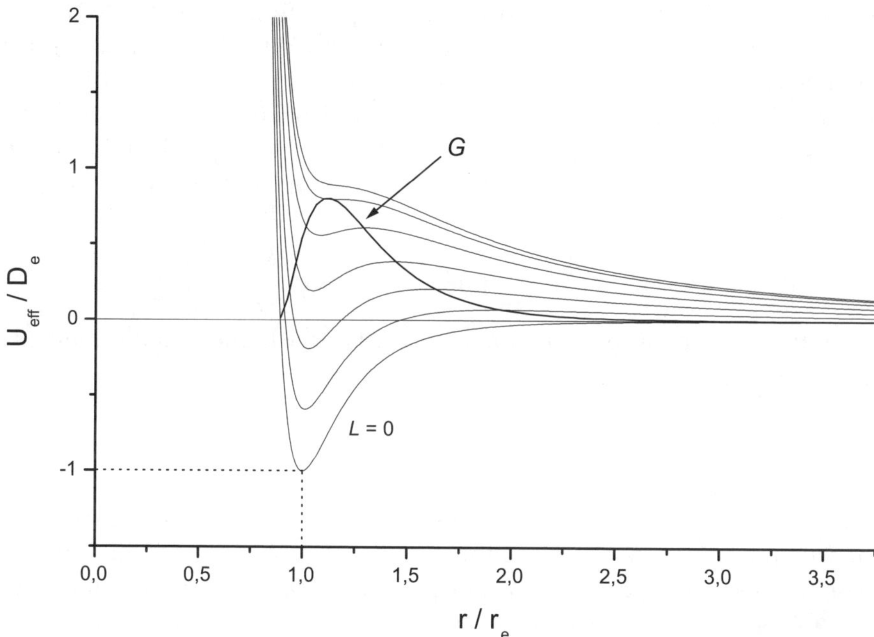

Figure 1. Model effective potential energy U_{eff} as a function of the end-over-end rotational quantum number L (only the lowest L value is labeled). The auxiliary function denoted G is also shown (see text).

Here R is the intermolecular separation, μ is the reduced mass, h is the Planck constant, c is the speed of light, and L is the orbital angular momentum quantum number.

In an anisotropic intermolecular potential the so-called Feshbach resonances can form. In this case a portion of the translational energy is temporarily transferred to excite the nondissociative internal degrees of freedom. The role of Feshbach resonances is clearly increasing in parallel with the increase in the number of internal degrees of freedom (see [4–6]). Let the total Hamiltonian of a pair (neglecting the movement of the center of mass) be written in the form:

$$H = U(R, \Omega) + E_{\text{tr}} + E_2 + E_{\text{i}}. \tag{4}$$

Here Ω stands for angular coordinates determining the spatial orientations of the two monomers, E_{tr} stands for the translational kinetic energy of the monomers moving along the line connecting their centers of masses, E_2 is the kinetic energy for the end-over-end rotation of the pair characterized by the overall angular momentum L, E_{i} is the kinetic energy of internal rotations (librations) of the monomers within the pair. It is assumed as first approximation that rotation of a pair against intermolecular axis does not lead to dissociation. With these assumptions the domain of true bound states is restricted to a set of canonical variables which leave the Hamiltonian (4) negative. This means that the sum of all kinetic energy terms in (4) is insufficient to overcome the potential energy needed to release the monomers trapped in the potential well:

$$E_{\mathrm{tr}} + E_2 + E_{\mathrm{i}} \leq -U(R, \Omega). \tag{5}$$

The total partition function of a system with the Hamiltonian (4) can be defined as:

$$Q = \int \exp\left(-\frac{H}{k\,T}\right)\, d\tau. \tag{6}$$

Here $d\tau$ defines elementary volume in phase space for all spatial coordinates and conjugated momenta. The truncated partition function Q_{b} for true bound states can be obtained from (6) provided the integration extends over the domain of states satisfying the condition (5). After integration over impulses, one obtains:

$$Q_{\mathrm{b}} = (k\,T)^{\frac{s+3}{2}} \int\limits_{\Omega} \int\limits_{\sigma}^{\infty} e^{-\frac{U}{kT}} \frac{\gamma\left(\frac{s+3}{2}, -\frac{U}{kT}\right)}{\Gamma\left(\frac{s+3}{2}\right)}\, r^2\, dr\, d\Omega. \tag{7}$$

Here s is the number of degrees of freedom defining relative orientations of the two monomers in a pair, $\sigma \equiv \sigma(\Omega)$ stands for the solution of the equation $U(R, \Omega) = 0$. The formulae (6, 7) make it possible to average any observable over either totally available or true dimeric domain of states.

More difficult is to define the boundary limiting the domain of quasibound states in the phase space. The definition is obvious for the domain of shape resonances as shown in Fig. 2 in case of structureless monomers where resonance states satisfy the condition $-U - E_2 < E_{\mathrm{tr}} < G_{\mathrm{h}} - U - E_2$. Here $G_{\mathrm{h}} = G_{\mathrm{h}}(E_2)$ is an auxiliary function describing the dependence of the barrier height (see Fig. 1) as a function of the end-over-end rotational energy E_2. A similar definition can be assumed to hold roughly for polyatomics irrespective to the energy E_{i} of internal movements. The corresponding partitioning in the $(E_2, E_{\mathrm{tr}}, E_{\mathrm{i}})$ coordinate space is schematically shown in Fig. 3.

2.2. From true bound states to free pair states

Partitioning the pair states in the phase space was carried out in a number of papers [1–8]. Obviously, true bound states always dominate

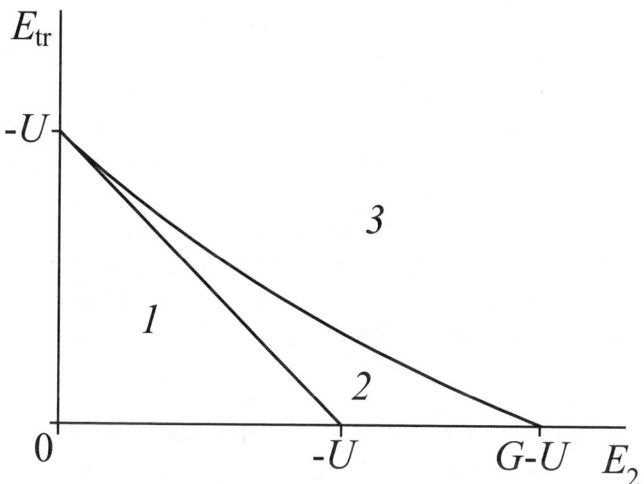

Figure 2.   Partitioning in the phase space of two isotropically interacting monomers into true bound (1), metastable (2), and free (3) domains of states.

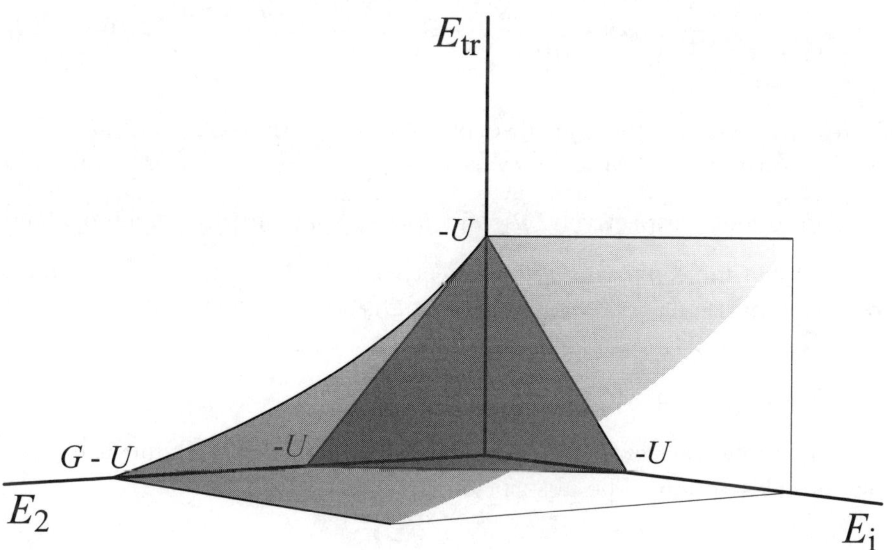

Figure 3.   Same as in Fig. 2 for anisotropically interacting monomers.

in the low temperature extreme. In the opposite extreme, corresponding to a heated gas, free pair states prevail. The fraction of metastable pair states has always a maximum at an intermediate temperature; this maximum appears as a result of the competition between true bound and free states. Starting from the low temperature regime quasibound states borrow their population by depleting true bound states. Then, at increasing temperature, the population is transferred inevitably from quasibound to free pair states. In case of shape resonances (i. e. for a spherically symmetrical interaction potential) the fraction of metastable states is shown to remain smaller than that of true bound states (see e. g. [7]). The anisotropy of interactions enhances significantly the domain of resonance states in the phase space. Typical temperature variations of the normalized contributions of the different pairs to CIA intensities are shown in Fig. 4 for some polyatomic dimers.

Let us look into more detail at how true bound and free pair states contribute to the spectral absorption coefficient. In the first extreme the associated species are assumed to be strong enough and many-particle states can be treated in terms of true bound states. The spectral absorption coefficient $\kappa(\nu)$ per unit length for a pure gas of the molecular species M then can be represented in terms of a sum of individual contributions of the various species in the mixture of polymers $(M)_i$:

$$\kappa(\nu) = \sum_i \frac{N_i}{V}\, \sigma_i(\nu) = \frac{\bar{\mu}}{\mu_1}\, \frac{N}{V}\, \frac{\sum_i x_i\, \sigma_i(\nu)}{\sum_i i\, x_i} = N_0\, \frac{\rho}{\mu_1}\, \sum_i \frac{\alpha_i}{i}\, \sigma_i(\nu). \qquad (8)$$

Here $\sigma_i(\nu)$ stands for the absorption cross-sections (in $cm^2\,molec^{-1}$) of the constituents $(M)_i$, $x_i = N_i/N$ and $\alpha_i = i x_i / \sum_j j x_j$ are mole and mass-fractions, respectively, μ_1 and $\bar{\mu} = \mu_1 \sum_i i x_i$ are the monomer and the average molecular masses, respectively, N_0 is the Avogadro's number. In a particular case of a dimerizing gas $(i = 2)$ one obtains

$$x_1 = \frac{2\,\alpha_1}{1 + \alpha_1}; \qquad x_2 = \frac{1 - \alpha_1}{1 + \alpha_1}; \qquad \bar{\mu} = \mu_1 \frac{2}{1 + \alpha_1}.$$

The mole fractions of monomers and dimers can be found using the law of mass-action. The absorption coefficient reads then as

$$\kappa(\nu) = \frac{N_0\,\rho}{\mu_1}\, \alpha_1\, \sigma_1(\nu) + \frac{N_0\,\rho}{\mu_2}\, \alpha_2\, \sigma_2(\nu)$$

$$= \frac{N_0\,\rho}{\mu_1} \left(\sigma_1(\nu) + \frac{\alpha_2}{2} \left(\sigma_2(\nu) - 2\sigma_1(\nu) \right) \right). \qquad (9)$$

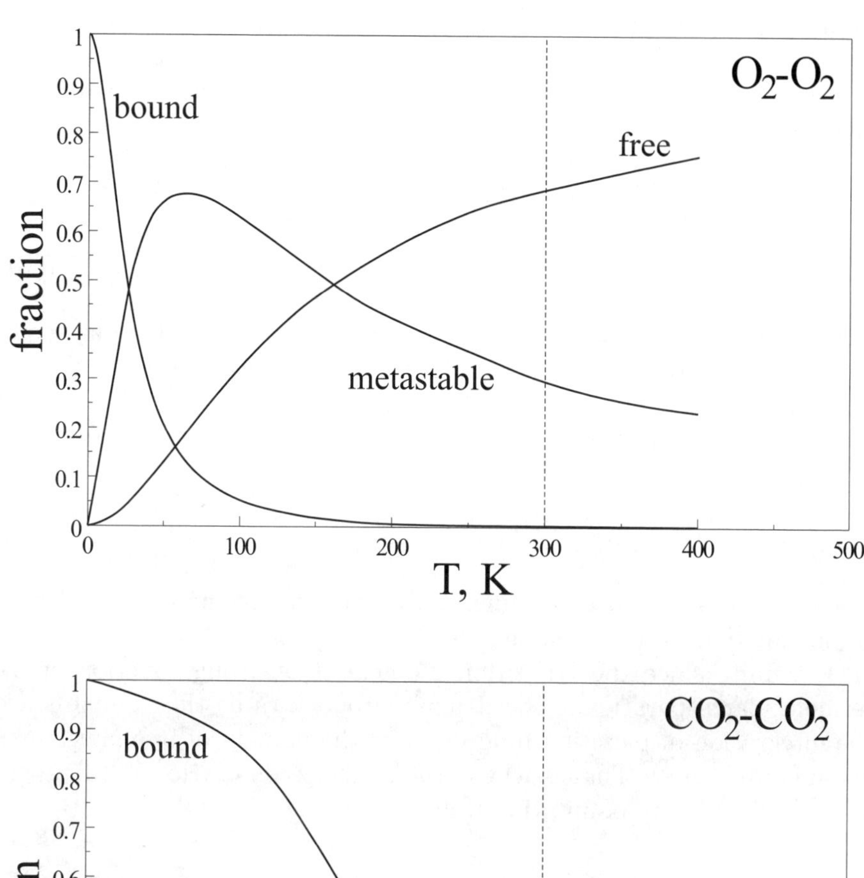

Figure 4. Partitioning of the normalized CIA intensity in the phase space of O_2 and CO_2 pairs [8].

Assuming the fraction of dimers is small and hence $\alpha_2 \sim \rho$, one arrives to an analogue of the virial expansion $\kappa(\nu) = k_1(\nu)\rho + k_2(\nu)\rho^2$, where the second virial $k_2(\nu)$ is proportional to the deviation of the dimer absorption cross-section from the sum of those characterizing two parent monomers.

In the second extreme, let us assume that free pair states dominate in the phase space. Then for the spectral absorption coefficient one has

$$\kappa(\nu) = \frac{N_1}{V}\,\sigma_1(\nu) + \frac{N_2}{V}\,\sigma_2(\nu). \tag{10}$$

The number of pairs N_2 which can be formed from N individual molecules is obviously $N_2 \approx N(N-1)/2 \approx N^2/2$. To calculate the absorption cross-section $\sigma_2(\nu)$ in this case one has to average the interaction induced transition moment squared $|\mu|^2$ over the domain of free pair states

$$<\sigma_2(\nu)>_{\mathrm{f}} = \frac{L_{\mathrm{f}}(\nu)}{Q_{\mathrm{f}}} \int_{\mathrm{f}} |\mu|^2 \, d\tau, \tag{11}$$

where $L_{\mathrm{f}}(\nu)$ is the spectral function, Q_{f} stands for the truncated partition function over the domain of free pair states. The numerator in (11) is finite since the transition moment is a strong function of intermolecular separation. The denominator varies as the volume since the integration of partition function over the domain of free pair states extends to infinity. That is why $\sigma_2(\nu)$ in the second term in (10) varies as V^{-1}. Finally on assumption that $N_1 \approx N$ one obtains

$$\kappa(\nu) = \frac{\rho}{\mu_1}\,\sigma_1(\nu) + \frac{\rho^2}{2\,\mu_1}\,<V\sigma_2(\nu)>_{\mathrm{f}}. \tag{12}$$

It is seen that free pairs provide a correction to the monomer absorption which is proportional to density squared. This correction term must be averaged over the domain of free pair states only. The use of simple Boltzmann exponential (or radial distribution function) in the averaging procedure is not recommended unless temperature is high enough to make free pair states dominant in the phase space.

2.3. Resonance states

The anisotropy of the intermolecular potential makes it possible for a pair of molecules to temporarily stabilize in the form of quasibound compound system or scattering resonance of the Feshbach type. The excess energy above the dissociation energy is distributed in this case among internal rotations of the monomers, librational vibrations of the dimer, or excitation of the intramolecular modes. A pair of monomers

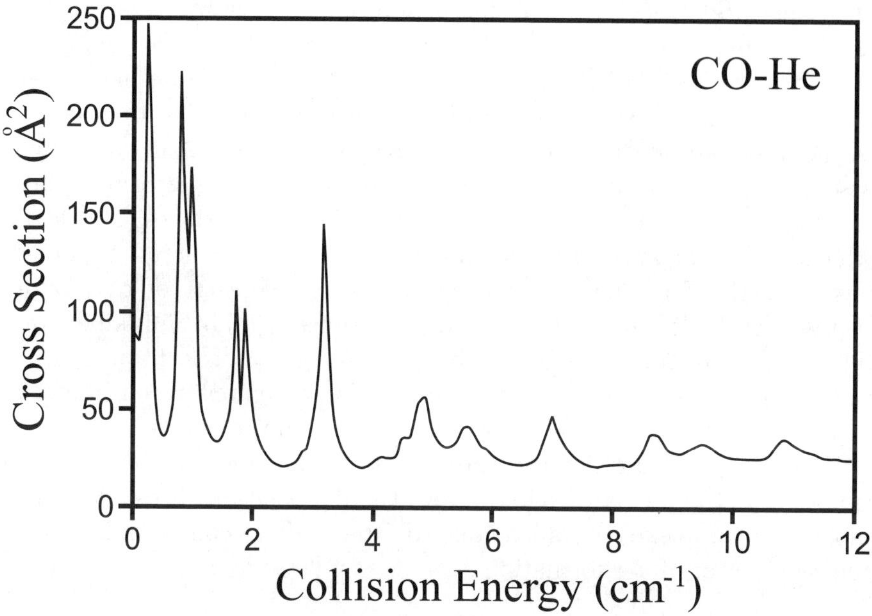

Figure 5.   Pressure broadening cross-section for collisions between CO and He at low energy (from Ref. [11]).

in a quasibound state predissociates by virtue of internal energy redistribution between the various modes. Direct spectroscopic observations of Feshbach resonances in molecular pairs are scarce (see e. g. [9]).

Shape resonances have usually very long lifetimes. In contrast the lifetimes of most of the Feshbach resonances are typically short. Trajectory calculations (see e. g. [10]) show that the average lifetime of the resonance states in collisions of CO_2 with Ar is of the order of 10^{-10}–10^{-11} s, corresponding to a broadening of about 0.3–3 cm^{-1} or even more. For molecules with dense rotational spectra the number of resonances is huge and one can expect many of them to have very short lifetimes. Hence, the spectroscopic features associated with these resonances are expected to be very broad and usually no detectable resonant structure appears in the spectrum. At low temperature Feshbach resonances give rise to sharp peaks in the energy dependent cross-section as displayed in Fig. 5. As temperature rises, however, this sharp structure washes out due to statistical averaging over thermally populated states. The appearance of resonances nicely explains the observed deviations in the line shift and line broadening cross-sections from semi-classical high temperature behavior. This has been demonstrated by

De Lucia and associates [11] for a number of colliding pairs in their elegant experiments at extremely low temperatures.

Quasibound pairs are apt to contribute to continuum-like absorption. The lack of distinct spectroscopic features precludes identification of the observed effect with individual resonant states. The only way to assess the role played by short-lived resonances is to perform sophisticated calculations using complete potential energy surfaces (PES) from which the importance of metastable states can be predicted unambiguous. Worth of noting is the need to use the concept of quasibound states in the calculations of partition functions, equilibrium, kinetic and relaxation constants for weakly interacting pairs. The importance of resonance states in these calculations has been demonstrated in a number of works [1, 2, 12]. Statistical theory of unimolecular decay (see e. g. [13]) aims at calculating the rate constants for spontaneous dissociation of a molecular pair excited above the dissociation threshold. These molecular complexes are analogous to intermediate compound states in the reaction of dimer formation.

3. How bimolecular states show up in absorption

3.1. Collision-induced absorption

Collision-induced absorption (CIA) provides a unique opportunity to study absorption by molecular pairs consisting of highly symmetrical monomers. In case of CIA the first linear term disappears from Eq. 2 due to selection rules prohibiting photon absorption in isolated dipoleless molecules. Hence, after removal of possible impurity contributions from the recorded spectrum, the observed absorption varies as the density squared, thus demonstrating the purely bimolecular nature of the effect.

One of the prototype systems for studies of CIA phenomena is molecular hydrogen (see e. g. [14, 15]). Spectroscopy of molecular hydrogen pairs is now a particularly well understood domain of CIA spectroscopy mainly due to the outstanding efforts undertaken by McKellar, Frommhold and others. In spite of the extreme weakness of van der Waals complexes formed by hydrogen molecules, these complexes are successfully characterized in much details through the study of their contribution to CIA. In molecular hydrogen the large spacing between rotational lines allows for observation of rotationally resolved interaction-induced transitions. The theory (see e. g. [16]) accounting for the formation of bound and quasibound pair states proves to be in excellent agreement with the observations. For molecules heavier

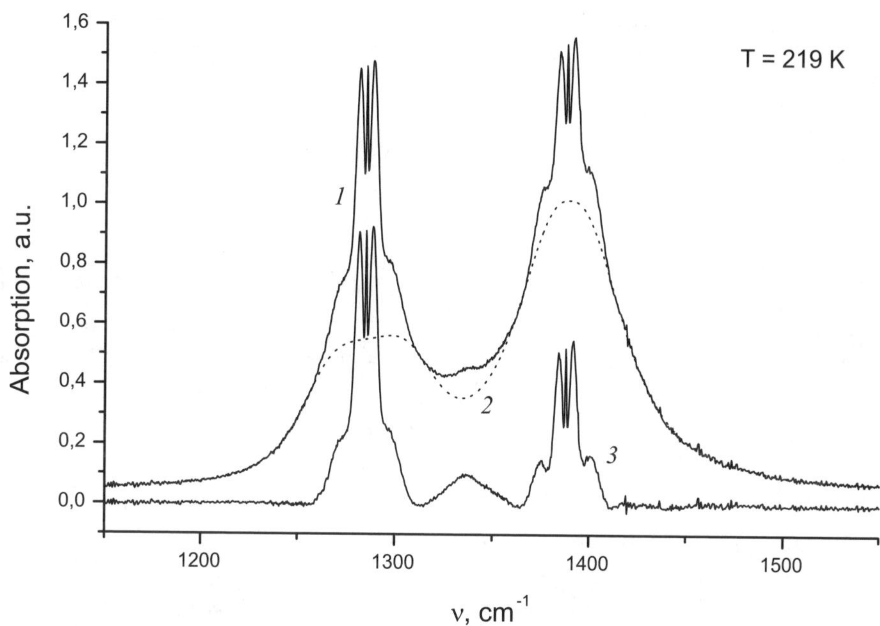

Figure 6. CIA spectrum in the region of the Fermi doublet of low-temperature CO_2 (from Ref. [21]). Total CIA (1), base profile (2), true dimer profile (3).

than molecular hydrogen one can barely expect to resolve the rotational structure in CIA spectra at densities typical to CIA experiments.

Recently, the use of Fourier transform techniques allowed for detection of signatures of fine structure in CIA spectra of N_2, O_2, and CO_2 [17–20]. These features appear mainly as P, Q, R-like structure atop the CIA rovibrational bands [20] along with the so-called "ripples" detected both in N_2 and O_2 fundamentals [17, 18] and in the rototranslational band of nitrogen [19]. Figure 6 shows the symmetric-top-like structure atop both members of the CO_2 Fermi doublet. After removing the structureless pedestal from the spectra, one can identify the residual absorption with the signatures of true dimers of the carbon dioxide.

This identification is strongly supported by the coincidence in the position of the Q-branches seen in induced infrared absorption bands with their counterpart in CARS spectra of CO_2 dimers probed in free jet expansion [22]. Normalizing the integrated dimer intensity by the total intensity of the CIA Fermi coupled bands one can trace the temperature variations of the fraction of true dimers as shown in Fig. 7. The solid line on this figure shows the result of a statistical physics calculation using Eqs. (3, 4) supported by a reliable multidimensional PES [23]. The agreement between theory and observation seems quite

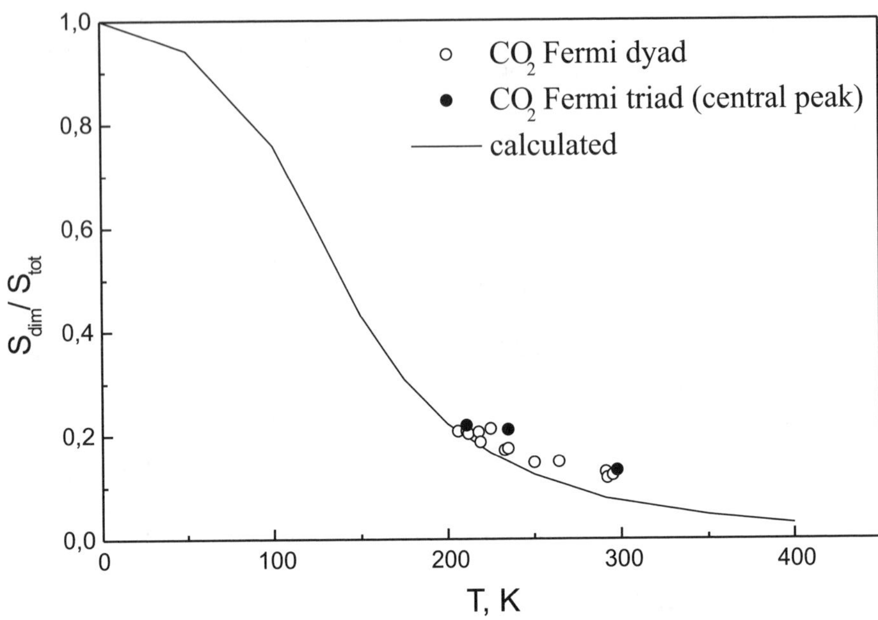

Figure 7.   Temperature variations of the normalized CIA intensity of true bound CO_2 dimers (from Ref. [21]).

reasonable. One can suggest therefore that the use of Eqs. $(3, 4)$ makes it possible to evaluate how much of true bound (dimeric) states are present among the totality of pair states. Very roughly, at any given temperature (irrespective to the gas density) the larger the number internal degrees of freedom are involved and the deeper the intermolecular potential, the higher the fraction of true bound and metastable states to the expense of free pair states. Note that the number density of dimers at equilibrium is governed by both temperature and pressure of a gas.

An intriguing property of the "ripples" in the CIA contours is the almost perfect match of the minima in absorption with location of quadrupole (or Raman) lines whereas the maxima of absorption appear between the lines. Several hypotheses were put forward to explain the nature of these ripples (see e. g. [24] and references therein). In [24, 25] they were suggested to arise from line-mixing effects of quadrupole induced lines.

Figures 8, 9 show that the model used is nicely reproducing these ripples both in far-infrared absorption in nitrogen and in mid-infrared CIA of oxygen. Note that these calculations match the observations with better accuracy at low rotational quantum numbers J; as J

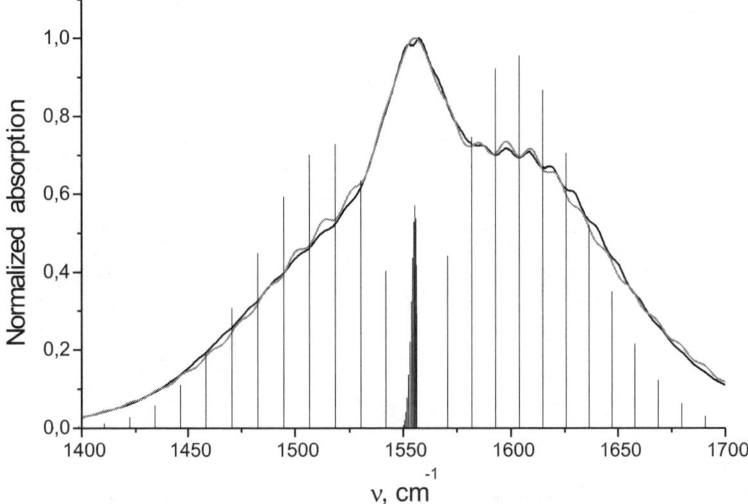

Figure 8. CIA profile in the region of the O_2 fundamental at $T = 296$ K. Heavier line shows normalized experimental profile from Maté *et al.* [18], lighter line refers to line-mixing model calculations from [24], vertical sticks stand for quadrupole induced lines.

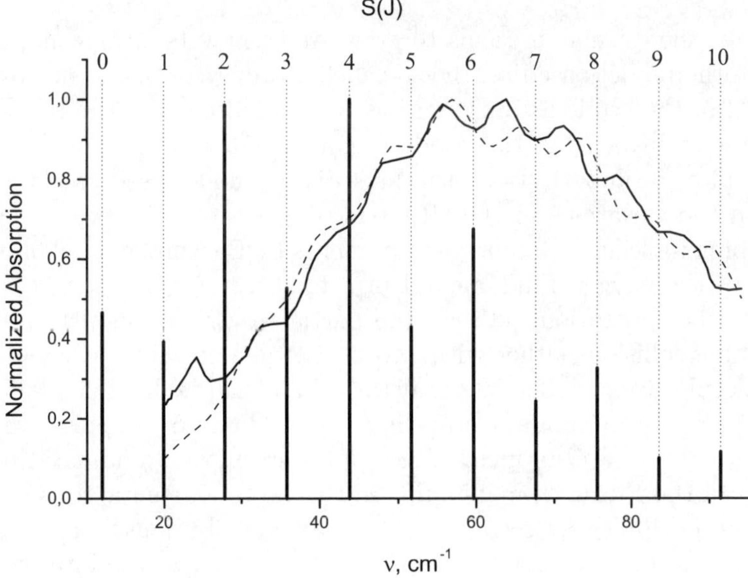

Figure 9. Collision-induced rototranslational band in N_2–Ar taken at $T = 89$ K. Solid line shows experimental profile from Wishnow *et al.* [19], dash line displays the result of line-mixing calculations.

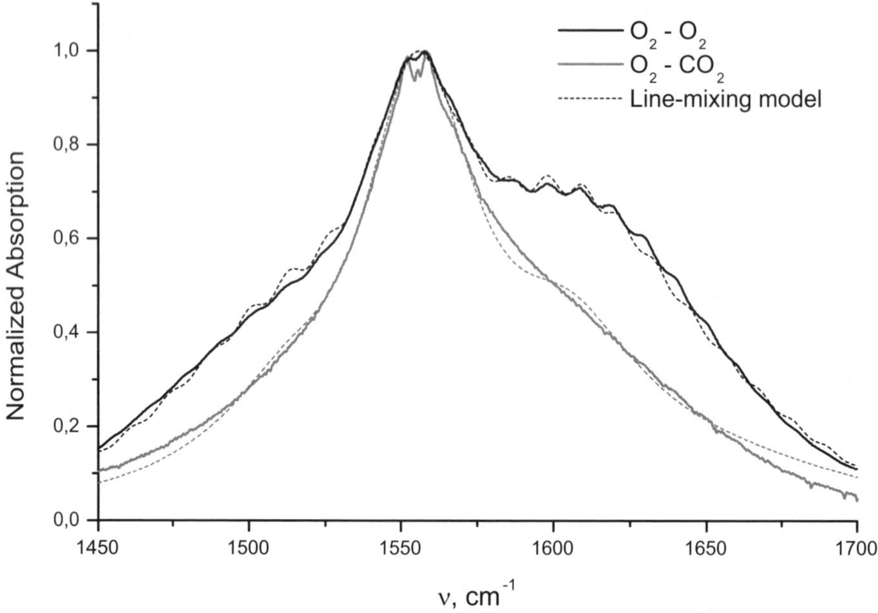

Figure 10.   CIA profiles in the region of the oxygen fundamental in pure O_2 and in a mixture of O_2 with CO_2 (from Ref. [26]).

increases, the mismatch seems to grow. This may be due to imperfections of the model based on the so-called "strong collision" approximation. Also, the crude assumption has been adopted that the relaxation parameter τ involved in this model is constant irrespective of J.

Despite its imperfections the line-mixing model used sheds a new light on the variations in the CIA bandshape caused by the change in perturbing molecule. Recent measurements by Baranov *et al.* [26] in the region of the oxygen fundamental in a mixture of CO_2 with O_2 make possible the separation between the partial contributions arising from perturbing collisions either with another oxygen or with a carbon dioxide molecule. Figure 10 shows that these bandshapes differ significantly. In the pure oxygen fundamental the distinct O and S branches are seen on both sides of a Q-branch. The ripples are most pronounced in the vicinity of the S-branch maximum. In the O_2 fundamental for O_2–CO_2 mixtures no ripples are seen and only weak contributions from O and S branches can be recognized. It looks as if the stronger the intermolecular perturbation, the more symmetrical the band profile. Modifications occurring in the overall bandshape can thus be considered in terms of a collapse of the rotational structure caused by line interference. Having adjusted three parameters in the line-mixing model, the calculated

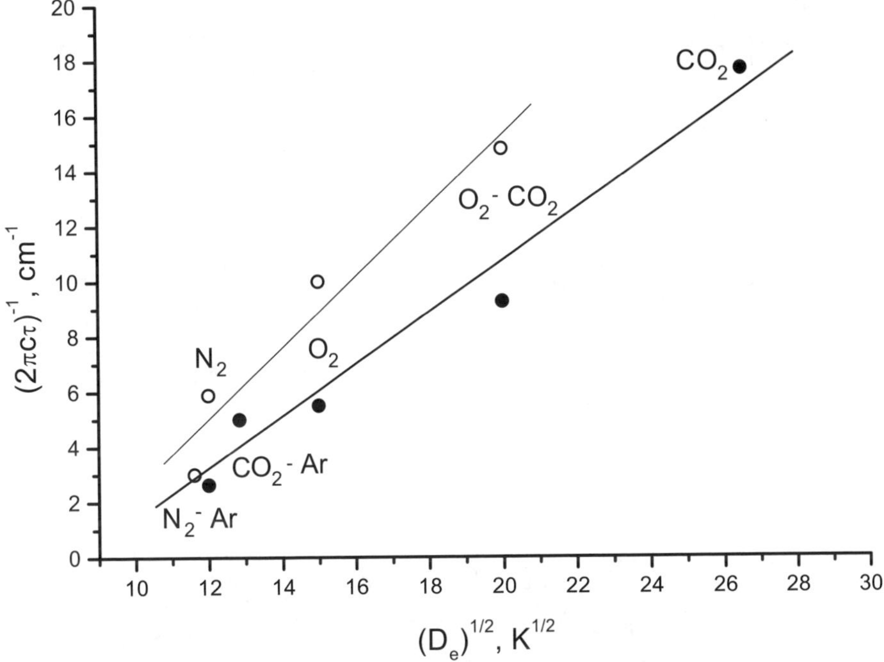

Figure 11. Reciprocal of the relaxation time vs square root of the dissociation energy for several molecular pairs. Solid circles refer to the fitting of the spectra taken near room temperature, open circles correspond to the low temperature spectra (77–220 K).

profiles can be fit to the observed ones with appreciable accuracy. The key parameter in our line-mixing model is the relaxation time τ. Figure 11 demonstrates an almost linear correlation between the reciprocal of τ and the strength of the intermolecular interaction. These observed variations (the stronger the interaction, the shorter τ) are thus supporting the adopted line-mixing model. It has to be recalled that all of the τ's in Fig. 11 were obtained by fitting the spectral profiles for the different molecular pairs.

3.2. Allowed absorption bands in dense gases

Various mechanisms have been proposed in the literature to explain how the spectral profiles in allowed monomer absorption bands are changing as the density of the gas is rising. These effects are line-mixing effect, finite duration of collisions, intercollisional correlation, molecular aggregation etc. The problem is how to rank these effects according their relative importance for any given combination of intermolecular

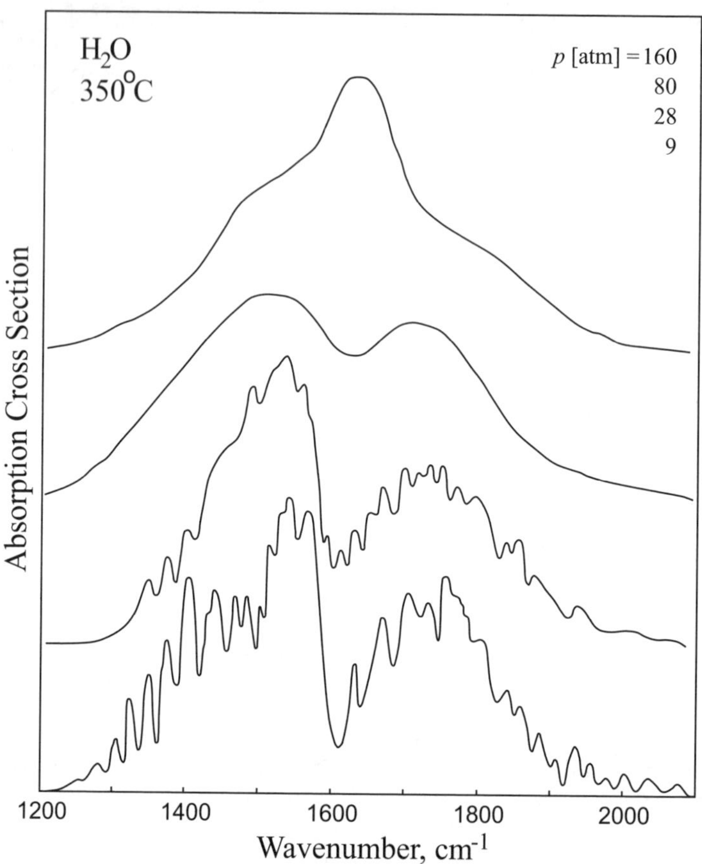

Figure 12.   Variations of the bending profile in pressurized water vapor [27].

PES, temperature, and density of the gas. Figure 12 displays typical variations occurring in a vibrational band profile as the density of the gas is increasing. This plot is taken from the data obtained by Vetrov [27] in the ν_2 bending region of high pressure water vapor. It is seen that a Q-branch-like maximum appears at high density in the trough between P and R branches and the total bandshape symmetrizes. Similar density evolution is found for various vibrational bands in CO, CO$_2$ and other species. For the CO fundamental and overtones and several allowed CO$_2$ bands such variations have been successfully reproduced in terms of pure line-mixing effects (see e.g. [28–30]) or line-mixing combined with molecular association [31–33]. Available mass action law estimates of the equilibrium number density of CO, CO$_2$ or H$_2$O dimers at relevant experimental conditions predict substantial molecular association (see e.g. [27, 32, 33]).

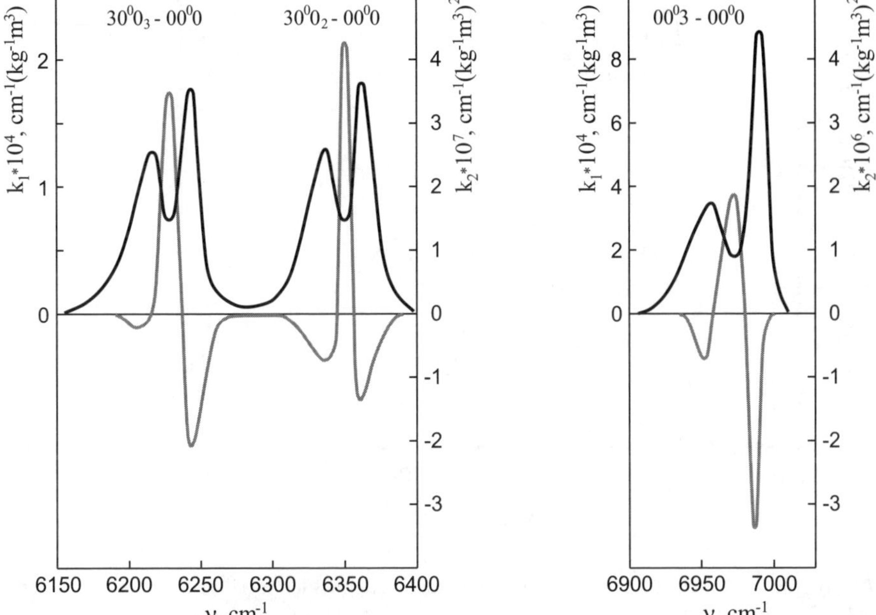

Figure 13. Coefficients $k_1(\nu)$ (heavier line) and $k_2(\nu)$ (lighter line) retrieved from various CO_2 absorption bands at elevated densities [34] (see Eq. (9) and below).

The presence of spectroscopically detectable amount of true bound dimers in CO_2 is obvious from the CIA spectra taken near room temperature as described above. These same dimers are expected to contribute significantly to the density evolution in allowed absorption band profiles and the spectral absorption coefficient (9) should apply. As the density increases there is a substantial decrease of the absorption in the vicinity of maxima of the monomer P and R branches, which is accompanied by an absorption increase in the region of the band center. The decomposition of the CO_2 spectral profiles performed by Adiks [34] reveals these expected variations as shown in Fig. 13. It is worth to mention that the dimer rotational structure condenses near the band center due to the much smaller rotational constants of the dimer in comparison to those of the monomer.

Line-mixing effects also favor an increase of absorption near the band center at the expense of a decrease of the absorption in the wings. Both effects appear to cause similar trends in the density evolution of allowed absorption profiles. Accounting for dimerization in a gas and correcting the monomer profile due to line-mixing effects seem to explain most of the observed density variations in spectral profiles of pressurized

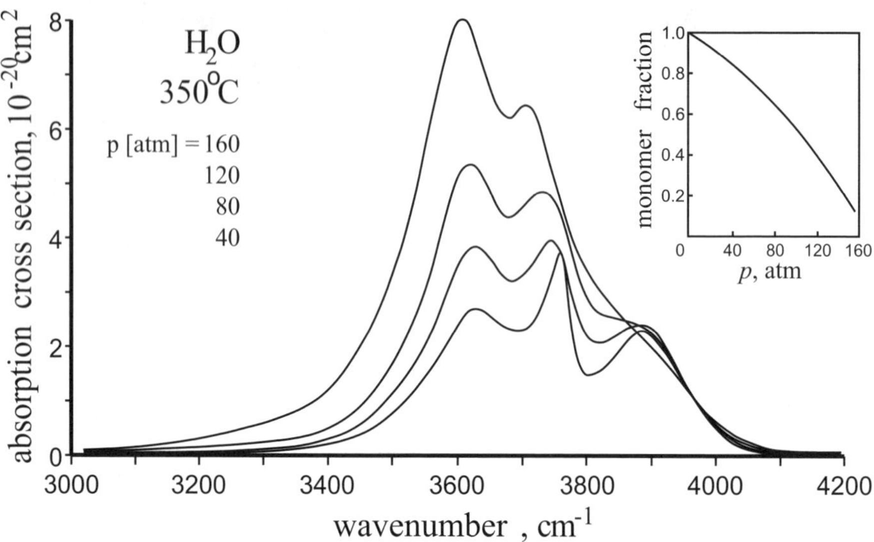

Figure 14.   Pressure evolution of the absorption profiles in the region of OH stretch [27]. The insert shows isothermal pressure variations of the monomer mole fraction obtained from decomposition of the spectral profiles.

CO and CO_2 [31–33] at least in the range of density where pair effects are expected to prevail. A satisfactory discrimination between the contributions from molecular association and from line-mixing effects is thus the result of detailed statistical physics considerations.

Association in water vapor is expected to reach high levels in the vicinity of the saturation line, especially near the critical point. Low resolution infrared absorption spectra taken by Vetrov [27] yielded useful information on the spectral manifestations of hydrogen bonding in pressurized water vapor. Figures 12, 14 give an idea of the modifications occurring to spectral profiles in the regions of bending and OH stretching vibrations. The distortions of spectral profiles for the OH stretch are related to hydrogen bonding which causes significant enhancement in intensity for the red-shifted OH-vibrations. Analysis of the spectral profiles showed that association of water molecules $(i > 2)$ generally exceeds the contribution of dimers in a superheated steam. The insert in Fig. 14 exemplifies the decrease of the monomers mole fraction following isothermal compression of water vapor at $T = 350^{\circ}$C. The use of such spectroscopic data can serve to verify the mass-action-law-based estimates of the equilibrium composition in associating water vapor.

3.3. Water vapor continuum

The nature of the excess or continuum absorption in water vapor is a matter of controversies in scientific literature since many decades (see e. g. [5, 35]). The theory of sub-lorentzian absorption in the far wings of monomer lines (see e. g. [36]) allows to reproduce effectively the observed continuum at least in the $10\,\mu$m window.

The monomer line wing theory as applied to the water vapor continuum absorption, leaves a number of questions open, however. First, the applicability of quasistatic lineshape theory seems highly questionable. Second, statistical averaging in this theory implies that free pairs dominate in room temperature water vapor whereas independent calculations show this is not the case. Third, the sensitivity of the final result to the chosen PES has not been examined yet in much detail. Fourth, the criterion governing the choice of cut-off in wavenumber of monomer lines contributing to continuum absorption at selected frequency is not obvious. Besides the monomer line wing model, the dimer hypothesis is vividly discussed in the literature for a long time. Very recently new indications on the possible water dimer involvement in continua absorption in the atmosphere appeared in the literature [37, 38].

Our consideration above pointed out the importance of bound and metastable pair states in various gas phase systems near room temperature. In case of room temperature carbon dioxide for instance, true dimers occupy about 15% of all available pair states, the result supported both by statistical physics partitioning in the phase space and by decomposition of CIA profile. Preliminary partitioning of the pair states in water vapor [2, 39] showed that the role of free pair states is almost negligible at near room temperature; metastable and true bound states should dominate instead. This is not surprising since the interaction between water molecules is at least three times stronger than that between carbon dioxide molecules. Additional degrees of freedom in water–water pairs as compared to carbon dioxide pairs are also expected to produce an enhanced contribution of bound or quasibound states. Moreover, available computations using the truncated partition function method [6, 39] showed that true bound and metastable states of a water dimer are almost equally populated around room temperature (see Fig. 15). A reliable theory of the water vapor continuum requires sophisticated partitioning in the phase space followed by the modeling of the spectral profiles for bound and quasibound dimers and statistical averaging over the ensemble of pair states.

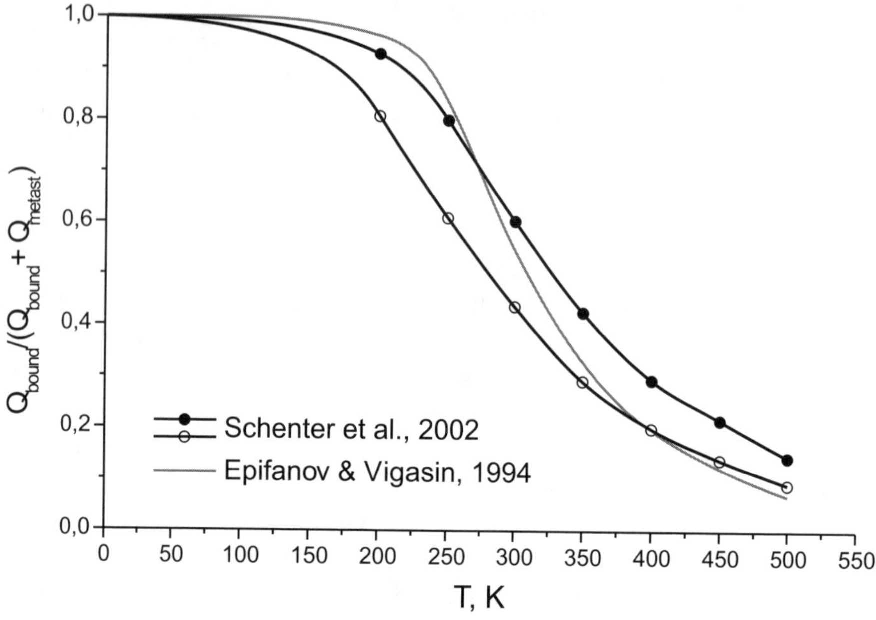

Figure 15.   Relative fraction of true bound and metastable dimers in water vapor. Lighter line is from [39], heavier lines refer to data from Ref. [6] (solid and open circles correspond to classical and quantum calculations, respectively).

4. Conclusions

The present paper demonstrates the importance of statistical physics partitioning in the phase space of interacting molecular pairs. True dimer, metastable, and free pair states can rigorously be defined in the phase space. Departures of the monomer line shape from conventional impact theory profile are due to free pair states mapping the repulsive branches of the intermolecular potential. The relevant statistical averaging has to extend over domain of free pair states in the phase space. True dimers are clearly seen in CIA spectra of carbon dioxide at room temperature and atmospheric pressure. This is an indication that the contribution from true bound and metastable dimeric states may not be neglected in the case of the water vapor continuum absorption in the atmosphere since association in water vapor is expected to be much stronger than that in carbon dioxide. In dense gases both molecular association and line-mixing effects play a role either in the regions of allowed or forbidden transitions. Line-mixing effects seem to be one of the major factors determining the shape of the usually structureless base profile of CIA bands. In pairs of molecules like N_2 or O_2 which are relatively weak perturbers and which have large spacings between quadrupole lines, line-mixing effects are responsible for the ripples observed. In molecules like CO_2, with a substantial quadrupole moment and a denser rotational structure, line-mixing at higher pressure results in narrowing and symmetrizing the base profile, an effect analogous to the collapse of the rotational structure in condensed fluids. To explain fully the nature of the water vapor continuum one has to start from sophisticated partitioning in the phase space using complete multidimensional PES. Spectroscopic observations have to be supported by independent statistical physics considerations in order to construct a reliable theoretical model for continuum absorption in atmospheric gases.

Acknowledgments

The author acknowledges partial support of this work from the Russian Foundation for Basic Researches through the Grant 02-05-64529.

References

[1] Stogryn, D. E. and Hirshfelder, J. O. (1959) Contribution of bound, metastable and free molecules to the second virial coefficients and some properties of double molecules, J. Chem. Phys., 31, 1531–1545.

[2] Vigasin, A. A. (1991) Bound, metastable and free states of bimolecular complexes, Infrared Phys., 32, 461–470.

[3] Hill, T. (1956) *Statistical Mechanics*, McGraw-Hill, New York.

[4] Vigasin, A. A. (1985) On the spectroscopic manifestations of weakly bound complexes in rarefied gases, Chem. Phys. Lett., 117, 85–88.

[5] Vigasin, A. A. (1998) Dimeric absorption in the atmosphere. In: A. A. Vigasin and Z. Slanina, editors. *Molecular Complexes in Earth's, Planetary, Cometary, and Interstellar Atmospheres*, World Scientific, Singapore, pp. 60–99.

[6] Schenter, G. K., Kathmann, S. M., and Garrett, B. C. (2002) Equilibrium constant for water dimerization: Analysis of the partition function for a weakly bound system, J. Phys. Chem. A, 106, 1557–1566.

[7] Levine, H. B. (1972) Spectroscopy of dimers, J. Chem. Phys., 56, 2455–2473.

[8] Epifanov, S. Yu. and Vigasin, A. A. (1994) Contribution of bound, metastable and free states of bimolecular complexes to collision-induced intensity of absorption, Chem. Phys. Lett., 225, 537–541.

[9] Pine, A. S., Lafferty, W. J., and Howard, B. J. (1984) Vibrational predissociation, tunneling, and rotational saturation in the HF and DF dimers, J. Chem. Phys., 81, 2939–2950.

[10] Ivanov, S. V. (2003) Trajectory study of CO_2–Ar and He collision complexes, this volume, pp. 49–63.

[11] Flatin, D. C., Goyette, T. M., Beaky, M. M., Ball, C. D., and De Lucia, F. C. (1999) Rotational state dependence of collision induced line broadening and shift at low temperatures, J. Chem. Phys., 110, 2087–2098.

[12] Sumpter, B. G., Thompson, D. L., and Noid, D. W. (1987) The effect of resonances on collisional energy transfer, J. Chem. Phys., 87, 1012–1021.

[13] Robinson, P. J. and Holbrook, K. A. (1972) *Unimolecular Reactions*, Wiley, New York.

[14] Frommhold, L. (1993) *Collision-induced Absorption in Gases*, Cambridge University Press, Cambridge.

[15] McKellar, A. R. W. (1996) High resolution infrared spectra of H_2–Ar, HD–Ar, and D_2–Ar van der Waals complexes between 160 and $8620\,cm^{-1}$, J. Chem. Phys., 105, 2628–2653.

[16] Dunker, A. M. and Gordon, R. G. (1978) Bound atom–diatomic molecule complexes. Anisotropic intermolecular potentials for the hydrogen–rare gas systems, J. Chem. Phys., 68, 700–725.

[17] McKellar, A. R. W. (1988) Infrared spectra of the $(N_2)_2$ and N_2–Ar van der Waals molecules, J. Chem. Phys., 88, 4190–4196.

[18] Maté, B., Lugez, C. L., Solodov, A. M., Fraser, G. T., and Lafferty, W. J. (2000) Investigation of the collision-induced absorption by O_2 near $6.4\,\mu m$ in pure O_2 and O_2/N_2 mixtures, J. Geophys. Research, 105, 22225–22230.

[19] Wishnow, E. H., Gush, H. P., and Ozier, I. (1996) Far-infrared spectrum of N_2 and N_2–noble gas mixtures near 80 K, J. Chem. Phys., 104, 3511–3516.

[20] Baranov, Yu. I. and Vigasin, A. A. (1999) Collision-induced absorption by CO_2 in the region of ν_1, $2\nu_2$, J. Molec. Spectrosc., 193, 319–325.

[21] Vigasin, A. A., Baranov, Y. I., and Chlenova, G. V. (2002) Temperature variations of the interaction induced absorption of CO_2 in the ν_1, $2\nu_2$ region: FTIR measurements and dimer contribution, J. Molec. Spectrosc., 213, 51–56.

[22] Huisken, F., Ramonat, L., Santos, J., Smirnov, V. V., Stelmakh, O. M., and Vigasin, A. A. (1997) High resolution CARS spectroscopy of small carbon dioxide clusters: investigation of the CO_2 dimer in the $\nu_1/2\nu_2$ Fermi dyad, J. Molec. Struct., 410/411, 47–50.

[23] Bouanich, J.-P. (1992) Site-site Lennard—Jones potential parameters for N_2, O_2, H_2, CO, and CO_2, JQSRT, 47, 243–250.

[24] Vigasin, A. A. (2000) Collision-induced absorption in the region of the O_2 fundamental: Bandshapes and dimeric features, J. Molec. Spectrosc., 202, 59–66.

[25] Vigasin, A. A. (1996) On the nature of collision-induced absorption in gaseous homonuclear diatomics, JQSRT, 56, 409–422.

[26] Baranov, Y. I., Vigasin, A. A., Lafferty, W. J., and Fraser, G. T. (2002) Continuum absorption in the region of the O_2 vibrational fundamental band in O_2 and CO_2 mixtures at temperatures ranging from 220 K to 296 K. In: Proc. 57th Ohio State University International Symposium on Molecular Spectroscopy, Columbus, Ohio, June 17–21.

[27] Vetrov, A. A. (1976) The Study of Absorption Coefficients and the Structure of Water Vapor at High Temperatures and Pressures. Ph. D. Thesis, Institute for High Temperatures, USSR Academy of Sciences, Moscow (in Russian).

[28] Filippov, N. N., Tonkov, M. V., Boulet, Ch., and Bouanich, J.-P. (1993) Analysis of line mixing in CO 2–0 band in high pressure nitrogen. In: Proc. High-Resolution Molecular Spectroscopy, SPIE Vol. 2205, pp. 328–331.

[29] Filippov, N. N., Bouanich, J.-P., Hartmann, J.-M., Ozanne, L., Boulet, C., Tonkov, M. V., Thibault, F., and Le Doucen, R. (1996) Line-mixing effects in the $3\nu_3$ badn of CO_2 perturbed by Ar, JQSRT, 55, 307–320.

[30] Hartmann, J.-M. and Boulet, C. (1991) Line mixing and finite duration of collision effects in pure CO_2 infrared spectra: Fitting and scaling analysis, J. Chem. Phys., 94, 6406–6419.

[31] Vigasin, A. A., Filippov, N. N., and Chlenova, G. V. (1992) Effect of the interference of spectral lines and van der Waals association of the molecules on the shape of the 2.0-μm band of compressed CO_2, Opt. Spectrosc. (USSR), 72, 56–59.

[32] Adiks, T. G., Tchlenova, G. V., and Vigasin, A. A. (1989) On the influence of van der Waals association on the IR absorption band shapes of the highly compressed carbon dioxide, Infrared Phys., 29, 575–582.

[33] Tchlenova, G. V., Vigasin, A. A., Bouanich, J.-P., and Boulet, C. (1993) The nature of the absorption bandshape density evolution for the first overtone of CO compressed by N_2, Infrared Phys., 34, 289–298.

[34] Adiks, T. G. (1982) Experimental Study of the CO_2 IR Absorption Spectra as Applied to the Windows of Transparency of Venusian Atmosphere. Ph. D. Thesis, Institute of Atmospheric Physics, USSR Academy of Sciences, Moscow (in Russian).

[35] Tobin, D. C., Strow, L. L., Lafferty, W. J., and Olson, B. (1996) Experimental investigations of the self- and N_2-broadened continuum within the ν_2 band of water vapor, Appl. Optics, 35, 4724–4734.

[36] Tipping, R. H. and Ma, Q. (1995) Theory of the water vapor continuum and validations, Atmospheric Research, 36, 69–94.

[37] Cormier, J. G., Ciurilo, R., and Drummond, J. R. (2002) Cavity ringdown spectroscopy measurements of the infrared water vapor continuum, J. Chem. Phys., 116, 1030–1034.

[38] Maurellis, A. N., Lang, R., Williams, J. E., van der Zande, W. J., Smith, K., Newnham, D. A., Tennyson, J., and Tolchenov, R. N. (2003) The impact of new water vapor spectroscopy on satellite retrievals, this volume, pp. 259–272.

[39] Epifanov, S. Yu. and Vigasin, A. A. (1997) Subdivision of the phase space for anistropically interacting water molecules, Molec. Phys., 90, 101–106.

TRAJECTORY STUDY OF
CO_2–Ar AND CO_2–He COLLISION COMPLEXES

S. V. Ivanov

Institute on Laser and Information Technologies,

Russian Academy of Sciences,

Pionerskaya 2, 142190, Troitsk, Moscow Region, Russia

Abstract   The formation of CO_2–Ar and CO_2–He collision complexes also called quasi-complexes (QC) or Feshbach-type resonances, is studied using the method of classical trajectories. Exact classical equations in body-fixed coordinates, *ab initio* interaction potentials and Monte-Carlo sampling of collision parameters are used in these computations. Statistical analysis is made for the parameters of QC formed in collisions. It is shown that QC can be both short-lived and very long-lived and are characterized by a variety of interparticle separations. Among the total number of collisions the fraction of QC increases rapidly with temperature decrease. The contribution of QC to CO_2 rotational inelastic cross sections evaluated at two temperatures (100 K and 240 K) was found to be small. At the same time, rotational relaxation via QC-forming collisions turned out to be much more efficient than via conventional inelastic collisions.

Keywords: collision complexes / classical trajectories / Monte-Carlo simulation

1. Introduction

Unconventional molecular absorbers of radiation in the atmosphere manifest themselves in different ways, e. g. by changing the rates of collisional energy transfer, by modifying the optical characteristics of the medium, etc. For instance, in an aircraft vortex wake, containing significant amount of H_2O and CO_2, various weakly bound species like H_2O–H_2O, H_2O–CO_2, H_2O–N_2 are expected to form. Either stable or metastable, these floppy complexes belong to the class of molecules with pronounced nonrigidity. Besides their fundamental interest, these molecular associates can in principle be used for remote sensing of the

C. Camy-Peyret and A.A. Vigasin (eds.),
Weakly Interacting Molecular Pairs: Unconventional Absorbers of Radiation in the Atmosphere, 49–64.
© 2003 *Kluwer Academic Publishers. Printed in the Netherlands.*

composition of gaseous mixtures. The conditions of formation, statistical and optical properties of weakly interacting molecular pairs are currently subject to extensive and sophisticated theoretical investigations.

Tightly bound complexes in the gas phase are usually formed in the course of stabilization of their precursors — quasibound complexes, i. e. quasi-complexes (QC) or scattering resonances. Generally, two types of quasibound complexes are distinguished: classically forbidden shape resonances (they arise from translational tunneling) and classically allowed Feshbach-type resonances. The latter can form provided that at least one of the colliding partners has internal degrees of freedom [1]. The simplest prototype system arises in case of a collision between atom and rigid diatomic characterized by two rotational degrees of freedom associated with only one classical angular coordinate. The resulting energetically unstable complex then undergoes stabilizing or destroying collision with a third body, otherwise it is subject to unimolecular decay (see e. g. [2–5]). The characterization of the birth and death of these unstable species is by no means straightforward. Present work aims at numerical investigation of the classical dynamics of formation and decay of QC by virtue of trajectory computer simulations. We restricted ourselves by consideration of the relatively simple atom–rigid diatom systems (Ar–CO_2 and He–CO_2) and have focused on the statistics of QC formation and on the computations of rotational energy relaxation cross sections.

2. Classical trajectory equations for atom-diatom collision

Let us consider the collision of structureless A particle (atom) with a rigid rotor BC (diatomic or linear polyatomic molecule). The system of coordinates $AXYZ$ (see Fig. 1) centered on A atom is laboratory system, while the system $Bxyz$ connected with BC molecule is body-fixed.

The z axis of $Bxyz$ system is always directed along the vector $\vec{R}$, connecting the center of masses of BC molecule with the origin of coordinates A in laboratory system. The center of the body-fixed system is positioned on one of the end atoms of a molecule. The length of $\vec{r}$ vector, directed from the center of $Bxyz$ system to another end atom (C), is constant since the molecule is assumed rigid. The elevation angle φ is the angle between the axis Bz and the YAZ plane. The plane in which the collisions start is XAZ. In the course of a collision the vector of relative velocity $\vec{v}_{rel}$ may change its direction, and the collision plane may turn by the angle θ. Exact classical equations of motion in the form of Hamilton equations describe the dynamics of the selected pair of particles. In the body-fixed coordinates these equations have the form [6]

$$\dot{R} = \frac{p_R}{\mu},$$

$$\dot{\theta} = \frac{(p_\theta + A)}{\mu R^2 \sin^2 \varphi},$$

$$\dot{\varphi} = \frac{(p_\varphi - J_y)}{\mu R^2},$$

$$\dot{p}_R = \frac{p_\theta^2 + A(A + 2p_\theta)}{\mu R^3 \sin^2 \varphi} + \frac{p_\phi^2 + J_y(J_y - 2p_\varphi)}{\mu R^3} - \frac{\partial V}{\partial R},$$

$$\dot{p}_\varphi = \frac{\cos\varphi \left(p_\theta^2 + A(A + 2p_\theta) \right)}{\mu R^2 \sin^3 \varphi} - \frac{B(A + p_\theta)}{\mu R^2 \sin^2 \varphi},$$

$$\dot{x} = \frac{p_x}{m} + \frac{y \cos\varphi (A + p_\theta)}{\mu R^2 \sin^2 \varphi} + \frac{z(J_y - p_\varphi)}{\mu R^2},$$

$$\dot{y} = \frac{p_y}{m} - \frac{(x \cos\varphi + z \sin\varphi)(A + p_\theta)}{\mu R^2 \sin^2 \varphi},$$

$$\dot{z} = \frac{p_z}{m} + \frac{y \sin\varphi (A + p_\theta)}{\mu R^2 \sin^2 \varphi} - \frac{x(J_y - p_\varphi)}{\mu R^2}, \tag{1}$$

$$\dot{p}_x = \frac{p_y \cos\varphi (A + p_\theta)}{\mu R^2 \sin^2 \varphi} + \frac{p_z(J_y - p_\varphi)}{\mu R^2} - \frac{\partial V}{\partial x},$$

$$\dot{p}_y = -\frac{(p_x \cos\varphi + p_z \sin\varphi)(A + p_\theta)}{\mu R^2 \sin^2 \varphi} - \frac{\partial V}{\partial y},$$

$$\dot{p}_z = \frac{p_y \sin\varphi (A + p_\theta)}{\mu R^2 \sin^2 \varphi} - \frac{p_x(J_y - p_\varphi)}{\mu R^2} - \frac{\partial V}{\partial z},$$

$$\vec{J} = \vec{r} \times \vec{p},$$

$$A = J_x \sin\varphi - J_z \cos\varphi,$$

$$r^2 = x^2 + y^2 + z^2,$$

$$B = J_x \cos\varphi + J_z \sin\varphi.$$

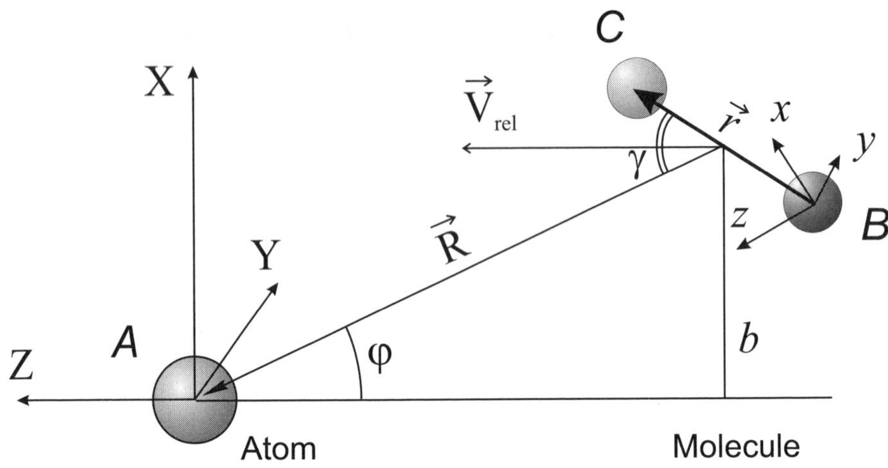

Figure 1.   Scheme of atom-diatom collision and coordinate systems used: laboratory $AXYZ$ and body-fixed $Bxyz$.

Here μ is the reduced mass of the A–BC system; m is the reduced mass of molecule BC; $V = V(\vec{R}, \vec{r})$ stands for the interaction potential between A and BC particles. The momentum p_θ in equations (1) is the integral of motion

$$p_\theta = \mu R^2 \dot{\theta} \sin^2 \varphi + m\Big(\big((x \cos \varphi + z \sin \varphi)^2 + y^2\big) \dot{\theta} - (y \cos \varphi)\, \dot{x}$$
$$+ (x \cos \varphi + z \sin \varphi)\, \dot{y} - (y \sin \varphi)\, \dot{z} + y\, (x \sin \varphi - z \cos \varphi)\, \dot{\varphi}\Big).$$

It was shown in [6] that while exact body-fixed equations are quite cumbersome, they are amenable to the introduction of simplifying approximations, e. g. planar or coplanar. For long-term numerical calculations using Monte-Carlo averaging over great number of trajectories these approximations are of extreme importance because they allow for significant reduction of the computational efforts keeping the calculations of rotational cross sections reasonably accurate. Despite notable computational inconvenience, we restricted ourselves hereafter to the solution of exact body-fixed equations since the accuracy of the use of planar trajectories in the problem of complex formation is difficult to assess.

3. Intermolecular interaction potential

The *ab initio* potential of Parker, Snow and Pack [7]

$$V(R, \gamma) = \sum_{n=0}^{10} V_n(R)\, P_n(\cos \gamma), \qquad n = 0,\, 2,\, 4 \ldots 10 \qquad (2)$$

was used for atom–linear polyatomic systems considered. Here $P_n(\cos \gamma)$ are Legendre polynomials; γ is the angle between the vectors $\vec{R}$ and $\vec{r}$; $\cos \gamma = z/r$. The radial coefficients $V_n(R)$ have the form

$$V_n(R) = \begin{cases} A_{n_1} \exp(A_{n_2} R + A_{n_3} R^2) \\ \quad - B_{n_1} \exp(B_{n_2} R + B_{n_3} R^2), & R \leq R_n, \\ A_{n_1} \exp(A_{n_2} R + A_{n_3} R^2) - \dfrac{C_6(n)}{R^6} - \dfrac{C_8(n)}{R^8}, & R \geq R_n. \end{cases} \qquad (3)$$

The values of the parameters A_{n_1}, A_{n_2}, A_{n_3}, B_{n_1}, B_{n_2}, B_{n_3}, $C_6(n)$, $C_8(n)$, and R_n for CO_2–He and CO_2–Ar were taken from Tables XI and XIII of Ref. [7].

4. Computational procedure

The computer code consists of several subroutines. The main program is driving most of the calculations, namely, the solution of the eleven dynamic equations (1) and the Monte-Carlo selection of initial conditions (collision parameters) according to the chosen distribution function (uniform, Maxwell-Boltzmann, delta-function). The distributions are sampled using standard Fortran 90 random numbers generator RAN and von Neumann procedure (method of exclusion [8]). Auxiliary subroutines are used for calculating the necessary molecular and potential parameters.

Numerical integration along trajectories is made by the methods of Adams-Moulton [9] and Gear [10] using the standard IMSL procedures. All the calculations are performed using double precision with a typical relative accuracy of 10^{-8} and variable integration step within fixed time-grid intervals of $\Delta t = 0.2 \times 10^{-13}$ s. The system of equations turns out to be "stiff" in particular cases (i. e. displaying pronounced difference in typical time-scales for different variables), and in this sense the possibility of switching from the Adams method ("nonstiff" problem) to that of Gear ("stiff" problem) is efficient. The Adams method is usually employed for the nearly straight part of each trajectory at $R > b_{\max}$ ($b_{\max}$ means the maximum impact parameter), while the Gear method is used elsewhere.

The calculations for CO_2–Ar and CO_2–He pairs were made using the *ab initio* potential of Ref. [7]. The chosen parameters of collisions were bracketed within the following ranges: impact parameter (0–12) Å, relative velocity $(0.01–3)\,v_p$; CO_2 rotation frequency $(0–8)\,\omega_p$. Here v_p and ω_p are the most probable Maxwell relative velocity of a pair and Boltzmann rotational frequency of CO_2, respectively. Initial intermolecular separation R_{max} was set equal 15 Å for most of the calculations. Initial orientation and direction of rotation for a CO_2 molecule were selected assuming uniform distribution. Computed trajectories were checked for numerical accuracy making use of back-integrations for selected trajectories and monitoring the total energy conservation. Semiclassical "quantization" procedure [11, 12] was applied to correlate the angular momentum quantum number J of the CO_2 molecule with its rotational frequency ω.

5. Results of simulation

Our calculations showed that in many cases the trajectories deviate strongly from the straight ones. Rotational frequency of a CO_2 molecule undergoes rather complicated variations as a function of collision parameters. In some cases initial and final rotational states of CO_2 (i.e. initial and final rotational quantum numbers J) do not change, despite significant fluctuations in rotational frequency in the course of a collision. Such collisions are classified as elastic or adiabatic. Other collision parameters give rise to initial and final rotational states which differ from each other. These result in inelastic or nonadiabatic collisions. Such collisions cause the transitions between rotational energy levels involved in formation of the spectral line shapes and in the energy relaxation processes.

In the course of a collision the formation of quasibound dimers can occur. The exchange between translational and rotational energy of particles results in regular oscillations in interparticle separation R and in rotational frequency ω of a CO_2 molecule. The lifetime of such quasi-complex can exceed the typical mean duration of a collision by orders of magnitude. Recall that bimolecular Feshbach-type resonances are unstable by definition. These QC are suitable for conversion into stable dimers, however, provided a third particle takes away the energy in excess of the dissociation energy of the complex.

Figure 2 shows some examples of temporal variations of the intermolecular separation R, the elevation angle φ, and the CO_2 rotational frequency ω in quasi-complexes.

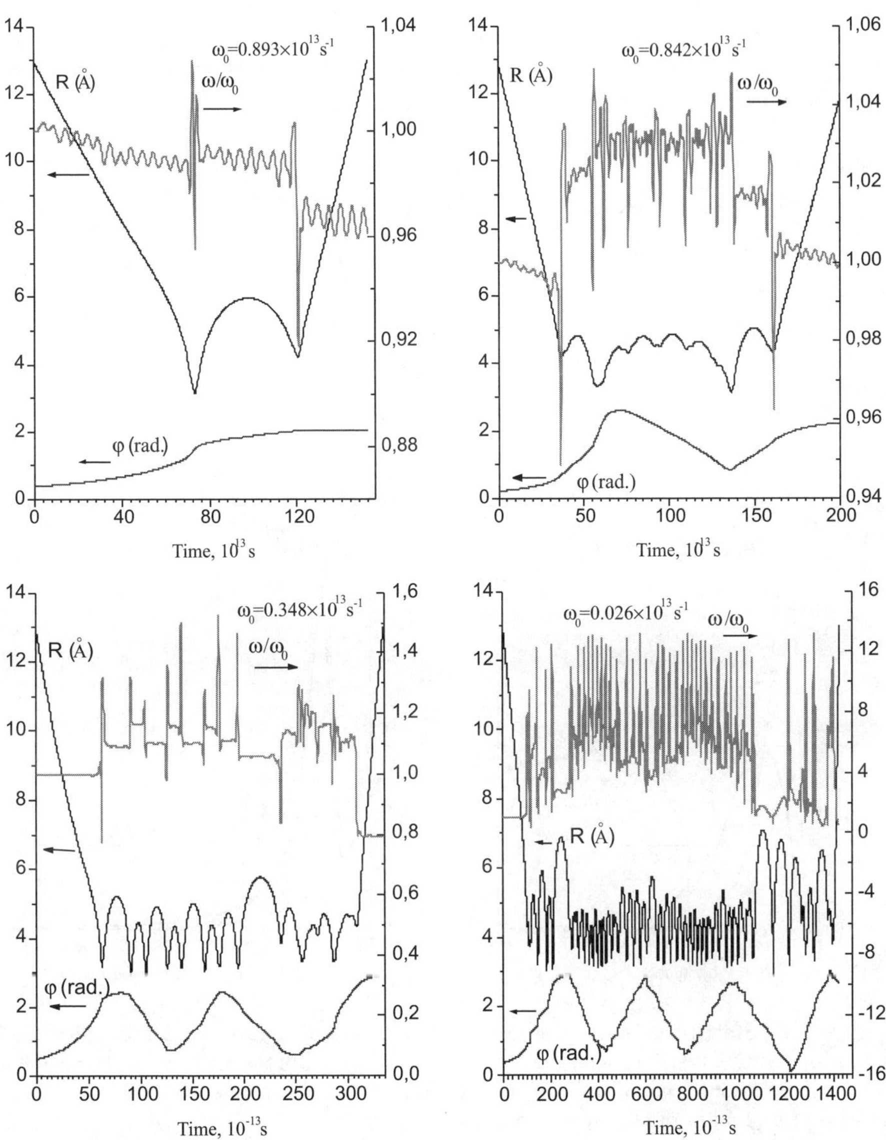

Figure 2. Examples of quasi-complexes formation in CO_2–Ar collisions. R is inter-particle distance, φ is elevation angle (see Fig. 1), ω_0 is initial rotation frequency of CO_2 molecule, $R_{max} = 13$ Å.

Table 1. Statistics of Ar–CO_2 and He–CO_2 collisions in Maxwell-Boltzmann equilibrium at $T = 240$ K. KC is the total number of collisions, KE is the number of elastic collisions, KQ is the number of quasi-complexes formed, KEQ is the number of elastic quasi-complexes, $f_E = $ KE/KC is the fraction of elastic collisions, $f_{QC} = $ KQ/KC is the fraction of quasi-complexes, $f_{EQC} = $ KEQ/KQ is the fraction of elastic quasi-complexes.

Parameter	Ar–CO_2	He–CO_2
KC	360443	2766670
KE	290729	2526670
KQ	11600	7900
KEQ	2847	2700
f_E	80.7%	91.3%
f_{QC}	3.2%	0.29%
f_{EQC}	24.5%	34.2%

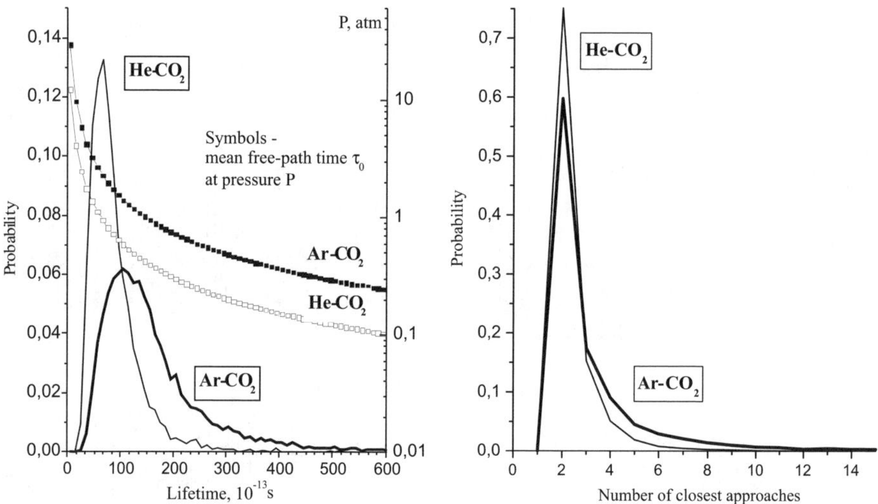

Figure 3. Statistical analysis of QC formation (see also Table 1). Probability distributions of QC lifetime (left) and number of closest approaches (right). $T = 240$ K. Maxwell-Boltzmann equilibrium. The step equals 10^{-12} s in case of a lifetime and 1 in case of a number of closest approaches.

In Figs. 3–5 and in Table 1 some results of the statistical analysis are presented for QC formed in CO_2–Ar and CO_2–He collisions.

The following probability distributions are plotted in Figs. 3–5: the lifetime of a complex, the number of the closest approaches (turning points) of particles, the center-of-mass separations for closest and

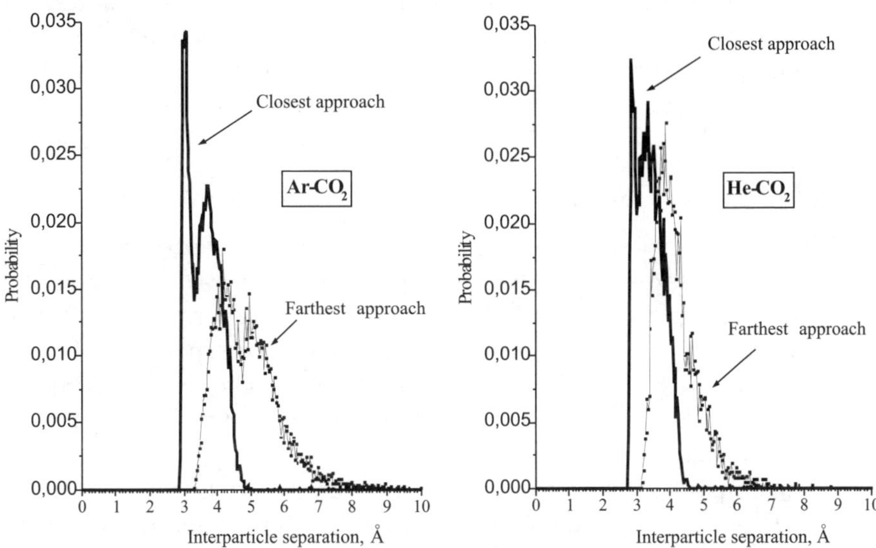

Figure 4. Statistical analysis of QC formation (see also Table 1). Probability distributions of atom-molecule center-of-masses separations in closest and farthest approaches. The step with respect to interparticle separation is 0.03 Å.

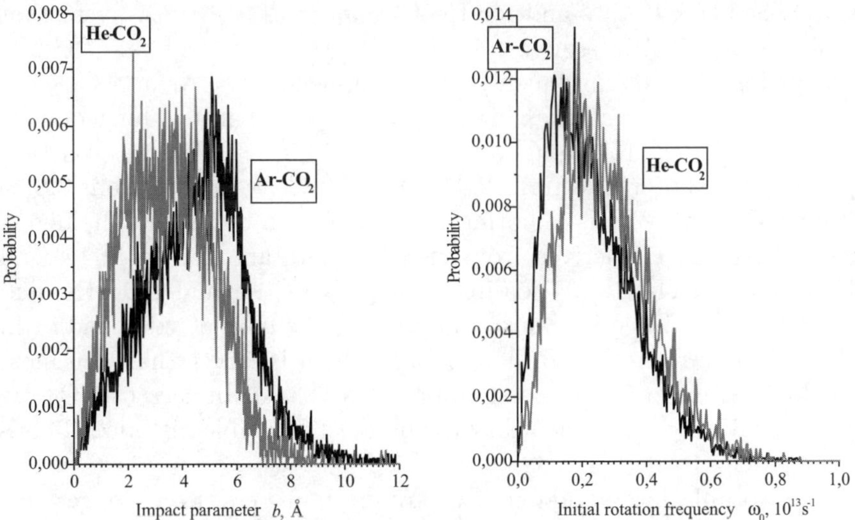

Figure 5. Statistical analysis of QC formation (see also Table 1). Probability distributions of impact parameter (left) and CO_2 initial rotational frequency (right) at which QC are formed. The step with respect to impact parameter is 0.024 Å, it amounts to $0.00344 \times 10^{13}\,\mathrm{s}^{-1}$ for rotational frequency. Most probable Boltzmann CO_2 rotation frequency is $\omega_p = 0.2 \times 10^{13}\,\mathrm{s}^{-1}$.

farthest approaches of particles, the impact parameter, the initial rotational frequency of CO_2. The probability for a parameter x (lifetime, number of turning points, etc.) is defined as $p_i = p(x_i) = N_i/KQ$, where N_i is the number of QC falling into the cell $[x_i, x_i + \Delta x]$ of the distribution, Δx is the step for parameter x and KQ is total number of formed complexes. The condition $\sum_{i=1}^{KQ} p_i = 1$ is obviously fulfilled. Figure 3 shows that QC can be classified in terms of short-lived or very long-lived complexes. Note that at $T = 240\,\mathrm{K}$ the average duration of a collision τ_c for Ar–CO_2 is $27 \times 10^{-13}\,\mathrm{s}$ and for He–CO_2 $\tau_c = 11 \times 10^{-13}\,\mathrm{s}$. The maxima in the QC lifetime distributions correspond to $4\tau_c$ for Ar–CO_2 and to $6\tau_c$ for He–CO_2, respectively. The average time τ_0 between the collisions is also plotted in Fig. 3 as a function of total pressure P.

The comparison of lifetime probability distributions with the function $\tau_0(P)$ allows for clarifying the following important question: in what range of gas pressures the formation of QC can be treated as bimolecular process? It is clear that when a QC lives longer than $\tau_0(P)$ it is on the average affected by a third particle. Hence, at any given pressure P, only QC having lifetimes $\tau < \tau_0(P)$ can be considered as truly bimolecular. Figure 3 demonstrates that for $P = 1\,\mathrm{atm}$ τ_0 for Ar–CO_2 is $147 \times 10^{-13}\,\mathrm{s}$ and for He–CO_2 $\tau_0 = 61 \times 10^{-13}\,\mathrm{s}$. Note that the above values of τ_0 almost coincide with the most probable QC lifetimes in Fig. 3 and, consequently, the long-lived CO_2–Ar and CO_2–He QC can hardly be treated as bimolecular even at $1\,\mathrm{atm}$. The geometry of a QC is subject to strong variations (see Fig. 4). The interparticle separations in QC range from 2.9 to $8\,\mathrm{\AA}$ for CO_2–Ar and from 2.8 to $7\,\mathrm{\AA}$ for CO_2–He. Worth of noting is that the van der Waals diameters for CO_2, Ar and He derived from viscosity data are $3.66\,\mathrm{\AA}$, $4.64\,\mathrm{\AA}$ and $2.19\,\mathrm{\AA}$, respectively [10]. The dimensions of CO_2–Ar and CO_2–He quasicomplexes can be less than or can exceed the sum of vdW radii. In fact for CO_2–Ar one has $R_{vdW} = 4.15\,\mathrm{\AA}$ which is larger than the most probable distance of the closest approach ($\approx 3\,\mathrm{\AA}$). In case of CO_2–He $R_{vdW} = 2.93\,\mathrm{\AA}$, which coincides with the most probable distance of closest approach.

The notable interaction in CO_2–Ar system persists at longer distances than that in CO_2–He. Therefore, more effective formation of QC can be expected for the former system. The calculations made for identical conditions (see Figs. 3–5) gave nearly the same fraction of elastic collisions (91% for CO_2–He and 81% for CO_2–Ar, see Table 1) but very different fractions of formed QC (0.3% for CO_2–He and 3.2% for CO_2–Ar). The most probable impact parameters for QC formation

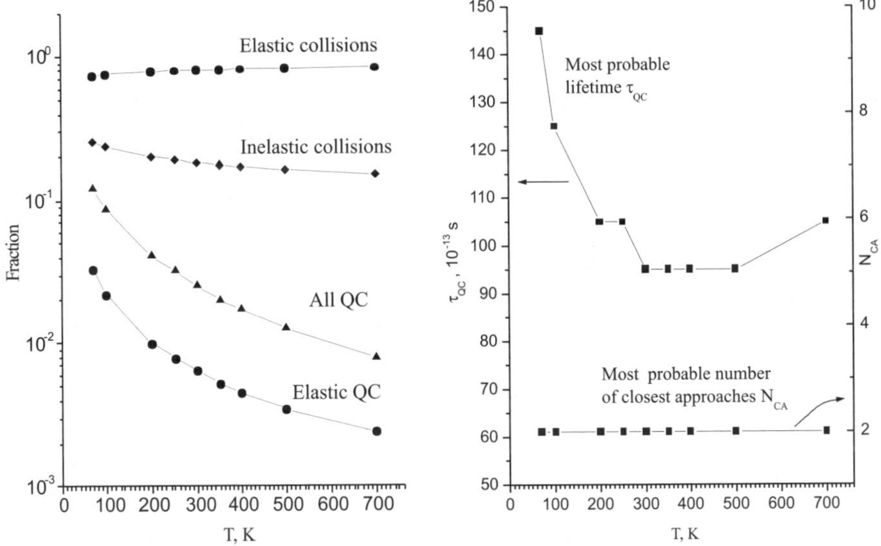

Figure 6. Temperature dependences of the fractions of Ar–CO_2 collisions of different types (left). Temperature dependencies of the most probable QC lifetime and number of closest approaches in QC (right). Maxwell-Boltzmann equilibrium.

are approximately $3\,\text{Å}$ for CO_2–He and $5\,\text{Å}$ for CO_2–Ar collisions (see Fig. 5). Figure 5 shows that the most probable initial rotation frequency of CO_2 for QC formation is close to Boltzmann value ω_p for the case of CO_2–He but is significantly smaller than that for CO_2–Ar system.

Figure 6 demonstrates some results of computations for CO_2–Ar collisions at different temperatures. It is seen that the fraction of elastic collisions rises slowly with temperature. This results obviously from the speeding-up of the CO_2 rotation which makes the interaction potential more "spherical" as well as from the shortening of the collision duration. Similar reasons explain the reduced QC formation resulting in the rapid decrease of the QC fraction as temperature is increasing. The calculations showed that the most probable lifetime depends distinctly on temperature (see Fig. 6). At any given temperature the number of the most probable approaches equals two.

Figures 7 and 8 display the cross sections for CO_2 rotational transitions in collisions with Ar and He atoms for two different temperatures.

The relevant statistics is collected in Table 2. It is clear from Figs. 7 and 8 that the contribution of inelastic QC in the total CO_2 rotational inelastic cross sections is relatively small. However, rotational relaxation for QC-forming collisions were found to be much greater than those for

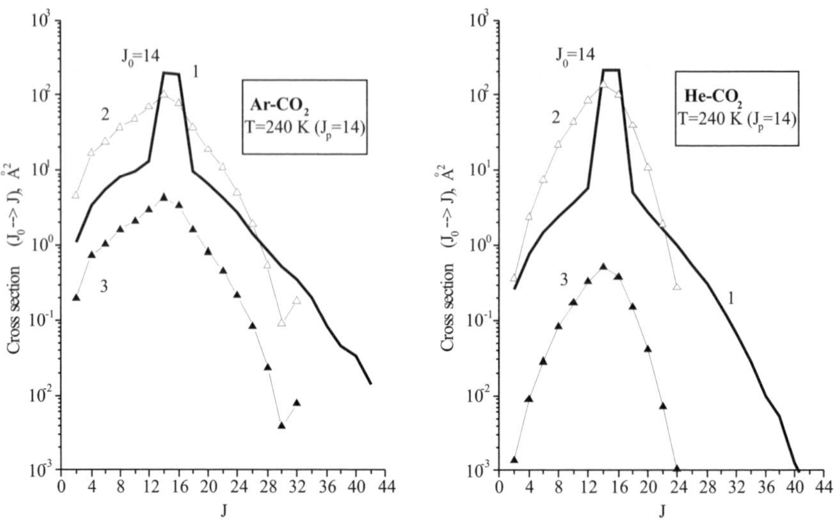

Figure 7. Cross sections for CO_2 rotational $J_0 \to J$ transitions in collisions with Ar and He atoms. Maxwell equilibrium at $T = 240\,\text{K}$. All collisions (1), QC-forming collisions only (2), and contribution of QC into total cross section (3), which is the curve (2) multiplied by the factor of (KQ-KEQ)/(KC-KE). Only even J values are allowed. J_p is most probable Boltzmann rotational quantum number corresponding to ω_p at temperature T.

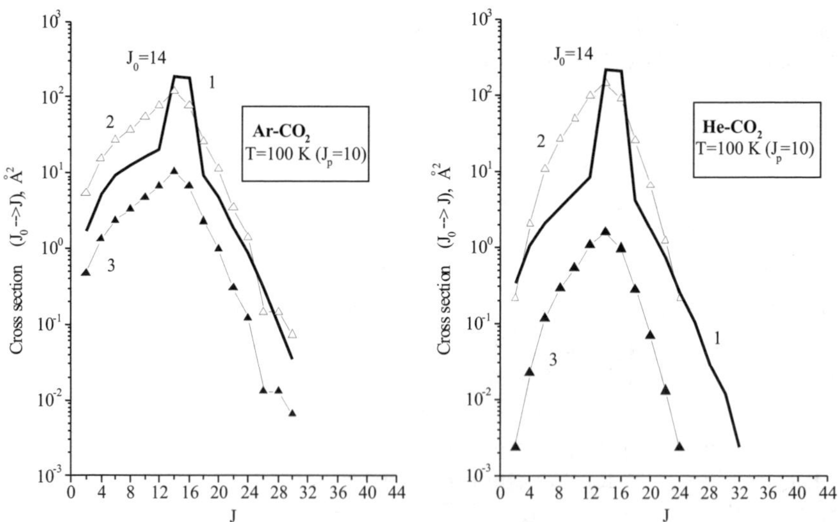

Figure 8. Cross sections for CO_2 rotational $J_0 \to J$ transitions in collisions with Ar and He atoms at $T = 100\,\text{K}$. All other parameters and designations are the same as in Fig. 7.

Table 2. Statistics of Ar–CO$_2$ and He–CO$_2$ collisions at two temperatures for the case of fixed CO$_2$ rotational quantum number $J_0 = 14$. Maxwell equilibrium. For other notations see the caption to Table 1.

Parameter	$T = 240\,\mathrm{K}$		$T = 100\,\mathrm{K}$	
	Ar–CO$_2$	He–CO$_2$	Ar–CO$_2$	He–CO$_2$
KC	161422	1755539	89950	778491
KE	69609	831497	38013	373620
KQ	5020	5000	6200	6440
KEQ	1107	1508	1635	2080
f_E	43.1%	47.4%	42.3%	48.0%
f_{QC}	3.1%	0.3%	6.9%	0.8%
f_{EQC}	22.1%	30.2%	26.4%	32.3%

conventional inelastic collisions. Because of the stronger interaction in CO$_2$–Ar system the collisions are more efficient for $|\Delta J| > 2$ transitions as compared to those in CO$_2$–He. The decrease of a gas temperature leads to small rise of cross-sections for $\Delta J < -2$ and to a significant drop of cross sections for $\Delta J > 2$, which prevents populating higher rotational levels. Note that rotational relaxation cross-sections in QC-forming collisions are quite stable to temperature changes for CO$_2$–He at all ΔJ and for CO$_2$–Ar if $\Delta J < -2$.

6. Conclusions

The main conclusions drawn from the above considerations can be formulated as follows.

- Quasibound collision complexes of CO$_2$–Ar and CO$_2$–He can be characterized by short, intermediate, or very long lifetimes. The lifetime of QC ranges from several typical duration of a collision τ_c to hundreds of τ_c. The most probable QC lifetimes amount to $4\tau_c$ for Ar–CO$_2$ and to $6\tau_c$ for He–CO$_2$, respectively;

- Mean free-path times τ_0 between Ar–CO$_2$ and He–CO$_2$ collisions at 1 atm are on the order of the most probable QC lifetimes. Consequently, the long-lived CO$_2$–Ar and CO$_2$–He QC can hardly be treated as bimolecular at pressures in excess of 1 atm;

- Collision complexes strongly vary in their geometrical dimensions. Typically the latter range from 2.9 to 8 Å for CO$_2$–Ar and from 2.8 to 7 Å for CO$_2$–He;

- The interaction between CO_2 and Ar is stronger and more efficient in QC formation than that between CO_2 and He. The fractions of QC formed in identical initial conditions are very different (at $T = 240\,K$ 3.2% for CO_2–Ar and 0.3% for CO_2–He);

- The fraction of QC in the total number of collisions rapidly decreases as the temperature increases. QC-forming collisions are mainly inelastic. In Maxwell-Boltzmann equilibrium at $240\,K$ the fraction of elastic QC (with respect to the total QC number) is only 25% for CO_2–Ar and 34% for CO_2–He;

- The contribution of QC to the total CO_2 rotational inelastic cross sections at $240\,K$ and $100\,K$ is relatively small. However, rotational relaxation via QC-forming collisions is much more efficient than via conventional inelastic collisions.

Acknowledgments

The author is greatly indebted to A. P. Gal'tsev who inspired this work for a long time. Special thanks to V. V. Tsukanov for computer consulting and hospitality. Particular thanks to A. A. Vigasin for keen interest and fruitful discussions. Partial support from the INTAS–Airbus Grant 1815 is gratefully acknowledged.

References

[1] Miller, W. (1971) Quasiclassical character of atomic and molecular collisions. In: B. Alder, S. Fernbach, and M. Rotenberg, editors. *Methods in Computational Physics. Advances in Research and Applications*, Vol. 10, Atomic and Molecular Scattering, Academic Press, New York and London.

[2] Gal'tsev, A. P. and Tsukanov, V. V. (1977) Investigation of spectral lines contour by numerical method, Optika i Spectroscopiya, 42, 1063–1069 (in Russian).

[3] Fitz, D. E. and Brumer, P. (1979) Geometric effects on complex formation in collinear atom-diatom collisions, J. Chem. Phys., 70, 5527–5533.

[4] Babcock, L. M. and Thompson, D. L. (1983) Dynamics of association and decay. A model study of $Cl^- + Cl_2 \rightleftarrows Cl_3^{-*}$ using quasiclassical trajectories, J. Chem. Phys., 78, 2394–2401.

[5] Marković, N. and Nordholm, S. (1989) Trajectory study of collision complex formation. Weak to strong coupling transition in atom–diatomic molecule collisions, Chem. Phys., 134, 69–84.

[6] Pattengill, M. D. (1977) On the use of body-fixed coordinates in classical scattering calculations: planar trajectory approximation for rotational excitation, J. Chem. Phys., 66, 5042–5045.

[7] Parker, G. A., Snow, R. L., and Pack, R. T. (1976) Intermolecular potential surfaces from electron gas methods. I. Angle and distance dependence of the He–CO_2 and Ar–CO_2 interactions, J. Chem. Phys., 64, 1668–1678.

[8] Bird, G. A. (1976) *Molecular Gas Dynamics*, Clarendon Press, Oxford.

[9] Korn, G. A. and Korn, T. M. (1968) *Mathematical Handbook for Scientists and Engineers*, McGraw-Hill Book Company, New York.

[10] Gear, C. W. (1971) *Numerical Initial Value Problem in Ordinary Differential Equations*, Englewood Cliffs: Prentice Hall, N.J.

[11] Pattengill, M. D. (1975) A comparison of classical trajectory and exact quantal cross sections for rotationally inelastic Ar–N_2 collisions, Chem. Phys. Lett., 36, 25–28.

[12] Nyeland, C. and Billing, G. D. (1978) Rotational relaxation of homonuclear diatomic molecules by classical trajectory computations, Chem. Phys., 30, 401–406.

THEORETICAL STUDY OF INTERACTION POTENTIAL AND PRESSURE BROADENING OF SPECTRAL LINES FOR THE He–CH$_3$F COMPLEX

B. Bussery-Honvault, J. Boissoles

Laboratoire PALMS, UMR CNRS 6627, Université de Rennes 1,
Campus de Beaulieu, 35042 Rennes Cedex, France

R. Moszynski

Department of Chemistry, University of Warsaw,
Pasteura 1, 02-093 Warsaw, Poland

Abstract   Pressure-broadening coefficients of spectral lines of CH$_3$F in a helium bath are evaluated by solving the close-coupling equations for the $J = 0$–1, $K = 0$ transition of CH$_3$F. For that, the energy surface of the interaction potential of the CH$_3$F–He complex has been calculated *ab initio* with the symmetry-adapted perturbation theory (SAPT), with CH$_3$F kept rigid. A global 3D fit on the *ab initio* points in Jacobi coordinates have been done for the dynamics. Good agreement is observed between present pressure broadening coefficients and experimental data of De Lucia and collaborators (J. Chem. Phys. 97 (1992) 4723).

Keywords: collisional broadening / potential energy surface / rotational spectra of dimer / methyl fluoride and helium / band shapes / linewidths / intermolecular interactions / symmetry-adapted perturbation theory (SAPT)

Over the past decade van der Waals complexes of methyl fluoride with model collisional partners, such as the rare gas atoms, attracted significant experimental and theoretical interest. Recent experiments were mainly focused on the collisional broadening of the rotational spectral lines of CH$_3$F perturbed by the helium gas. Consequently, pressure broadening and shifting coefficients for this system are known for a few

65

C. Camy-Peyret and A.A. Vigasin (eds.),
Weakly Interacting Molecular Pairs: Unconventional Absorbers of Radiation in the Atmosphere, 65–71.
© 2003 *Kluwer Academic Publishers. Printed in the Netherlands.*

spectroscopic transitions [1–6] Theoretical modelling of the line broadening coefficients for He–CH_3F are scarce, and limited to simple models [6, 7] with parameters fitted to the observed spectral parameters. Such a theoretical approach, although useful to predict band shapes and linewidths, does not give any information on the interactions in the He–CH_3F complex that are probed by the experiment. In the present communication we report *ab initio* calculations of the intermolecular interactions in the He–CH_3F complex and the pressure broadening coefficients of the microwave lines of CH_3F in the helium bath at various temperatures.

Our calculations of the potential energy surface for the He–CH_3F complex employ the symmetry-adapted perturbation theory (SAPT) [8, 9], and follow the approach introduced and tested in previous papers (see e. g. Ref. [10]). The interaction energy was computed from the following expression:

$$E_{\text{int}}^{\text{SAPT}} = E_{\text{elst}}^{(1)} + E_{\text{ind}}^{(2)} + E_{\text{disp}}^{(2)} + E_{\text{exch}}, \tag{1}$$

where the consecutive terms on the r. h. s. of Eq. (1) denote the electrostatic, induction, dispersion, and exchange energies, respectively. See Ref. [12] for a more detailed description of the computational approach. The SAPT calculations were done with the program SAPT [11].

The intermolecular potential energy surface for the He–CH_3F system, where CH_3F is kept rigid, can be naturally described in the Jacobi coordinates (R, ϑ, ϕ), where R is the distance from the center of mass of CH_3F to the He atom, ϑ is the angle between the C–F axis and the vector pointing from the center of mass of CH_3F to He, and ϕ is the rotation angle around the C–F axis. The angle $\vartheta = 0°$ corresponds to the linear geometry H_3CF–He with the He atom lying on the F side of the C–F axis, while the $\phi = 0°$ geometry corresponds to a structure with the He atom lying in the plane defined by the H, C, and F atoms. In all calculations the CH_3F molecule was kept at its experimental equilibrium geometry [13].

We performed a global 3D fit of the 225 computed points to the following expression:

$$E_{\text{int}}(R, \vartheta, \phi) = \sum_{L=0}^{14} \sum_{M=0, 3, \dots}^{L} v_{LM}(R) C_L^M(\vartheta, \phi), \tag{2}$$

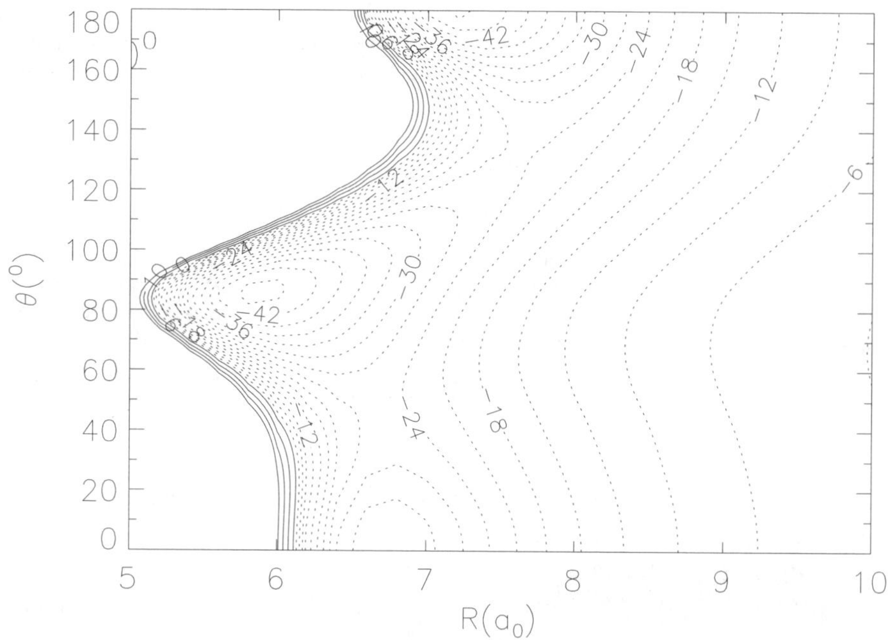

Figure 1. Contour plot (in cm^{-1}) of the He–CH$_3$F potential energy surface as a function of R and ϑ at $\phi = 60°$, passing through the global minimum. Contours are spaced by $3\,cm^{-1}$ with negative values labeled by dotted lines.

where $C_L^M(\vartheta, \phi)$ denotes the tesseral harmonics [14], and the radial coefficients $v_{LM}(R)$ are given by:

$$v_{LM}(R) = A_{LM} \exp\left(-\alpha_{LM} R - \beta_{LM} R^2\right) - \sum_{n=6}^{10} \sum_{LM} \frac{C_n^{LM}}{R^n} f_n(\widetilde{\beta}, R). \quad (3)$$

The long-range coefficients C_n^{LM} corresponding to the multipole-expanded induction and dispersion energies [14] were computed independently, at the same level of theory and with the same basis sets, by means of the POLCOR package [15]. Other parameters appearing in Eq. (3) were fitted to the computed points. The damping function $f_n(\widetilde{\beta}, R)$ was assumed in the Tang-Toennies form Ref. [16]. The main features of the potential energy surface can be observed in Fig. 1, showing a contour plot of the 3D surface at $\phi = 60°$. An inspection of this figure shows that the computed energy surface has a global minimum of $-48.9\,cm^{-1}$ at the center of mass separation of $7.22\,$bohr with the helium atom lying along the C–F bond on the hydrogen's side (i.e. $\vartheta = 180°$). In addition two local minima were found. They correspond

Table 1. The minima and saddle points of the He–CH$_3$F interaction potential, and the electrostatic, induction, dispersion, and exchange contributions (in cm^{-1}) to the interaction energies at these points.

Extrema	R, a_0	ϑ, °	ϕ, °	$E^{(1)}_{\text{elst}}$	$E^{(2)}_{\text{ind}}$	$E^{(2)}_{\text{disp}}$	E_{exch}	$E^{\text{SAPT}}_{\text{int}}$
Global minimum	7.22	180.0	–	−13.5	−5.5	−100.4	70.5	−48.9
Local minimum (I)	5.87	86.0	60	−11.6	−5.2	−89.8	61.9	−44.7
Local minimum (II)	6.73	0.0	–	−8.4	−6.24	−62.5	44.9	−32.2
Saddle point (I)	7.59	132.6	60	−6.1	−2.3	−51.6	34.8	−25.1
Saddle point (II)	6.69	40.6	0	−6.1	−4.7	−49.4	35.2	−25.0
Saddle point (III)	6.00	77.0	20	−10.7	−5.6	−84.1	58.1	−42.4

to geometries of the complex with the He atom lying along the C–F bond on the fluorine side (i. e. $\vartheta = 0°$) and with the He atom almost perpendicular to the C–F bond ($\vartheta = 86°$) in the middle of two C–H planes (i. e. $\phi = 60°$, see Table 1). It is worth noting that the global minimum corresponds to a geometry of the complex with the maximal (in the absolute value) attraction. All these minima are separated by small barriers whose locations are given in Table 1 as well.

The first two saddle points of Table 1 correspond to potential barriers when going from the global minimum to the first local minimum and from the first local minimum to the second one. The third saddle point corresponds to potential barrier when the dimer is in the first local minimum with the He atom rotating around the C–F bond. For all these geometries the dispersion energy and exchange-repulsion term represent the two major contributions to the interaction energy. Their sum reproduces over 60–70% of the total interaction energies. So, the dispersion interaction together with the valence-repulsion govern the most important features of the potential energy surface. Other terms, i. e. the attractive electrostatics and induction contribute around 30–40% of the total interaction energies.

Once the potential energy surface is obtained one can proceed to the next step of our study, and compute the pressure broadening coefficients. If a thermal (Maxwellian) distribution for the translational motions of the collisional partners is assumed, the pressure broadening coefficients γ_0 can be expressed as:

$$\gamma_0 = \frac{<v\,\sigma_{ii,ff}>}{2\pi c\,k_{\text{B}}\,T}\,, \tag{4}$$

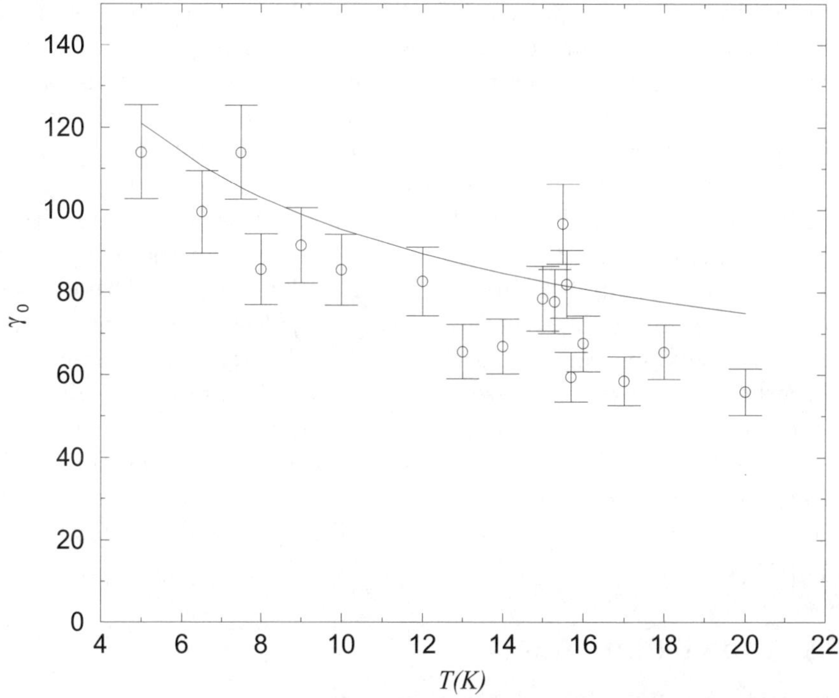

Figure 2. Comparison of the theoretical and experimental low-temperature pressure broadening coefficients for the $(j,k) = (0,0) \rightarrow (1,0)$ microwave transition of CH_3F in the helium bath. The experimental data are taken from Ref. [3].

where the average $< \cdots >$ over the relative velocity distribution is defined as:

$$<v\,\sigma_{ii,ff}> = \left(\frac{8\,k_B\,T}{\pi\mu}\right)^{1/2} \int_0^\infty d\left(\frac{E_{CM}}{k_B T}\right) \frac{E_{CM}}{k_B T}\, e^{-E_{CM}/k_B T} \sigma_{ii,ff}(E_{CM}), \quad (5)$$

where v denotes the relative velocity, $\sigma_{ii,ff}$ is the state-to-state integral generalized (pressure broadening) cross section, k_B is the Boltzmann constant, μ is the reduced mass, c is the speed of light, and E_{CM} stands for the collisional energy. The generalized cross sections $\sigma_{ii,ff}$ are computed directly from the S matrix elements [17], which in turn are obtained by solving the close-coupling (CC) equations [17] for He–CH_3F. Finally, the pressure broadening coefficients γ_0 are obtained from the Boltzmann average according to Eq. (5). The MOLSCAT system of codes [18] was applied to determine the solutions of the close-coupling equations. The results of the close-coupling calculations of the pressure broadening coefficients for the $(j,k) = (0,0) \rightarrow (1,0)$ transition of CH_3F

perturbed by helium are presented in Fig. 2. Also reported in this figure are the experimental data of Willey *et al.* [3]. An inspection of Fig. 2 shows that the agreement between the theory and experiment is rather good. For several temperatures the theoretical result lies well within the experimental error bars. However, the experimental data show oscillations that are not seen on the theoretical curve. In Ref. [4] these oscillations were attributed to Feshbach resonant states of the He–CH$_3$F complex. Our generalized cross sections as functions of the energy also show oscillations that can directly be related to the He–CH$_3$F resonances, but the resonant structure of the $\sigma_{ii,ff}(E_{\mathrm{CM}})$ curve disappears after the Boltzmann average. At this point of this study it is difficult to say whether the existing disagreement between the present results and the experimental data [4] is due to some inaccuracies in the calculations or in the measurements. A more detailed investigation of the He–CH$_3$F pressure broadening data for other spectroscopic lines and for a wider range of temperatures is necessary, and will be reported in a forthcoming paper [12].

Acknowledgments

This work was supported by the Polish Scientific Research Concil (KBN) within the grant 4 T09A 071 22.

References

[1] Willey, D. R., Goyette, T. M., Ebenstein, W. L., Bittner, D. N., and De Lucia, F. C. (1989) Collisionally cooled spectroscopy: Pressure broadening below 5 K, J. Chem. Phys., 91, 122–125.

[2] Willey, D. R., Bittner, D. N., and De Lucia, F. C. (1989) Very low temperature spectroscopy: The Helium pressure broadening coefficients below 4.3 K for the higher lying states of CH$_3$F, J. Molec. Spectrosc., 133, 182–192.

[3] Willey, D. R., Choong, V.-E., Goodelle, J. P., and Ross, K. A. (1992) Collisional cooling between 5 and 20 K: Low-temperature helium pressure broadening of CH$_3$F, J. Chem. Phys., 97, 4723–4726.

[4] Beaky, M. M., Flatin, D. C., Holton, J. J., Goyette, T. M., and De Lucia, F. C. (1995) Hydrogen and helium pressure broadening of CH$_3$F between 1 K and 600 K, J. Molec. Struct., 352/353, 245–251.

[5] Grigoriev, I. M., Bouanich, J. P., Blanquet, G., Walrand, J., and Lepère, M. (1997) Diode-laser measurements of He-broadening coefficients in the ν_6 Band of CH$_3$F, J. Molec. Spectrosc., 186, 48–53.

[6] Grigoriev, I. M., Le Doucen, R., Benidar, A., Filippov, N. N., and Tonkov, M. V. (1997) Line-mixing effects in the ν_3 parallel absorption band of CH$_3$F perturbed by rare gases, J. Quant. Spectrosc. Radiat. Transfer, 58, 287–299.

[7] Thibault, F., Boissoles, J., Grigoriev, I. M., Filippov, N. N., and Tonkov, M. V. (1999) Line mixing effects in the ν_3 band of CH_3F in helium: Experimental band shapes and ECS analysis, Eur. Phys. J. **D**, 6, 343–353.

[8] Jeziorski, B., Moszynski, R., and Szalewicz, K. (1994) Perturbation theory approach to intermolecular potential energy surfaces of van der Waals complexes, Chem. Rev., 94, 1887–1930.

[9] Moszynski, R., Wormer, P. E. S., and van der Avoird, A. (2000) Symmetry adapted perturbation theory applied to the computation of intermolecular forces. In: *Computational Molecular Spectroscopy*, edited by P. R. Bunker and P. Jensen, Wiley, New York, pp. 69–108.

[10] Korona, T., Moszynski, R., Thibault, F., Launay, J.-M., Bussery-Honvault, B., Boissoles, J., and Wormer, P. E. S. (2001) Spectroscopic, collisional, and thermodynamic properties of the He–CO_2 complex from an *ab initio* potential: theoretical predictions and confrontation with the experimental data, J. Chem. Phys., 115, 3074–3084.

[11] Jeziorski, B., Moszynski, R., Ratkiewicz, R. A., Rybak, S., Szalewicz, K., and Williams, H. L. (1993) SAPT: a programm for many-body symmetry adapted perturbation theory calculations of intermolecular interaction energies. In: *Methods and Techniques in Computational Chemistry: METECC94*, edited by E. Clementi, STEF, Cagliari, Vol. B, pp. 79–129.

[12] Bussery-Honvault, B., Boissoles, J., and Moszynski, R. to be published.

[13] Demaison, J., Breidung, J., Thiel, W., and Papoušek, D. (1999) The equilibrium structure of methyl fluoride, Struct. Chem., 10, 129–133.

[14] Bulski, M., Wormer, P. E. S., and van der Avoird, A. (1991) *Ab initio* potential energy surfaces of Ar–NH_3 for different NH_3 umbrella angles, J. Chem. Phys., 94, 491–500.

[15] Wormer, P. E. S. and Hettema, H. (1992) POLCOR *package*, University of Nijmegen, The Netherlands.

[16] Tang, K. T. and Toennies, J. P. (1984) An improved simple model for the van der Waals potential based on universal damping functions for the dispersion coefficients, J. Chem. Phys., 80, 3726–3741.

[17] Green, S. (1979) Rotational excitation of symmetric top molecules by collisions with atoms. II. Infinite order sudden approximation, J. Chem. Phys., 70, 816–829.

[18] Hutson, J. M. and Green, S. (1994) MOLSCAT *computer code, version 14*, distributed by Collaborative Computational Project No. 6 of the Science and Engineering Research Council (UK).

VARIATIONAL SOLUTION OF ANHARMONIC VIBRATIONAL PROBLEMS FOR POLYATOMICS AND MOLECULAR PAIRS

A. I. Pavlyuchko,* B. S. Orlinson

Volgograd State Technical University,
Lenina 28, 400131, Volgograd, Russia

A. A. Vigasin

Wave Research Center,
General Physics Institute, Russian Academy of Sciences,
Vavilova 38, Moscow 119991, Russia

Abstract   This paper illustrates the use of computer codes for the variational solution of anharmonic vibrational-rotational problems for polyatomic molecules and molecular pairs. The computational procedure starts from internal curvilinear coordinates and the Rayleigh-Ritz variational method. A mixed Morse-harmonic basis, a purely harmonic basis, and an extended harmonic basis can be used. Various techniques for taking into account kinematic and dynamic anharmonicity are used.

Keywords: polyatomic molecules / molecular pairs / anharmonic vibrational motions / variational methods / kinematic and dynamic anharmonicity / curvilinear coordinates / computer codes

1. General outline

This paper aims at developing variational methods and computer codes for calculating anharmonic vibrations in polyatomic molecules. This problem is written using internal curvilinear coordinates and solved by the Rayleigh-Ritz variational method [1] characteristic for our approach to anharmonic vibrational problem.

---

*Corresponding author: `pavl@vstu.ru`

C. Camy-Peyret and A.A. Vigasin (eds.),
Weakly Interacting Molecular Pairs: Unconventional Absorbers of Radiation in the Atmosphere, 73–82.
© 2003 *Kluwer Academic Publishers. Printed in the Netherlands.*

The vibrational Hamiltonian is assumed to have the form

$$\hat{H}_v = -\frac{\hbar^2}{2} \sum_{i,j} \frac{\partial}{\partial q_i} \tau_{ij}(q) \frac{\partial}{\partial q_j} + T(q) + V(q),$$

where q_i are valence vibrational coordinates, $\tau_{ij}(q)$ are the elements of the kinematic coefficients matrix, $T(q)$ is the so-called nondifferential kinematic operator, and $V(q)$ is the potential energy operator. The dependence of the $\tau_{ij}(q)$ kinematic coefficients and the $T(q)$ nondifferential kinematic operator on valence coordinates q results in the so-called kinematic anharmonicity. Various techniques [1] for taking into account kinematic anharmonicity have been used. It was shown in [1] that the contribution of the $T(q)$ nondifferential kinematic operator to the values of the vibrational energy levels is small and can therefore be ignored.

The presence of terms higher than quadratic in the expansion of $V(q)$, the potential energy operator, with respect to vibrational coordinates gives rise to the so-called dynamic anharmonicity. For polyatomic molecules and molecular pairs, the potential energy operator can be written as a sum of Morse functions for terminal XH bonds r_i and of a fourth-order polynomial for bending and skeletal vibrations q_i

$$V(q) = \frac{1}{2} \sum_{i,j} D_{ij}\, x_i\, x_j + \frac{1}{6} \sum_{i,j,k} D_{ijk}\, x_i\, x_j\, x_k + \frac{1}{24} \sum_{i,j,k,l} D_{ijkl}\, x_i\, x_j\, x_k\, x_l,$$

where

$$x_i = \begin{cases} 1 - e^{-\alpha_i r_i} \\ q_i \end{cases}.$$

The variational problem can then be solved most efficiently if the variational functions χ are represented [1] in terms of the products

$$\chi_{kn} = \prod_i \phi_{k_i}(r_i) \prod_s \psi_{n_s}(Q_s),$$

where $\phi_{k_i}(r_i)$ are the eigenfunctions of the Morse oscillator for terminal bond vibrations, and $\psi_{n_s}(Q_s)$ are the harmonic oscillator functions pertaining to skeletal and bending modes. The Q_s coordinates for skeletal and bending modes are introduced as a linear combination of valence vibrational coordinates diagonalizing the corresponding harmonic component of the total Hamiltonian.

Intensities of vibrational transitions are calculated using valence-optical scheme, which presupposes the dipole moment of a molecule be expanded as a sum of dipole moments characteristic to individual bonds

$$\vec{\mu}(q) = \vec{\mu}(0) + \sum_{i,j} \left(\mu_i \left(\frac{\partial \vec{e}_i(q)}{\partial q_j} \right)_0 + \frac{\partial \mu_i}{\partial q_j} \vec{e}_i(0) \right)$$
$$+ \frac{1}{2} \sum_{i,j,k} \left(\mu_i \left(\frac{\partial^2 \vec{e}_i(q)}{\partial q_j \, \partial q_k} \right)_0 + \frac{\partial \mu_i}{\partial q_j} \left(\frac{\partial \vec{e}_i(q)}{\partial q_j} \right)_0 + \frac{\partial^2 \mu_i}{\partial q_j \, \partial q_k} \vec{e}_i(0) \right),$$

where $\vec{e}_i(q)$ and $\left(\frac{\partial \vec{e}_i(q)}{\partial q_j} \right)_0$, $\left(\frac{\partial^2 \vec{e}_i(q)}{\partial q_j \partial q_k} \right)_0$ are, respectively, directing vectors for valence bonds and their derivatives over valence coordinates at equilibrium and μ_i, $\frac{\partial \mu_i}{\partial q_j}$, $\frac{\partial^2 \mu_i}{\partial q_j \partial q_k}$ are the electro-optical parameters.

In our computer codes [1], a mixed Morse-harmonic basis, a purely harmonic basis, and an extended harmonic basis can be used. The number of atoms in a molecule as well as the number of valence vibrational coordinates and basis functions are only limited by the available memory. At present, we are in a position to solve vibrational problems for molecules containing several dozens of atoms, several hundreds of valence vibrational coordinates, and several dozen thousands of variational functions [1]. The capture of the molecule structure, the data input, and the analysis of results obtained are performed with the use of dedicated graphical user interface.

The programs also offer the possibility of calculating the rotational structure of the different vibrational energy levels. These calculations are necessary for the estimation of IR absorption band intensities of vibrational transitions and for the simulation of the corresponding spectral profiles. The starting approximation for the potential and electro-optical functions can be found from *ab initio* calculations which can be refined then by solving the inverse mechanical and electro-optical problems making use of the available experimental vibrational frequencies and absorption band intensities.

In [1], direct and inverse mechanical problems are solved variationally for the molecules H_2O, H_2Se, H_2S, NO_2, O_3, SO_2, ClO_2, HCN, C_2H_2, CO_2, N_2O, NH_3, H_2CO, and C_2H_4 and their isotopomers with the use of large mixed Morse—anharmonic basis sets, which ensured accurate calculations of their vibrational energy levels and potential function parameters. This allowed us to compare the calculated bond dissociation energies with the experimental bond rupture energies in the gas phase.

Table 1.　Electro-optical parameters for the CO_2 monomer and $(CO_2)_2$ dimer.

Parameter*	*Ab initio* calc. 6-31G(2d,2p)/MP2		Empirical value	Units
	CO_2	$(CO_2)_2$	$(CO_2)_2$	
μ_{r_i}	-1.3325	-1.5129	-1.5129	D
μ_{r_o}	-1.3325	-1.2316	-1.2316	D
$\partial\mu_{r_i}/\partial r_i$	-2.4736	-2.5410	-2.3700	D/Å
$\partial\mu_{r_o}/\partial r_i$	2.4736	2.5410	2.3700	D/Å
$\partial\mu_{r_i}/\partial r_o$	2.4736	2.3822	2.5700	D/Å
$\partial\mu_{r_o}/\partial r_o$	-2.4736	-2.3822	-2.5700	D/Å
$\partial^2\mu_{r_i}/\partial r_i^2$	4.4950	4.2500	4.0000	D/Å^2
$\partial^2\mu_{r_o}/\partial r_i^2$	-4.4950	-4.2500	-4.0000	D/Å^2
$\partial^2\mu_{r_i}/\partial r_o^2$	-4.4950	-4.3645	-5.0000	D/Å^2
$\partial^2\mu_{r_o}/\partial r_o^2$	4.4950	4.3645	5.0000	D/Å^2

*The bonds r_i and r_o look inside and outside a dimer, respectively.

Useful general conclusions can be drawn from our anharmonic vibrational calculations which are briefly summarized below:

- The calculated dissociation energy of a bond correlates well with estimates drawn from experiments in the gas phase. As one could expect, the former are somewhat larger than the latter;

- The Morse function for XH bonds differs somewhat from the real potential function at energies in the vicinity of dissociation and leads to an overestimate of the calculated energy values. The difference between the calculated CH bond dissociation energies and the energies of the CH bond in the gas phase does not change as function of a molecule containing these bonds;

- Independent solution of the inverse spectral problem for molecules containing CH bonds gives rise to the same value of parameter α describing the "half-width" of the Morse potential reasonably well;

- For all the molecules listed above, the anharmonic character of stretching vibrations up to overtone energies of about one half of the dissociation energy is well described by a single Morse function;

- The anharmonic character of all molecular vibrations is mainly determined by deviation of the well shape from harmonic one for the stretching vibrations and by the kinematic anharmonicity for bending vibrations. The anharmonic character of bending vibrations i. e. the difference between the shape of the corresponding potential well and that of an harmonic potential, is not pronounced in most cases, as is evidenced by the values obtained for the coefficients of the potential function;

- For all the molecules listed above, the potential energy function was derived from experimental data pertaining to low-level vibrational transitions (not higher than second overtones and combination tones) for one or two isotopomers. These functions containing a minimum set of anharmonic parameters describe a good number of high-lying (i. e. having quantum numbers corresponding to one half of the potential well depth) vibrational transitions (virtually the complete observed spectrum) in a satisfactory manner. The only exceptions are HCN and C_2H_2, for which these functions failed to describe highly excited bendings.

Using these computer codes we have simulated the IR absorption spectra for the $(CO_2)_2$ dimer and for two structural isomers of CO_2–H_2O complexes at near room temperature. Morse oscillator functions were used for terminal (CO and HO) stretches, and harmonic oscillator functions were adopted for the skeletal ($C\cdots C$, $C\cdots O$ and $H\cdots O$ bond) and bending modes. The total basis set consisted of 12870 wave functions (either harmonic or Morse oscillator) for the $(CO_2)_2$ dimer, 6435, and 12870 wave functions for CO_2–H_2O complexes in C_s and C_{2v} geometries, respectively.

2. Case study of the carbon dioxide complexes

2.1. The Fermi dyad and the Fermi triad regions in a $(CO_2)_2$ dimer

Starting approximation for potential energy function of a dimer was obtained from *ab initio* calculations at 6-31G(2d,2p)/MP2 level through the use of GAMESS package [2]. The obtained force field was then scaled so that an agreement is reached between the calculated and observed fundamentals and overtones for isolated CO_2 molecules. Table 1 lists main electrooptical parameters for the $(CO_2)_2$ dimer used in the variational anharmonic calculations. Also given are these parameters for an individual CO_2. At the first stage the electro-optical parameters were just taken from *ab initio* calculations and no adjustment of these parameters have been made. The comparison of calculated intensity distribution in the regions of the Fermi doublet and the Fermi triplet with that observed in CIA[1] (or $(CO_2)_2$ dimer, see below) spectra revealed notable difference in both spectral ranges. It was interesting therefore to proceed to the next stage and to solve an inverse problem in order to realize, how deep the changes in electro-optics have to be, which would result in a qualitative agreement between the calculated and the observed intensities.

Worth of noting is that the reason for the absorption intensity raise in the region of dipole forbidden CO_2 transitions consists of non-equivalence of the two C=O bonds in each CO_2 unit in a dimer. Consequently the C=O bond dipole moments and their first and second derivatives with respect to stretching coordinates become different for C=O bonds in the monomers forming a dimer. Pure *ab initio* calculation returns the first derivatives larger (by about 4% of its mean value) and the second derivatives smaller (by about 2%) for the outer bonds having larger dipole moments. The solution of inverse problem leads to both derivatives smaller — for the first derivative by about 4% and for the second derivative by about 11%. Figures 1, 2 demonstrate that these modifications are sufficient to obtain qualitative agreement with the observations. Lighter lines in the bottom part of these figures show the result of calculations employing pure *ab initio* force and electro-optical fields. Heavier lines on these figures relate to the use of adjusted molecular parameters as described above.

Note that experimental spectral curves shown in Figs. 1, 2 have somewhat different sense. In Fig. 1 the so-called 'dimer' spectrum is shown which was obtained from the measured CIA profile after subtraction of the structureless pedestals from two main components of the Fermi coupled bands. In Fig. 2 the CIA profile is depicted which has been obtained according the procedure the details of which can be found in [3].

Interestingly, the use of empirically adjusted parameters (see above) instead of pure *ab initio* resulted in predicted intensity of the central peak between main Fermi doublet components much weaker (by an order of magnitude) than in our previous analysis [5]. This seems reasonable, however, since in the course of retrieval of the 'dimer' spectrum in [5] (see also Fig. 1) we were unable to remove the underlying CIA contribution from the central peak profile and its seeming intensity was thus significantly overestimated.

---

[1]Hereafter CIA means collision-induced absorption.

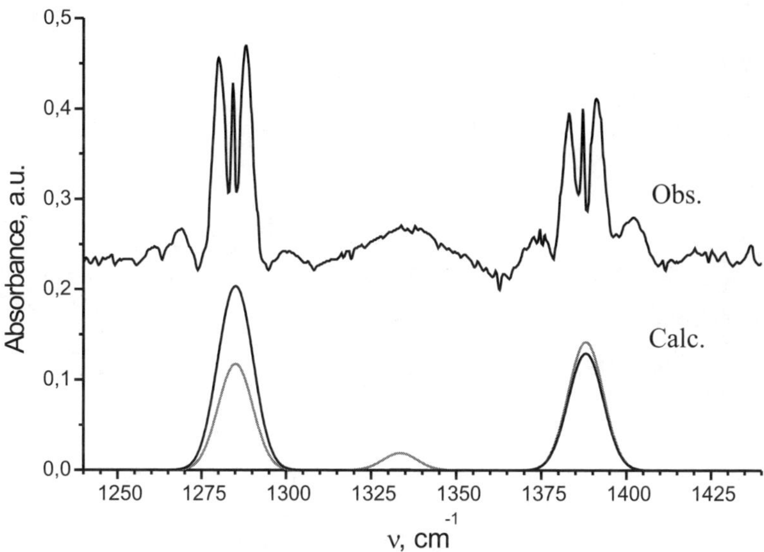

Figure 1. Calculated and experimental [3] $(CO_2)_2$ dimer absorption in the ν_1, $2\nu_2$ region.

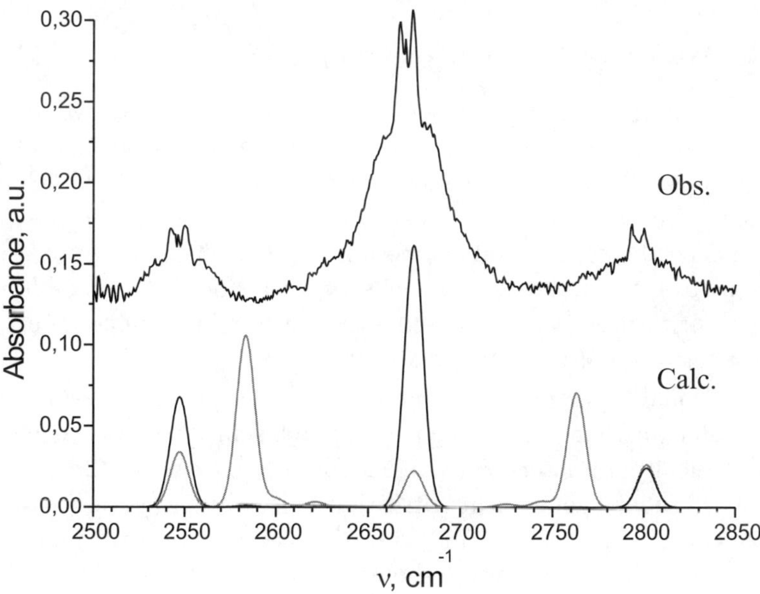

Figure 2. Calculated and experimental [4] $(CO_2)_2$ dimer absorption in the $2\nu_1$, $\nu_1 + \nu_2$, $4\nu_2$ region.

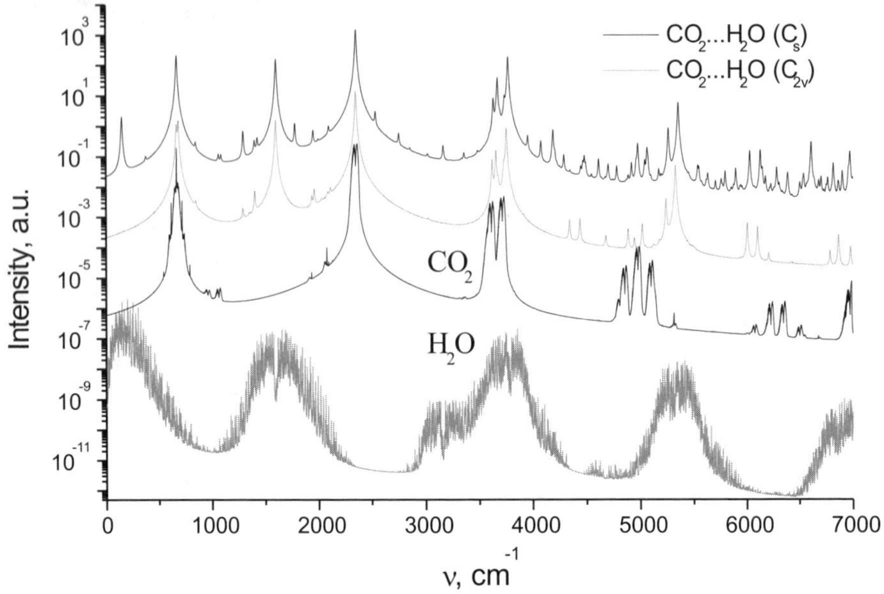

Figure 3. Simulated illustrative spectra of the CO$_2$–H$_2$O complexes (upper traces) and of the CO$_2$ and H$_2$O monomers (lower traces). The vertical scales and the broadening are arbitrary for all the spectral curves shown.

2.2. Water–carbon dioxide complexes

Anharmonic vibrational calculations were made for two possible structural isomers of H$_2$O–CO$_2$ complexes belonging to C$_{2v}$ and C$_s$ point groups. The former isomer with a van der Waals C$\cdots$O intermolecular bond was proven to form in the low-temperature gas phase and is indeed the ground state structure (see e. g. [6]). The latter isomer with a hydrogen bond was never observed, although its possible spectroscopic signatures were supposed either in the liquid phase spectra or in the spectra of matrix trapped species (see e. g. [7, 8][2]).

Vibrational frequencies and intensities were calculated for both isomers for all transitions extending to about 7000 cm^{-1}. The force field and electro-optical parameters were taken from *ab initio* 6-31G(2d,2p)/MP2 without subsequent empirical correction since the measurements of vibrational origins are scarce and vibrational intensities are unavailable for these complexes. Figure 3 shows an overview spectra for the two

---

[2]Other available matrix data [9,10] give no indications on the presence of a hydrogen-bonded isomer.

isomers. Also shown are the spectra of parent CO_2 and H_2O molecules simulated using HITRAN data base [11].

It is seen that in most cases the vibrational bands for the complexes in study are masked by those of pure CO_2 and H_2O. The calculated absorption spectrum of the C_s isomer is richer than that of the C_{2v} isomer. Weak high lying overtone bands having appreciable intensity, e. g. the combination transition $\nu_{OH}^s + \nu_{CO}^{as}$ in the vicinity of $\sim 6000\,\mathrm{cm}^{-1}$, fall occasionally in the regions of relative transparency of both isolated monomers and can therefore be discriminated from the monomer absorption.

Acknowledgments

This work was financially supported in part by the Russian Foundation for Basic Researches, Projects 02-05-64529 and 02-03-32416. A. A. V. acknowledges partial support from the INTAS–Airbus Grant 1815.

References

[1] Gribov, L. A. and Pavlyuchko, A. I. (1998) *Variational Methods for Solving Anharmonic Problems in the Theory of Vibrational Spectra of Molecules*, Nauka, Moscow (in Russian).

[2] Schmidt, M. W., Baldridge, K. K., Boatz, J. A., Elbert, S. T., Gordon, M. S., Jensen, J. H., Koseki, S., Matsunaga, N., Nguyen, K. A., Su, S. J., Windus, T. L., together with Dupuis, M., and Montgomery, J. A. (1993) GAMESS version from Iowa State University, J. Comput. Chem., 14, 1347–1363.

[3] Vigasin, A. A., Baranov, Y. I., and Chlenova, G. V. (2002) Temperature variations of the interaction induced absorption of CO_2 in the ν_1, $2\nu_2$ region: FTIR measurements and dimer contribution, J. Molec. Spectrosc., 213, 51–56.

[4] Baranov, Y. I., Fraser, G. T., Lafferty, W. J., and Vigasin, A. A. (2003) Collision-induced absorption in the CO_2 Fermi triad for temperatures from 211 K to 296 K, this volume, 149–158.

[5] Vigasin, A. A., Huisken, F., Pavlyuchko, A. I., Ramonat, L., and Tarakanova, E. G. (2001) Identification of the $(CO_2)_2$ dimer vibrations in the ν_1, $2\nu_2$ region: Anharmonic variational calculations, J. Mol. Spectrosc., 209, 81–87.

[6] Peterson, K. I. and Klemperer, W. (1984) Structure and internal rotation of H_2O–CO_2, HDO–CO_2 and D_2O–CO_2 van der Waals complexes, J. Chem. Phys., 80(6), 2439–2445.

[7] Vigasin, A. A., Adiks, T. G., Tarakanova, E. G., and Yukhnevich, G. V. (1994) Simultaneous infrared absorption in a mixture of CO_2 and H_2O: The role of hydrogen-bonded aggregates, JQSRT, 52(3/4), 295–301.

[8] Tso Tai-Ly and Lee, E. K. C. (1985) Role of hydrogen bonding studied by the FTIR spectroscopy of the matrix isolated molecular complexes and dimers of H_2O, H_2O–CO_2, H_2O–CO, H_2O_2–nCO in solid O_2 at 12–17 K, J. Phys. Chem., 89(9), 1612–1618.

[9] Fredin, L., Nelander, B., and Ribbegård, G. (1975) A matrix isolation study of the interaction between water and carbon dioxide, Chemica Scripta, 7, 11–13.

[10] Svensson, T., Nelander, B., and Karlström, G. (2001) The CO_2 complexes with HOO and HO in argon matrices, Chemical Physics, 265, 323–333.

[11] Rothman, L. S., Rinsland, C. P., Goldman, A., Massie, S. T., Edwards, D. P., Flaud, J.-M., Perrin, A., Camy-Peyret, C., Dana, V., Mandin, J.-Y., Schroeder, J., McCann, A., Gamache, R. R., Wattson, R. B., Yoshino, K., Chance, K. V., Jucks, K. W., Brown, L. R., Nemtchinov, V., and Varanasi, P. (1998) The HITRAN molecular spectroscopic database and HAWKS (HITRAN atmospheric workstation): 1996 edition, JQSRT, 60, 665–710.

INTERFERENCE EFFECTS
IN THE INFRARED SPECTRUM OF HD;
ATMOSPHERIC IMPLICATIONS

G. C. Tabisz

Department of Physics and Astronomy, University of Manitoba,
Winnipeg MB R3T 2N2, Canada

Abstract   Interference between the amplitudes of electric-dipole allowed and collision induced transitions affects the total intensity and lineshape of the infrared spectrum of gaseous HD. Experimental and theoretical aspects as well as implications for the spectra of Saturn and Jupiter are discussed. The anisotropy of the interaction between HD and a perturber permits collisional propagation to occur during the collision.

Keywords: HD / infrared / interference / collision-induced / collisional propagation / Saturn / Jupiter / spectral line shape / D to H ratio

1. Introduction

The theory of spectral line broadening is described in a voluminous literature, the greatest part of which concerns electric dipole transitions. The lineshape accompanying transitions which proceed by collision-induced electric dipole has received appreciable, but less, attention. Usually allowed lines have an intensity which is sufficient to dominate collision-induced effects and the two types need not be considered for the same transition. In the infrared spectrum of the hydrogen isotopomer HD, the small permanent dipole of the molecule (10^{-4}–10^{-3} D) yields allowed transitions comparable in strength to the collision-induced transitions and an observable in interference effect occurs between them. In this paper both experimental and theoretical aspects of the effect are described and possible atmospheric applications indicated.

83

C. Camy-Peyret and A.A. Vigasin (eds.),
Weakly Interacting Molecular Pairs: Unconventional Absorbers of Radiation in the Atmosphere, 83–92.
© 2003 *Kluwer Academic Publishers. Printed in the Netherlands.*

2. Interference

Coherence, which is a necessary ingredient for the occurrence of interference, is a manifestation of indistinguishability. In the HD problem, the amplitudes of two electric dipole transitions between the same initial and final states can interfere despite vastly different origins. The phenomenon was first identified by McKellar [1] and by Reddy and collaborators [2]. The first thorough theoretical description was given by Herman, Tipping and Poll [3].

In the simplest scenario, if interference is to occur, both the allowed and induced moments, must be of the same symmetry, that is, give rise to the same selection rules. The induced moment in the HD–X collision pair, where X is a perturber, can contain a component of the required symmetry that does not depend on the orientation of the perturber nor of the intermolecular axis [3]. This component lies in the direction of the internuclear axis of HD, either parallel or antiparallel to the allowed moment and the result is, respectively, positive or negative interference. The induced moment dies away quickly as it depends strongly on the magnitude of the intermolecular distance r, usually exponentially or as r^{-n}. A subsequent collision can induce another dipole with a component in the same direction. The allowed dipole transition is perturbed by collisions and undergoes collision-broadening. The phase associated with the induced transition is locked to that of the allowed transition and the loss of phase correlation in time for both is caused by collision-broadening type encounters. The resulting lineshape is narrow with a width characteristic of collision-broadening rather than of the extremely broad profile that accompanies purely collision-induced transitions.

3. Experimental evidence

The following will briefly summarize data taken at the University of Manitoba. In the far-infrared spectrum, the sharp lines sit on a broad collision-induced background (Fig. 1). This background is removed numerically to yield the individual $R_0(J)$ lines. In the dense gases, unlike the case of an ordinary allowed transition, the integrated intensity of a line is found to be a function of perturber density, increasing or decreasing as the density increases [4, 5]. The profile is nearly lorentzian at low density but an asymmetry develops as the line broadens. This asymmetry is slight for purely rotational transitions. For vibration-rotation transitions, the asymmetry is extremely pronounced at high

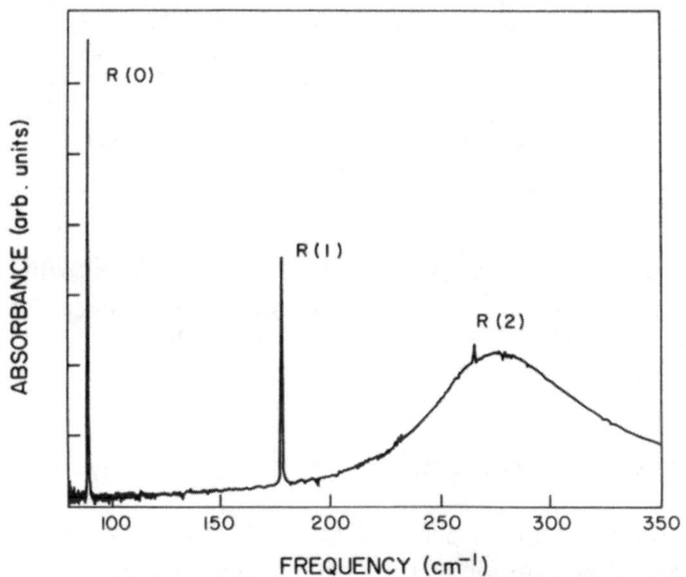

Figure 1.　　The far infrared spectrum of HD at 77 K.

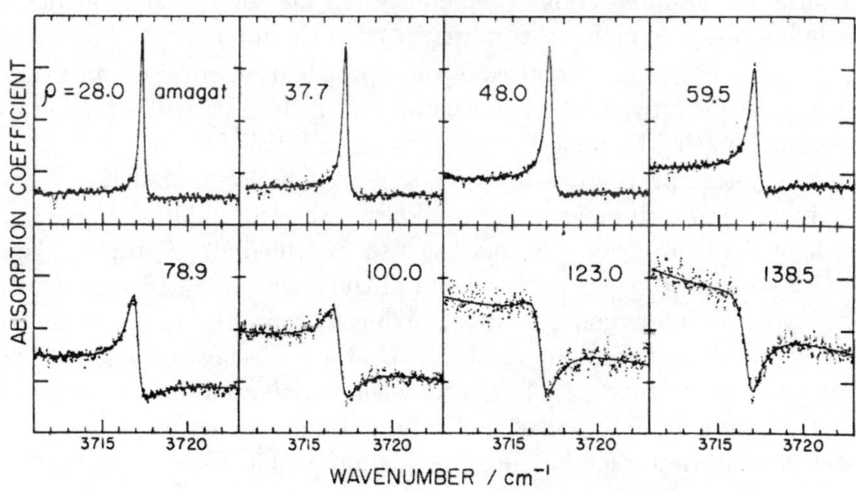

Figure 2.　　The $R_1(0)$ line of HD at 77 K as a function of amagat density [6].

density (150 amagats) as the now famous results of Rich and McKellar [6] show; the absorption line begins to resemble an emission feature at the highest density studied (Fig. 2).

Table 1.   Allowed dipole transition elements for $J \to J+1$ (10^{-4} D).

	$J=0$	$J=1$	$J=2$	$J=3$
experiment	8.10(15)	8.04(12)	8.16(12)	8.55(42)
ab initio calc [16]	8.36	8.38	8.39	8.41

From data for the integrated absorption coefficient the allowed dipole matrix elements for the isolated molecule have been determined (Table 1) [4,5]. They are about 5% less than the values given by *ab initio* calculations.

4. Collisional propagation

Gao, Tabisz, Trippenbach and Cooper [7] have developed a general theoretical description of the phenomenon. It allows for collisional propagation of the HD molecule, among rotational states, during inelastic collisions with a perturber with widely spaced internal energy levels. The thereby necessary inclusion of an anisotropic intermolecular potential in the interaction scheme also permits several components of the induced dipole moment to participate in the interference process.

To appreciate the role of collisional propagation consider an example (Fig. 3) for a transition between an initial state J_g and a final state J_e. In one path, the molecule starts in an arbitrary state i at $t=-\infty$ and the system propagates to $t=t_0$, while the molecule undergoes J or m mixing. At t_0, there is an interaction between an induced dipole and the light field, in general connecting two intermediate J states, 2 and 3. The system propagates to the end of the collision with the molecule again possibly undergoing J or m mixing to leave the molecule in the excited state e at $t=\infty$. In another path, the molecule propagates from state i at $t=-\infty$ to state g at $t=\infty$, where an allowed dipole transition between states g and e occurs. The amplitudes for these two possible paths can interfere. The mixing effect is important if $(\omega_{eg} - \omega_{32})\tau_c \lesssim 1$, where τ_c is the collision duration.

The formalism has formed the basis for the calculation of spectral lineshape parameters for pure rotational [8,9] and vibrational-rotational [10] transitions in HD–He and HD–Ar. In order to perform these calculations, three types of quantities are needed:

- ■ allowed dipole transition elements, either measured or calculated;

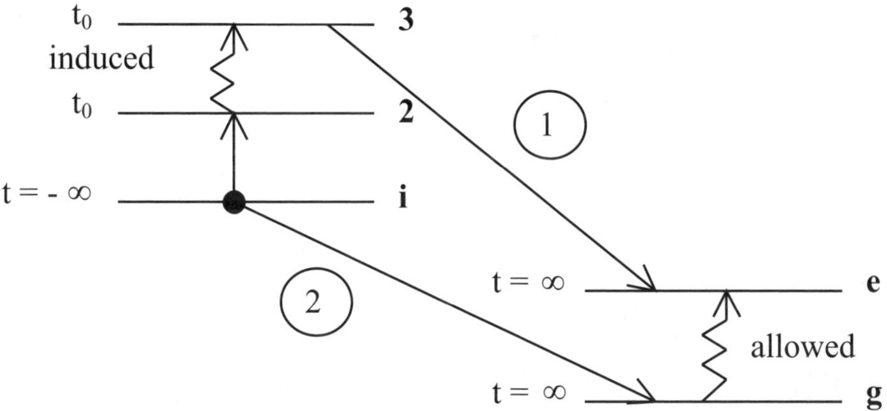

Figure 3. The interference between allowed and induced transitions.

- induced dipole transition elements as calculated by Borysow, Frommhold and Meyer [11];

- intermolecular potentials for H_2–He [12] and HD–Ar [13]. A shift of the origin to the center of mass of the HD molecule yields the HD–X potential; an approximate vibrational dependence was assigned to the potentials.

Calculations were performed in the classical path and impact approximations.

The expression for the absorption coefficient $\alpha(\omega)$ at frequency ω is given below:

$$\alpha(\omega) = \rho_R \left(\frac{4\pi^2 \omega}{3\hbar c}\right) \left(1 - \epsilon^{-\beta\hbar\omega}\right) P(J_g) |\mu_R|^2$$

$$\times \left(\frac{B\rho_x/\pi}{(\Delta - S\rho_x)^2 + (B\rho_x)^2} \left(1 + a\rho_x + b\rho_x^2\right) \right.$$

$$\left. - \frac{(\Delta - S\rho_x)/\pi}{(\Delta - S\rho_x)^2 + (B\rho_x)^2} \left(c\rho_x + d\rho_x^2\right) \right). \qquad (1)$$

Here ρ_R and ρ_x are the HD and perturber densities; $P(J_g)$ is the Boltzmann distribution for the ground state J_g; μ_R is the reduced matrix element for allowed HD dipoles μ_A, i.e. $\langle J_e \| \mu_A \| J_g \rangle$. The total lineshape is a sum of lorentzian and anomalous dispersion profiles whose relative strengths are given by the quantities a, b, c and d, which depend on matrix elements of the induced dipole moment divided by μ_R. B is the

Table 2. Broadening coefficient B, $10^{-2}\,\mathrm{cm}^{-1}$ amagat.

	temperature, K	experiment	theory
HD–He			
$R_0(0)$	77	0.15	0.31
			0.18 [14]
$R_0(2)$	195	0.41	0.48
			0.26
HD–Ar			
$R_0(0)$	195	1.4	2.2
$R_0(2)$	77	0.28	0.32

line broadening coefficient, S is the frequency shift coefficient and Δ is the detuning $(\omega - \omega_{eg})$. The leading interference parameter a affects the strength of the symmetric lorentzian component. The asymmetry of the profile is controlled by c which determines the magnitude of the anomalous dispersing component. The quantities b and d play analogous roles to a and c but in successive collisions and are important only at high density.

5. Calculations

Now let us consider some results for purely rotational transitions which demonstrate that inelastic collisions are crucial to the effect.

First the broadening coefficients, obtained with the full calculation which includes inelastic collisions, are presented in Table 2. For the interference parameter a, large differences occur between parameters calculated with the full formalism compared to those for which only elastic collisions are considered (Table 3). In particular, a pronounced J-dependence appears. The results are very sensitive to the potential in the region near the repulsive wall. A quantum treatment may be needed; a classical trajectory may not be adequate to describe the important close inelastic collisions. For the asymmetry parameter c, the same trends are evident. Notice that experimentally its sign may change from line to line and the calculation is capable of providing agreement.

In the case of vibration-rotation transitions, collisional propagation can occur among levels in the ground vibrational state and among levels in the excited vibrational state [10]. (The collision duration and experimental temperatures mitigate against propagation between the

Table 3. Interference parameter a, 10^{-3} amagat^{-1}.

	temperature, K	experiment	theory	
			full calc.	elastic coll.
		HD–He		
$R_0(0)$	77	6.0(16)	2.99	4.33
$R_0(2)$	77	4.4(15)	6.47	4.34
$R_1(0)$	77	−1.44(2)	−1.83	
		HD–Ar		
$R_0(1)$	295	1.8(3)	3.22	8.92
$R_0(2)$	295	6.1(2)	7.80	8.92

two manifolds.) Comparison with experiment is possible only for and of HD–He at 77 K. Agreement for a is good; c on the other hand is too small (Table 4). The calculation does not predict the large asymmetry observed [6]. There have been other recent calculations of these quantities. Gustafsson and Frommhold [14] have performed a fully quantum mechanical close coupled calculation on the HD–He complex at 77 K. Lewis and Herman [15] have used a fully quantum mechanical perturbation theory approach. The close coupled results agree better than do ours with the measured broadening coefficients and asymmetry parameters for both purely rotational and vibration-rotational lines. The necessity of the quantum mechanical approach is thereby confirmed. In defence of the classical path approach, it must be said that for the case where comparison is made, the $J=0$ to $J=1$ transition at low temperature, the semiclassical approximation is least valid because of its overestimation of the inelastic energy transfer rates. The failure of the perturbation theory to account for the large line asymmetry perhaps reflects the fact that it does not include all collisional propagation effects.

6. Dipole moment components

Investigation of the contribution of various individual components of the induced dipole moment revealed that different components play major roles in the intensity interference and line asymmetry aspects of the phenomenon [9]. For example, the so-called $A_1(10)$ component is the one which lies along the internuclear axis of HD and has the same symmetry as the allowed moment. As expected it is by far the largest single contributor to the interference parameter a and makes a non-negligible contribution to the asymmetry parameter c. It is completely negligible to the purely induced spectrum. Mainly short range

Table 4. Asymmetry parameter c, 10^{-3} amagat^{-1}.

	temperature, K	experiment	theory	
			full calc.	elastic coll.
		HD–He		
$R_0(0)$	77	−3.0(6)	−1.2 −2.18 [14]	
$R_0(0)$	195	−3.4(24)	−1.5	−0.67
$R_0(2)$	195	−1.8(7)	−0.31	-1×10^{-3}
$R_1(0)$	77	11.8	5.33 9.8 [14] 2.2 [15]	
		HD–Ar		
$R_0(1)$	195	−1.6(2)	−0.21	−0.04
$R_0(2)$	195	4.06(30)	1.3	-1.7×10^{-3}

anisotropic components also contribute to a and act to reduce the magnitude of the interference. As more induced dipole components are implicated in the process, the distinguishability of the two possible paths becomes greater (Section 2). Long range multipole components contribute to the asymmetry; these are the most important components for the purely induced spectrum. The intensity effect is not sensitive to the vibrational dependence of the potential but the line asymmetry is. Collisional interference seems to consist of two distinct effects which are complementary sources of information on intermolecular interactions.

7. Planetary atmospheres

Awareness of the interference effect may be valuable for the interpretation of the HD spectrum from planetary atmospheres. Professor Gary Davis from the University of Saskatchewan and his collaborators have used the high resolution mode of the Long Wavelength Spectrometer on the Infrared Space Observatory launched by the European Space Agency. They have observed the $R_0(0)$ and $R_0(1)$ lines of the purely rotational spectrum of HD in the atmospheres of Jupiter and Saturn at a resolving power of 10^4 [17].

Armed with knowledge of the lineshape parameters from the laboratory work, they use a multilayer atmospheric spectral model to invert the equations of radiative transfer to infer the HD abundance. From it they obtain the D/H ratio in these atmospheres, a number with cosmological implications. The primordial D/H ratio is a key measure of the baryon density of the universe. The D/H ratio of the universe decreases over time through stellar fusion processes and is not regenerated by any known process. Measurement of the ratio in the atmospheres of Jupiter and Saturn gives its value at the epoque of the formation of the solar system and supplies an important point on the D/H time curve. Their work has not yet been published.

8. Summary and needs for future

A considerable amount of experimental data on HD–X systems as function of temperature and density have been accumulated. A general theory of the interference effect has been developed which demonstrates the importance of collisional propagation in the process. Classical trajectories may be inadequate for calculation in many situations. Among the data available are HD–HD, HD–H_2 and HD–N_2 spectra which are important for atmospheric applications. These however cannot be analyzed with the present theoretical formalism because of the existence in each case of relevant closely spaced internal energy levels in the perturbing molecule. The present formalism needs to be extended to treat such systems. The line asymmetry is not fully understood, particularly at high density, and needs more investigation.

References

[1] McKellar, A. R. W. (1973) Intensities and the Fano line shape in the infrared spectrum of HD, Can. J. Phys., 51, 389–397.

[2] Poll, J. D., Tipping, R. H., Prasad R. D. G., and Reddy, S. P. (1976) Intracollisional interference in the spectrum of HD mixed with rare gases, Phys. Rev. Lett., 36, 248–251.

[3] Herman R., Tipping, R. II., and Poll, J. D. (1979) Shape of the R and P lines in the fundamental band of gaseous HD, Phys. Rev. A, 20, 2006–2012.

[4] Drakopoulos, P. G. and Tabisz, G. C. (1987) Collisional interference in the foreign-gas-perturbed far-infrared rotational spectrum of HD, Phys. Rev. A, 36, 5566–5574.

[5] Lu, Z., Tabisz, G. C., and Ulivi, L. (1993) Temperature dependence of the pure rotational band of HD: Interference, widths and shifts, Phys. Rev. A, 47, 1159–1173.

[6] Rich, N. and McKellar, A. R. W. (1983) Interference effects in the spectrum of HD: The fundamental band of pure HD, Can. J. Phys., 61, 1648–1654.

[7] Gao, B., Tabisz, G. C., Trippenbach, M., and Cooper, J. (1991) Spectral line shape arising from collisional interference between electric-dipole allowed and collision-induced transitions, Phys. Rev. A, 44, 7379–7391.

[8] Gao, B., Tabisz, G. C., and Cooper, J. (1991) Rotational spectrum of HD perturbed by He or Ar gases: The effects of rotationally inelastic collisions on the interference between allowed and collision-induced components, Phys. Rev. A, 46, 5781–5788.

[9] McQuarrie, B., Tabisz, G. C., Gao, B., and Cooper, J. (1995) Role of the induced dipole moment in the collisional interference in the pure rotational spectrum of HD–He and HD–Ar, Phys. Rev. A, 52, 1976–1981.

[10] McQuarrie, B. and Tabisz, G. C. (1996) Collisional interference in the infrared spectrum of HD: Calculation of the line shape of vibrational transitions for HD–He, J. Molec. Liq., 70, 159–168.

[11] Borysow, A., Frommhold, L., and Meyer, W. (1988) Dipoles induced by the interactions of HD with He, Ar, H_2 or HD, J. Chem. Phys., 88, 4855–4860.

[12] Mulder, F., van der Avoird, A., and Wormer, P. E. S. (1979) Anisotropy of long range interactions between linear molecules H_2–H_2 and H_2–He, Molec. Phys., 37, 159–180.

[13] LeRoy, R. J. and Hutson, J. M. (1987) Improved potential energy surfaces for the interaction of H_2 with Ar, Kr and Xe, J. Chem. Phys., 86, 837–853.

[14] Gustafsson, M. and Frommhold, L. (2001) Intracollisional interference of R Lines of HD in mixtures of Deuterium hydride and Helium gas, Phys. Rev. A, 63, 052514-1–052514-6.

[15] Herman, R. M. and Lewis, J. C. (2001) Scalar collisional interference parameters for the HD $R_1(0)$ and $R_1(1)$ lines in mixtures with He. In: J. Seidel, editor. *Spectral Line Shapes*, Vol. 10, AIP, New York, pp. 457–459.

[16] Wolniewicz, L. (1975) On the computation of dipole transitions in the HD molecule, Can. J. Phys., 54, 672–679.

[17] Davis, G. D., private communication.

COLLISION-INDUCED ABSORPTION IN DIPOLAR MOLECULE — HOMONUCLEAR DIATOMIC PAIRS

A. Brown, R. H. Tipping

Department of Physics and Astronomy, University of Alabama,
Tuscaloosa, AL 35487-0324, U.S.A.

Abstract   Theoretical expressions for the collision-induced absorption spectra of a dipolar and a homonuclear diatomic pair are presented. For the dipolar species, the line strengths and transition frequencies are taken from the HITRAN database, whereas for the homonuclear molecule, the polarizability matrix elements and spectroscopic parameters are obtained from the literature. As specific examples, we consider the double fundamental vibrational transition in CO_2–N_2, and the H_2O–N_2 transition in the region of the nitrogen fundamental. For the former pair, the theoretical results are in good agreement with experimental data. For the latter pair, we show that the collision-induced absorption for high humidity conditions is larger than that of N_2–N_2 pairs above $2500\,cm^{-1}$, but for lower wavenumbers, it is much weaker than the measured foreign continuum resulting from the far wings of allowed transitions.

Keywords: collision-induced absorption / double transitions / dipole–induced dipole absorption

1. Introduction

The existence of transient dipoles during binary collisions gives rise to collision-induced absorption (CIA); for an excellent review see Ref. [1]. There are three primary ways in which CIA differs from allowed absorption. The CIA spectrum and the integrated intensity are proportional to the square of the density for pure gases and to the product of the densities for gas mixtures. The integrated intensity of the CIA band depends on temperature; it decreases with increasing temperature. CIA line widths are large, typically 10–$50\,cm^{-1}$, and, thus, CIA absorption bands are relatively featureless. Most experimental and the-

93

C. Camy-Peyret and A.A. Vigasin (eds.),
Weakly Interacting Molecular Pairs: Unconventional Absorbers of Radiation in the Atmosphere, 93–99.
© 2003 *Kluwer Academic Publishers. Printed in the Netherlands.*

oretical studies of CIA have been made for homonuclear diatomic pairs [2–4], e. g. N_2-N_2, but CIA also occurs in pairs containing a dipolar molecule. Here we briefly review an efficient method for computing the CIA profile for dipolar molecule-homonuclear diatomic pairs on the basis of quantum line shapes, an isotropic potential, and dipole-induced dipole functions. For a more detailed explanation see Refs. [5, 6]. The line strengths and positions of the vibration-rotation transitions are treated using the HITRAN database [7] for the dipolar molecule and explicitly for the homonuclear diatomic. Results are presented for CO_2–N_2 and H_2O–N_2.

2. Theory

For X–N_2 (X = H_2O or CO_2) pairs, the absorption coefficient, $\alpha(\omega)$, divided by the product of the densities in amagats can be written as [8]

$$A(\omega, T) = \frac{\alpha(\omega)}{\rho_X \, \rho_{N_2}} = \frac{4\pi^2}{3} \, \alpha_F \, n_0^2 \, a_0^5 \, \omega \sum_{n, J_1, J_1', J_2, J_2'} S_n(T) \, L_n(\omega - \omega_{if}). \quad (1)$$

The strengths, $S_n(T)$, of the n components contributing to the absorption can be expressed as

$$S_n = P_{J_1}(T) \, C(J_1 \, \lambda_1 \, J_1')^2 \, P_{J_2}(T) \, C(J_2 \, \lambda_2 \, J_2')^2 \, <B_n(R)^2>. \quad (2)$$

In Eq. (2), P_J are normalized Boltzmann factors, the C's are Clebsch-Gordan coefficients, and

$$\begin{aligned}
<B_n(R)^2> &= \langle v_1 \, v_2 \, J_1 \, J_2 \mid A_\Lambda(\lambda_1 \, \lambda_2 \, L; r_1 \, r_2 \, R) \mid v_1' \, v_2' \, J_1' \, J_2' \rangle^2 \\
&\approx \langle v_1 \, J_1 \mid a_\Lambda^{(1)}(\lambda_1; r_1) \mid v_1' \, J_1' \rangle^2 \\
&\quad \times \langle v_2 \, J_2 = 0 \mid a_\Lambda^{(2)}(\lambda_2; r_2) \mid v_2' \, J_2' = 0 \rangle^2 \, I.
\end{aligned} \quad (3)$$

We have neglected the slight dependence of the N_2 matrix elements on the rotational quantum numbers. The angular brackets in Eq. (2) denote an average over the separation between centers of mass R as calculated using well-known results [2–4]. Explicit results for the dipole-induced dipole coefficients A are listed in Table 1. The key to the computational method is that $B_n(R)$, and subsequently $S_n(T)$, can be factored into three components depending upon molecule 1 (H_2O or CO_2), molecule 2 (N_2), and an integral

$$I = \int_0^\infty R^{-4} \, e^{-V_{iso}(R)/kT} \, dR \quad (4)$$

Table 1. The dipole–induced dipole coefficients $A_\Lambda(\lambda_1\,\lambda_2\,L; r_1\,r_2\,R)$.

λ_1	λ_2	Λ	L	$A_\Lambda(\lambda_1\,\lambda_2\,L; r_1\,r_2\,R)$
1	0	1	2	$-\sqrt{2}\,(\mu(r_1)\,\alpha(r_2))\,R^{-3}$
1	2	1	2	$(1/15)\,(\mu(r_1)\,\gamma(r_2))\,R^{-3}$
1	2	1	2	$(1/\sqrt{15})\,(\mu(r_1)\,\gamma(r_2))\,R^{-3}$
1	2	3	2	$(14/5\sqrt{21})\,(\mu(r_1)\,\gamma(r_2))\,R^{-3}$

over the separation between the two molecules and depending on the isotropic potential $V_{\text{iso}}(R)$. In Eq. (1), $L_n(\delta\omega)$ is a normalized line shape for component n. Therefore, as in the homonuclear diatomic cases [2–4], through independent knowledge of the dipole and polarizability parameters, line positions, line shapes, and the isotropic potential, we can calculate the CIA spectrum.

3. Results and discussion

For the isotropic and anisotropic polarizability matrix elements of N_2, the values [4] (in $a_0{}^3$) $\alpha_{00} = 11.74$, $\alpha_{01} = 0.365$, $\gamma_{00} = 4.75$, and $\gamma_{01} = 0.438$, have been used. The use of the "00" or "01" component depends upon whether N_2 makes a pure rotational or vibration-rotational transition. The dipole moments for CO_2 and H_2O are obtained from the HITRAN database [7]. Lennard—Jones isotropic potentials are used for the CO_2–N_2 and H_2O–N_2 interactions, where the LJ parameters are $\sigma = 6.88\,a_0$ and $\epsilon/k = 128\,\text{K}$ and $6.746\,a_0$ and $207\,\text{K}$, respectively. The line positions are obtained by treating the energies of the N_2 lines explicitly and by utilizing the transition frequencies for the dipolar molecule directly from the HITRAN database [7]. For the line shape, $L_n(\omega)$, we utilize the BC model [9] parameterized for N_2 [10]. Since these line shapes are for quadrupole-induced dipoles rather than dipole-induced dipoles, the line width is treated as an adjustable parameter.

3.1. CO_2–N_2

Figure 1 illustrates the theoretical [5, 6] and the experimental [11] collision-induced double transition $CO_2(v_3{=}1) + N_2(v{=}1) \leftarrow CO_2(v_3{=}0) + N_2(v{=}0)$ spectra at 296 K for four product densities. Good agreement between theory and experiment is obtained by decreasing the N_2 line widths by approximately 50% from their values for N_2–N_2 CIA. This comparison allows a reasonable estimate of the line shape to use

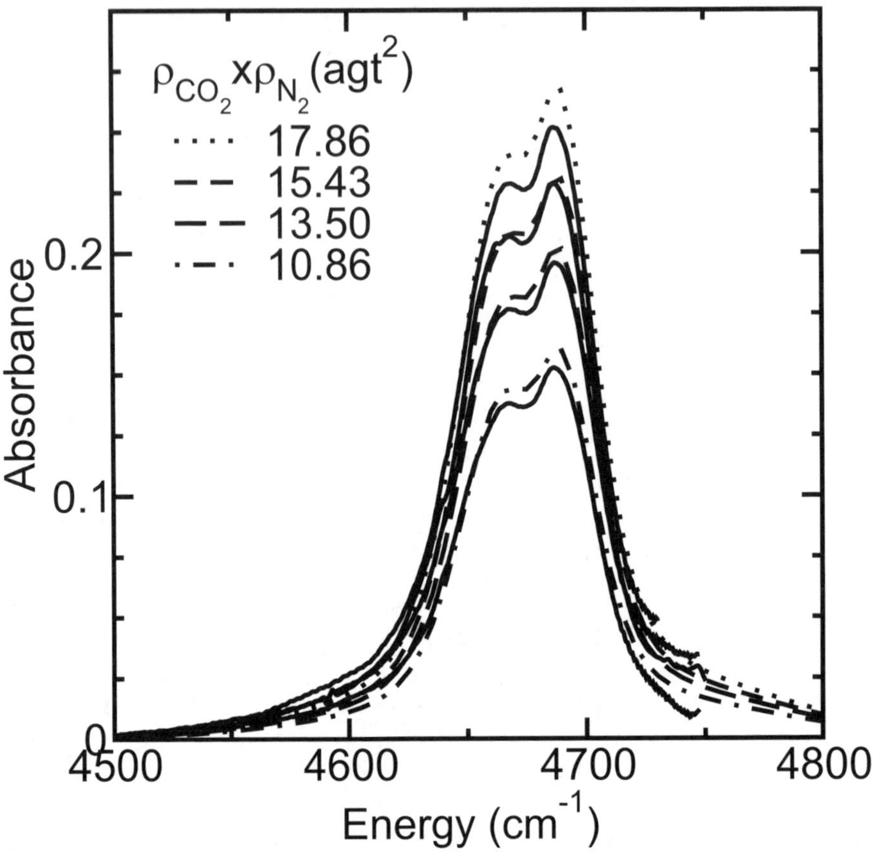

Figure 1. Comparison of the theoretical (dashed lines) and the experimental (solid lines) collision-induced double transition $CO_2(v_3{=}1) + N_2(v{=}1)$ $\leftarrow CO_2(v_3{=}0) + N_2(v{=}0)$ absorption spectra for four product densities.

for the calculation of spectra for other dipole-induced dipole processes involving N_2 for which experimental results are not available.

3.2. H_2O–N_2

Figure 2 illustrates the theoretical CIA spectra for H_2O–N_2 and N_2–N_2 [4] in the vicinity of the N_2 fundamental band. In addition to the results for equal numbers of H_2O and N_2 molecules, the CIA spectra are shown for 5% H_2O in N_2 (corresponding to approximately 100% relative humidity) and 1% H_2O in N_2.

The H_2O–N_2 CIA can be important beyond $2500\,\mathrm{cm}^{-1}$, as seen in the inset of Fig. 2, and its relative importance will be strongly humidity dependent. The theoretical model is predictive as functions of

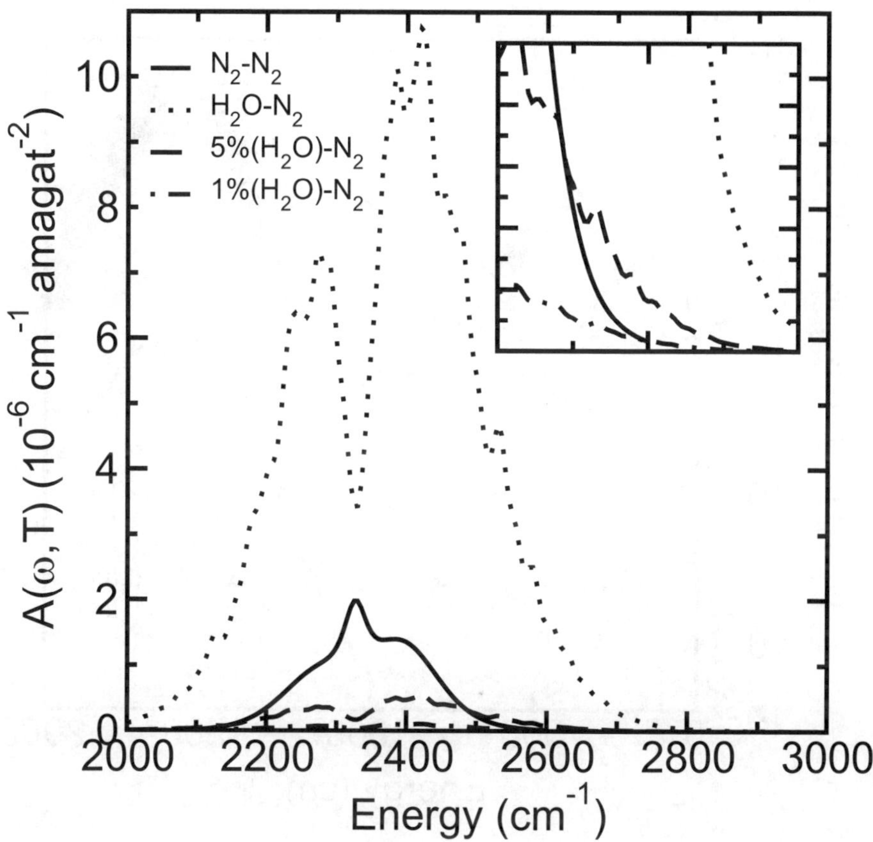

Figure 2. The H_2O–N_2 (dotted line) and N_2–N_2 (solid line) CIA spectra in the region of the N_2 fundamental. For comparison, H_2O–N_2 spectra are shown for 5% H_2O in N_2 (dashed line) and 1% H_2O in N_2 (dashed dot line). Inset is the region 2400–3000 cm^{-1} (y-scale is 0–0.5 × 10^{-6} cm^{-1} amagat^{-2}).

both water density (humidity) and temperature. Figure 3 plots the theoretical CIA spectrum for H_2O–N_2 in the region 0–2000 cm^{-1}, i.e. when N_2 makes a pure rotational transition. The experimental measurements [12–15] of the foreign (N_2) continuum arising from the far-wing absorption within this same spectral region are also illustrated. It is clear that within the present model, which contains no adjustable parameters, the contribution of CIA from H_2O–N_2 pairs is negligible.

4. Conclusions

From these results, the utility of HITRAN for determining CIA spectra involving dipole-induced dipole processes is evident.

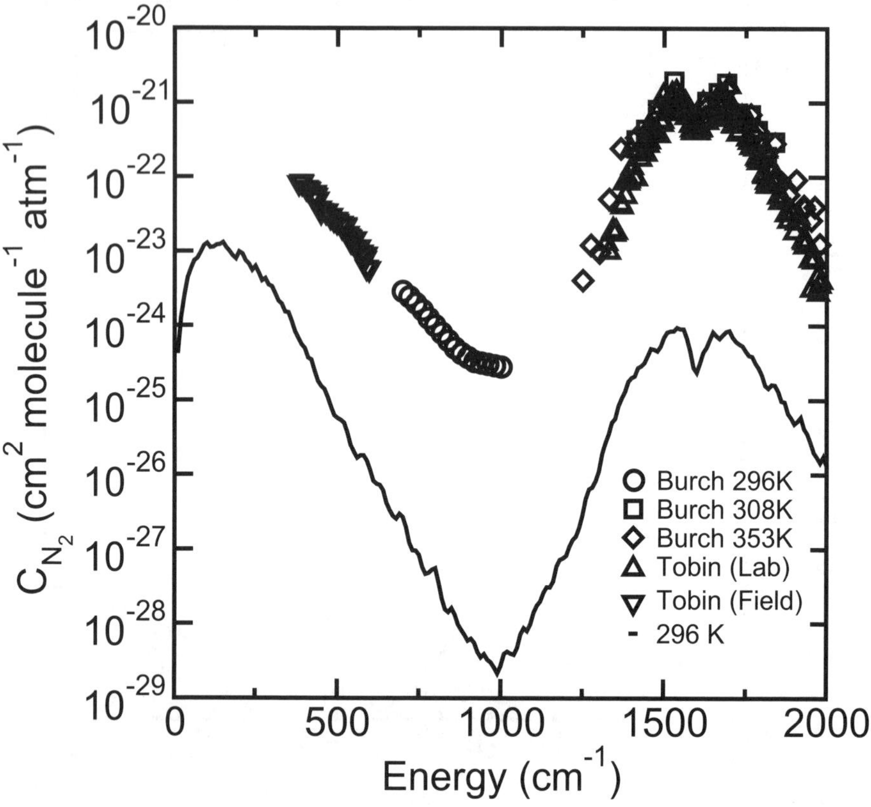

Figure 3.   Comparison of theoretical results for H_2O–N_2 CIA with experimentally measured absorptions in H_2O–N_2 mixtures for energies below the N_2 fundamental.

Acknowledgments

The authors would like to thank B. Maté, G. T. Fraser and W. J. Lafferty for providing their CO_2–N_2 experimental data and D. C. Tobin for providing his H_2O–N_2 experimental data.

References

[1] Frommhold, L. (1993) *Collision-induced Absorption in Gases*, Cambridge University Press, Cambridge.

[2] Moreau, G., Boissoles, J., Le Doucen, R., Boulet, C., Tipping, R. H., and Ma, Q. (2001) Experimental and theoretical study of the collision-induced fundamental absorption spectra of N_2–O_2 and O_2–N_2 pairs, J. Quant. Spectrosc. Radiat. Transfer, 69, 245–256.

[3] Moreau, G., Boissoles, J., Boulet, C., Tipping, R. H., and Ma, Q. (2000) Theoretical study of the collision-induced fundamental absorption spectra of O_2–O_2 pairs for temperatures between 193 and 273 K, J. Quant. Spectrosc. Radiat. Transfer, 64, 87–107.

[4] Boissoles, J., Tipping, R. H., and Boulet, C. (1994) Theoretical study of the collision-induced fundamental absorption spectra of N_2–N_2 pairs for temperatures between 77 and 297 K, J. Quant. Spectrosc. Radiat. Transfer, 51, 615–627.

[5] Brown, A., Tipping, R. H., and Maté, B. (2000) Theoretical study of the collision-induced double transition $CO_2(v_3{=}1)$ + $N_2(v{=}1)$ ← $CO_2(v_3{=}0)$ + $N_2(v{=}0)$ at 296 K, J. Mol. Spectrosc., 204, 153–158.

[6] Brown, A. and Tipping, R. H., (2001) Theoretical study of collision-induced double transitions in CO_2–X_2 (X_2 = H_2, N_2, and O_2) pairs, J. Mol. Spectrosc., 205, 319–322.

[7] Rothman, L. S., Rinsland, C. P., Goldman, A., Massie, S. T., Edwards, D. P., Flaud, J.-M., Perrin, A., Camy-Peyret, C., Dana, V., Mandin, J.-Y., Schroeder, J., McCann, A., Gamache, R. R., Wattson, R. B., Yoshino, K., Chance, K., Jucks, K., Brown, L. R., Nemtchinov, V., and Varanasi, P. (1998) The HITRAN molecular spectroscopic database and HAWKS (HITRAN Atmospheric Workstation): 1996 edition, J. Quant. Spectrosc. Radiat. Transfer, 60, 665–710.

[8] Poll, J. D. and Hunt, J. L. (1976) On the moments of the pressure-induced spectra of gases, Can. J. Phys., 54, 461–470.

[9] Birnbaum, G. and Cohen, E. R. (1976) Theory of line shape in pressure-induced absorption, Can. J. Phys., 54, 593–602.

[10] Borysow, A. and Frommhold, L. (1986) Collision-induced rototranslational absorption spectra of N_2–N_2 pairs for temperatures from 50 to 300 K, Astrophys. J., 311, 1043–1057.

[11] Maté, B., Fraser, G. T., and Lafferty, W. J. (2000) Intensity of the simultaneous vibrational absorption $CO_2(v_3{=}1)$ + $N_2(v{=}1)$ ← $CO_2(v_3{=}0)$ + $N_2(v{=}0)$ at $4680\,cm^{-1}$, J. Mol. Spectrosc., 201, 175–177.

[12] Tobin, D. C., Strow, L. L., Lafferty, W. J., and Olson, W. B. (1996) Experimental investigation of the self- and N_2 broadened continuum within the ν_2 band of water vapor, Appl. Opt., 35, 4724–4734.

[13] Tobin, D. C., Best, F. A., Brown, P. D., Clough, S. A., Dedecker, R. G., Ellingson, R. G., Garcia, R. K., Howell, H. B., Knuteson, R. O., Mlawer, E. J., Revercomb, H. E., Short, J. F., van Delst, P. F., and Walden, V. P. (1999) Downwelling spectral radiance observations at the SHEBA ice station: Water vapor continuum measurements from 17–26 μm, J. Geophys. Res., Atmospheres, 104, 2081–2092.

[14] Burch, D. E. (1981) Continuum absorption by H_2O, AFGL-TR-81-0300.

[15] Burch, D. E. and Alt, R. L. (1984) Continuum absorption by H_2O in the 700–1200 cm^{-1} and 2400–2800 cm^{-1} windows, AFGL-TR-84-0128.

CLUSTERING, SATURATED VAPORS, AND THE ATMOSPHERE: THE $(H_2O)_n$, H_2O-N_2, AND H_2O-O_2 CASES

Z. Slanina

Institute of Chemistry, Academia Sinica, 128 Yen-Chiu-Yuan Rd., Sec. 2,
Nankang, Taipei 11529, Taiwan — R.O.C., and
Institute for Molecular Science, Okazaki 444-8585, Japan

F. Uhlík

School of Science, Charles University,
128 43 Prague 2, Czech Republic

Abstract   Recent heuristic computational findings of temperature increase of clustering degree in several quite different saturated vapors are analyzed. A thermodynamic proof is presented, showing that this event should be rather common. Illustrations are primarily based on saturated steam and also on carbon vapor. Consequences for the atmosphere are discussed. Some preliminary $(U)MP2\,FC/6\text{-}311G^{**}$ results on the H_2O-N_2 and H_2O-O_2 hetero-dimers are reported. The G1 calculations for the $(O_2)_2$ system are presented for the first time too.

1. Introduction

Studies of atomic and molecular clusters have developed into a considerably broad field (see, e. g. [1–3]), covering aggregates of very different types. Clusters of basically any species can be considered — cf. for example the He dimer [4]. Although their experimental generation is frequently rather far from thermodynamic equilibrium, computational modeling of clusters has mostly worked — owing to obvious reasons — with dimers under the equilibrium conditions.

Water clusters have been studied very vigorously [5–7]. In order to estimate possible effects of water clusters of any dimension, a thermodynamic extrapolation scheme has been applied [8, 9]. Relative populations of water clusters in saturated steam (over ice or liquid water) have

C. Camy-Peyret and A.A. Vigasin (eds.),
Weakly Interacting Molecular Pairs: Unconventional Absorbers of Radiation in the Atmosphere, 101–110.
© 2003 *Kluwer Academic Publishers. Printed in the Netherlands.*

also been computed [10–12] based on the observed saturation pressure. It has been found in the model studies that the clustering degree increases with increasing temperature. It may be a surprising result but in fact it can easily be rationalized. While the equilibrium constants for cluster formation decrease with temperature, the saturated pressure increases. It is just the competition between these two terms which decides the final temperature behavior.

As several computational approximations have to be used with the water clusters, some simplesystems have also been investigated like dimerization in the argon saturated vapor [13] with the dimerization equilibrium constant from advanced evaluations [14]. A temperature increase of the dimeric fraction has again been found. A similar result could be found [15] for Mg vapor. The treatment has also been applied [16–18] to topical system of carbon clusters in order to explain a paradox that C_{60} is more populated than C_{70} in spite of the fact that the former has higher relative energy than the latter [19–23]. Carbon saturated pressure is not well known and has to be estimated owing to high temperatures involved, however, the temperature increase of the clustering degree results from the computational modeling, too.

In this report, the interesting computational finding is further analyzed without any reference to a particular potential function or approximation, i. e. from thermodynamic point of view. A simple thermodynamic criterion for the temperature enhancement of clustering degree is suggested and relation to the atmosphere discussed.

2. Simple dimerization case

Let us consider a substance A that is in both condensed (liquid) and gas phase and those two phases are in thermodynamic gas-liquid equilibrium. Moreover, in the gas phase a dimerization equilibrium exists:

$$2A(g) = A_2(g) \tag{1}$$

that is described by the equilibrium constant:

$$K_{p,2} = \frac{p_{A_2}}{p_A^2}, \tag{2}$$

in the terms of partial pressures of the components, p_A and p_{A_2}. Let P is the pressure of the saturated vapor, then the dimer equilibrium mole fraction x_2 is given [8, 9]:

$$x_2 = \frac{2PK_{p,2} + 1 - (4PK_{p,2} + 1)^{1/2}}{2PK_{p,2}}. \tag{3}$$

Temperature dependence of the dimerization equilibrium constant $K_{p,2}$ is described by the van't Hoff equation:

$$\frac{d\ln K_{p,2}}{dT} = \frac{\Delta H_2^0}{RT^2},\tag{4}$$

where ΔH_2^0 stands for the (negative) standard change of enthalpy. Let us further suppose that the saturated pressure P obeys the Clausius-Clapeyron equation:

$$\frac{d\ln P}{dT} = \frac{\Delta H_{\mathrm{vap}}}{RT^2},\tag{5}$$

where ΔH_{vap} symbolizes the (positive) enthalpy of vaporization.

Temperature derivative dx_2/dT shows the temperature evolution of the dimer equilibrium mole fraction. If it is positive, the dimerization degree x_2 will be increasing with temperature. In order to derive the temperature derivative dx_2/dT, it is more convenient to start from the quadratic equation that gives solution (3):

$$\Xi\, x_2^2 - (1 + 2\,\Xi)\, x_2 + \Xi = 0,\tag{6}$$

where $\Xi = PK_{p,2}$. It follows from the temperature differentiation of Eq. (6):

$$\frac{dx_2}{dT} = \frac{(x_2 - 1)^2}{2\,\Xi\,(1 - x_2) + 1}\frac{d\Xi}{dT}.\tag{7}$$

Hence, it turns out that the sign of the dx_2/dT derivative is determined by the sign of the $d\Xi/dT$ derivative. The last mentioned term reads:

$$\frac{d\Xi}{dT} = \frac{\Xi}{RT^2}\left(\Delta H_2^0 + \Delta H_{\mathrm{vap}}\right).\tag{8}$$

Consequently, the $(\Delta H_2^0 + \Delta H_{\mathrm{vap}})$ term is decisive. As long as $\Delta H_{\mathrm{vap}} > |\Delta H_2^0|$, the temperature derivative dx_2/dT is positive and the x_2 term increases with temperature, i.e. the dimeric mole fraction in the saturated vapor increases with temperature. Table 1 compares the ΔH_2^0 and ΔH_{vap} terms for water as an illustration.

3. Clusters of any dimension

The case of monomers and dimers only is a simplification, in a more general scheme one should allow for clusters A_i of any dimension:

$$i\mathrm{A}(g) = \mathrm{A}_i(g)\ (i = 2,\,\ldots),\tag{9}$$

Table 1. Comparison of the $\frac{\Delta H_i^0}{i-1}$ and ΔH_{vap} terms (kcal/mol) for some illustrative examples.

System	T, K	i	$\dfrac{\Delta H_i^0}{i-1}$	ΔH_{vap}
H_2O	298.15	2	-4.3^a	10.5^b
C	3000	2	-144.2^b	170.2^c
C	5000	2	-143.4^b	167.5^c
C	298.15	60	-163.6^d	171.3^c
C	298.15	70	-164.3^d	171.3^c

[a] An average value from Ref. [12]
[b] Ref. [30]
[c] Estimated from the data for C(solid), C(gas) in Ref. [30]
[d] Refs. [24, 28]

their formation being described by equilibrium constants:

$$K_{p,i} = \frac{p_{A_i}}{p_A{}^i}, \quad (i = 2, \ldots). \tag{10}$$

Temperature dependence of the dimerization equilibrium constant $K_{p,i}$ is also described by the van't Hoff equation:

$$\frac{d \ln K_{p,i}}{dT} = \frac{\Delta H_i^0}{RT^2}, \quad (i = 2, \ldots). \tag{11}$$

If we want to compare the ΔH_i^0 values, it is convenient to consider the $(1/i)\Delta H_i^0$ terms as it has been customary with carbon clusters. For the carbon clusters either atomization heats per carbon atom [24] (if they are taken so that they have negative values, they have the meaning of the $(1/i)\Delta H_i^0$ terms) or heats of formation per carbon atom [25] are considered. The two relative terms differ just by a constant term. Already the first-ever computed heats of formation per atom for a few smaller carbon clusters revealed [26, 27] a fast smooth decrease of the $(1/i)\Delta H_i^0$ term (cf. also Table 1).

The total pressure P will be the result of all the partial pressures derived from Eqs. (10):

$$P = p_A + \sum_{i=2}^{\infty} p_A{}^i K_{p,i}. \tag{12}$$

We can also switch to the monomeric molar fraction x_1:

$$1 = x_1 + \sum_{i=2}^{\infty} x_1{}^i P^{i-1} K_{p,i}. \tag{13}$$

This algebraic equation cannot be solved in a closed form (in contrast to the simple dimerization case, Eq. (3)). Nevertheless, the temperature derivative dx_1/dT can still be evaluated:

$$-\frac{dx_1}{dT}\left(1+\sum_{i=2}^{\infty} i\,\xi_i\right) = \sum_{i=2}^{\infty} x_1\,\xi_i\,\frac{\Delta H_i^0 + (i-1)\,\Delta H_{\text{vap}}}{R\,T^2}\,, \qquad (14)$$

where the ξ_i terms stand for expressions:

$$\xi_i = {x_1}^{i-1}\,P^{\,i-1}\,K_{p,i}. \qquad (15)$$

From Eq. (14) a simple sufficient (though not necessary) condition can be derived for temperature decrease of x_1 in the saturated vapor (i.e. for temperature increase of the clustering degree $1-x_1$). As long as $\Delta H_{\text{vap}} > \frac{|\Delta H_i^0|}{i-1}$ for every $i \geq 2$, the temperature derivative dx_1/dT is negative and thus, the clustering degree increases with temperature in the saturated vapor. Obviously, in the case $i=2$ this rule is reduced to that formulated previously for the simple dimerization case.

Interestingly enough, while with carbon clusters the energy terms divided by the number of carbon atoms have so far been used [19, 25], for the rule on clustering degree a bit different terms are relevant, i.e. those produced through division by $i-1$, not by i. Table 1 documents, on the best available energy data for higher carbon clusters [24, 28, 29], that for C_{60} and C_{70} the condition is indeed valid (though there are only rough estimates [30] of ΔH_{vap} available for graphite). It is also possible that in fullerene synthesis a supersaturated rather than saturated vapor is present. It should also be analyzed to what degree all the clusters involved can effectively be treated as ideal gases. Anyhow, for an experimental verification of the predicted temperature increase of the clustering degree in the saturated vapor, steam should definitely be more suitable system [31, 32].

4. H_2O–N_2 and H_2O–O_2 hetero-dimers

The importance of molecular complexes for spectral records and for thermophysical properties of the atmosphere has clearly been established [33–35]. There is a particular sub-group in the set of molecular complexes relevant to the atmosphere — the complexes with water, both water clusters and hetero-clusters with water as one component. According to the previous section, contents of molecular complexes in saturated or nearly-saturated vapors increase with temperature.

The atmosphere can also be close to its saturation with water. Hence, the contents of water dimers or even of water-containing hetero-dimers can actually increase with the atmospheric warming-up (strictly speaking, contents of water clusters should increase under near-saturation conditions — the case of hetero-clusters is not covered by the above treatment and should be given a special reasoning). In addition to that, those water-containing species are also green-house effect agents, somewhat different from free water. Hence, it make sense to clarify if the water-containing complexes themselves could — if increased — contribute to some additional warming-up. This hypothesis can only be studied computationally at present. Anyhow, it would be interesting to estimate what kind of contribution into the green-house effect can be expected from water-containing homo- and hetero-dimers.

There are two particularly interesting water-containing hetero-dimers, viz. H_2O–N_2 and H_2O–O_2. In order to investigate their temperature development in a mixture of air and saturated steam, we have been performing *ab initio* computations with the Gaussian program package [36]. The computations are carried out at the second order Møller-Plesset (MP2) perturbation treatment with the frozen-core option (MP2 FC) in the standard 6-311G** basis set, i.e. the MP2 FC/6-311G** or UMP2 FC/6-311G** treatment for the H_2O–N_2 and H_2O–O_2 species, respectively. In both cases, three stationary points have been located. If we, somewhat arbitrarily, define hydrogen bonds as H–O or H–N long-range linkages shorter than 3 Å, the stationary points can be described as exhibiting two or one hydrogen bond.

In the H_2O–N_2 case, the lowest-energy structure is non-planar with two hydrogen bonds 2.76 and 2.80 Å long. Its MP2=FC/6-311G** dimerization potential energy is $\Delta E = -1.62$ kcal/mol. It is followed by a planar structure with one hydrogen bond of 2.49 Å and dimerization energy of $\Delta E = -1.59$ kcal/mol. The third lowest structure is a planar species with two hydrogen bonds of 2.68 and 2.87 Å and a stabilization energy slightly less than -1.59 kcal/mol.

Decrease in the potential energy upon dimerization is by about 0.5 kcal/mol smaller for the H_2O–O_2 species. The lowest-energy structure is also non-planar in this case, it has two hydrogen bonds of 2.75 and 2.87 Å, and the related UMP2 FC/6-311G** dimerization potential energy is $\Delta E = -0.96$ kcal/mol. The second lowest isomer is a planar structure with two hydrogen bonds 2.65 and 2.68 Å long, stabilized by -0.91 kcal/mol. Finally, the third species is a C_{2v} planar form with two hydrogen bonds 2.72 Å long and the dimerization energy $\Delta E = -0.81$ kcal/mol.

The listed dimerization potential energy terms are without the BSSE correction [37]. It still seems that the physical significancy of the correction is not clarified [4]. In order to avoid the BSSE correction, one can think on application of some version of the developing G-theories [38–40]. The G1 theory [38] applies the full fourth-order Møller-Plesset (MP4) treatment with the 6-311G**, 6-311+G**, and 6-311G(2df,p) basis sets. In addition, a quadratic configuration interaction with all singles and doubles (QCISD) and with a quasi-perturbative treatment of higher excitations (QCISD(T)) is applied with the 6-311G** basis set. The effects of polarization space extension and diffuse functions are then assumed to be additive at the MP4 level, and a correction formula for further basis set incompleteness is applied. The G-theories are aiming [38–40] at a precise description of thermochemistry, however, they have not been tested for description of weak molecular complexes.

In order to see performance of the new tool, we have applied the G1-theory [38] to the three structures considered for $(O_2)_2$ in our previous evaluations [41]. The T-shape C_{2v}, rhomboid C_{2h}, and linear $D_{\infty h}$ exhibit the G1 dimerization energies of -0.62, -0.51, and $-0.48\,$kcal/mol, respectively. In order to get through the computations, the frozen-core approached had to be applied, though not supposed in the original G1 scheme [38]. The potential-energy terms should however be further corrected for the zero-point vibrational energy and for temperature effects in order to be compared with experiment [42]. Orlando *et al.* studied [42] the temperature dependence of collision-induced absorption by oxygen over the temperature range 225–356 K. They concluded a heat of formation ΔH_{f} for $(O_2)_2$ as $-1.1 \pm 0.5\,$kcal/mol. Long and Ewing reported [43, 44] a ΔE_{f} term of $-0.53\,$kcal/mol near 100 K which corresponds [42] to ΔH_{f} of about $-0.75\,$kcal/mol (for a wider, diversified perspective of tetraoxygen, cf. also papers [45–50]). In overall, the G1 results seem to be in an encouraging agreement with the observed values. However, from a practical point of view, the H_2O–N_2 system should be given preference in submission to still higher levels of theory (also because it is free of open-shell difficulties).

Acknowledgments

The reported research has been supported by the National Science Council, Taiwan, Republic of China and by the Japan Society for the Promotion of Science. In its initial stage, the study had been supported by the Alexander von Humboldt-Stiftung.

References

[1] Slanina, Z. (1986) *Contemporary Theory of Chemical Isomerism*, D. Reidel Publ. Comp., Dordrecht.

[2] Castleman, A. W. and Wei, S. (1994) Cluster reactions, Annu. Rev. Phys. Chem., 45, 685–719.

[3] Castleman, A. W. and Bowen, K. H. (1996) Clusters: Structure, energetics, and dynamics of intermediate states of matter, J. Phys. Chem., 100, 12911–12944.

[4] Gutowski, M., van Duijneveldt-van de Rijdt, J. G. C. M., van Lenthe, J. H., and van Duijneveldt, F. B. (1993) Accuracy of the Boys and Bernardi function counterpoise method, J. Chem. Phys., 98, 4728–4737.

[5] Clementi, E. (1976) *Determination of Liquid Water Structure, Coordination Numbers for Ions and Solvation for Biological Molecules*, Springer-Verlag, Berlin.

[6] Bondybey, V. E. and Beyer, M. K. (2002) How many molecules make a solution? Intl. Rev. Phys. Chem., 21, 277–306.

[7] van Duijneveldt-van de Rijdt, J. G. C. M. and van Duijneveldt, F. B. (1999) Interaction optimized basis sets for correlated *ab initio* calculations on the water dimer, J. Chem. Phys., 111, 3812–3819.

[8] Slanina, Z. (1987) A cluster model of the gas-phase water, Thermochim. Acta, 114, 273–280.

[9] Slanina, Z. (1990) Computational studies of water clusters: temperature, pressure and saturation effects on cluster fractions within the RRHO MCY-B/EPEN steam, J. Mol. Struct., 273, 81–92.

[10] Slanina, Z. (1987) Pressure enhancement of gas-phase water-oligomer populations: a RRHO MCY-B/EPEN computational study, Z. Phys. D, 5, 181–186.

[11] Slanina, Z. (1987) Temperature enhancement of water-oligomer populations in saturated aqueous vapor, Thermochim. Acta, 116, 161–169.

[12] Slanina, Z. (1991) A comparative study of the water-dimer gas-phase thermodynamics in the BJH- and MCYL-type flexible potentials, Chem. Phys., 150, 321–329.

[13] Slanina, Z. (1992) A useful test of temperature dependency of cluster populations under saturation conditions: the Ar_2 case, Thermochim. Acta, 209, 1–6.

[14] Dardi, P. S. and Dahler, J. S. (1990) Equilibrium constants for the formation of van der Waals dimers—calculations for Ar–Ar and Mg–Mg, J. Chem. Phys., 93, 3562–3572.

[15] Slanina, Z. (1992) On temperature increase of the Mg_2 mole fraction in saturated magnesium vapor, Thermochim. Acta, 207, 9–13.

[16] Slanina, Z., Rudziński, J. M., and Ōsawa, E. (1987) $C_{60}(g)$ and $C_{70}(g)$: a computational study of pressure and temperature dependence of their populations, Carbon, 25, 747–750.

[17] Slanina, Z., Rudziński, J. M., and Ōsawa, E. (1987) $C_{60}(g)$, $C_{70}(g)$, saturated carbon vapor and increase of cluster populations with temperature: a combined AM1 quantum-chemical and statistical-mechanical study, Collect. Czech. Chem. Commun., 52, 2381–2838.

[18] Slanina, Z., Zhao, X., Kurita, N., Gotoh, H., Uhlík, F., Rudziński, J. M., Lee, K. H., and Adamowicz, L. (2001) Computing the relative gas-phase populations

of C_{60} and C_{70}: beyond the traditional $\Delta H^0_{f,298}$ scale, J. Mol. Graphics Mod., 19, 216–221.

[19] Curl, R. F. (1993) On the formation of the fullerenes, Phil. Trans. Roy. Soc. London A, 343, 19–32.

[20] Cioslowski, J. (1995) *Electronic Structure Calculations on Fullerenes and Their Derivatives*, Oxford University Press, Oxford.

[21] Scuseria, G. E. (1995) Theoretical studies of fullerenes. In: *Modern Electronic Structure Theory, part I*, Yarkony, D. R., editor. World Scientific Publ., Singapore, pp. 279–310.

[22] Slanina, Z., Lee, S.-L., and Yu, C.-H. (1996) Computations in treating fullerenes and carbon aggregates, Rev. Comput. Chem., 8, 1–62.

[23] Slanina, Z., Zhao, X., and Ōsawa, E. (1999) Relative stabilities of isomeric fullerenes, Advan. Strain. Inter. Org. Mol., 7, 185–235.

[24] Slanina, Z. and Ōsawa, E. (1997) Average bond dissociation energies of fullerene, Fullerene Sci. Technol., 5, 167–175.

[25] Sun, M.-L., Slanina, Z. and Lee, S.-L. (1995) Square/hexagon route towards the boron–nitrogen clusters, Chem. Phys. Lett., 233, 279–283.

[26] Slanina, Z. (1974) Ph. D. Thesis, Czech. Acad. Sci., Prague.

[27] Slanina, Z. (1975) Remark on the present applicability of quantum chemistry to the calculations of equilibrium and rate constants of chemical reactions, Radiochem. Radioanal. Lett., 22, 291–298.

[28] Kiyobayashi, T. and Sakiyama, M. (1993) Combustion calorimetric studies on C_{60} and C_{70}, Fullerene Sci. Technol., 1, 269–273.

[29] Beckhaus, H.-D., Verevkin, S., Rüchardt, C., Diederich, F., Thilgen, C., ter Meer, H.-U., Mohn, H., and Müller, W. (1994) C_{70} is more stable than C_{60}: experimental determination of the heat of formation of C_{70}, Angew. Chem., Int. Ed. Engl., 33, 996–998.

[30] *JANAF Thermochemical Tables*, (1971, 1985) Natl. Bur. Stand., Washington, D.C.

[31] Walrafen, G., Fisher, M. R., Hokmabadi, M. S., and Yang, W. H. (1986) Temperature dependence of low- and high-frequency Raman scattering from liquid water, J. Chem. Phys., 85, 6970–6971.

[32] Fujita, Y. and Ikawa, S. (1989) Effect of temperature on the hydrogen bond distribution in water as studied by infrared spectra, Chem. Phys. Lett., 159, 184–188.

[33] Vigasin, A. A. (2000) Water vapor continuous absorption in various mixtures: possible role of weakly bound complexes, J. Quant. Spectr. Radiat. Transf., 64, 25–40.

[34] Vaida, V., Headrick, J. E. (2000) Physicochemical properties of hydrated complexes in the Earth's atmosphere, J. Phys. Chem. A, 104, 5401–5412.

[35] Slanina, Z., Uhlík, F., Saito, A. T., and Ōsawa, E. (2001) Computing molecular species of atmospheric significancy in earth's and other atmospheres, Phys. Chem. Earth C, 26, 505–511.

[36] Frisch, M. J., Trucks, G. W., Schlegel, H. B., Scuseria, G. E., Robb, M. A., Cheeseman, J. R., Zakrzewski, V. G., Montgomery Jr., J. A., Stratmann, R. E., Burant, J. C., Dapprich, S., Millam, J. M., Daniels, A. D., Kudin, K. N., Strain, M. C., Farkas, O., Tomasi, J., Barone, V., Cossi, M., Cammi, R., Mennucci, B.,

Pomelli, C., Adamo, C., Clifford, S., Ochterski, J., Petersson, G. A., Ayala, P. Y., Cui, Q., Morokuma, K., Malick, D. K., Rabuck, A. D., Raghavachari, K., Foresman, J. B., Cioslowski, J., Ortiz, J. V., Stefanov, B. B., Liu, G., Liashenko, A., Piskorz, P., Komaromi, I., Gomperts, R., Martin, R. L., Fox, D. J., Keith, T., Al-Laham, M. A., Peng, C. Y., Nanayakkara, A., Gonzalez, C., Challacombe, M., Gill, P. M. W., Johnson, B., Chen, W., Wong, M. W., Andres, J. L., Gonzalez, C., Head-Gordon, M., Replogle, E. S. and Pople, J. A. (1998) *Gaussian 98, Revision A.7*, Gaussian, Inc., Pittsburgh, PA.

[37] Boys, S. F. and Bernardi, F. (1970) The calculation of small molecular interactions by the difference of separate total energies. Some procedures with reduced errors, Mol. Phys., 19, 553–568.

[38] Curtiss, L. A., Jones, C., Trucks, G. W., Raghavachari, K., and Pople, J. A. (1990) Gaussian-1 theory of molecular-energies for 2nd-row compounds, J. Chem. Phys., 93, 2537–2545.

[39] Curtiss, L. A., Raghavachari, K., Trucks, G. W., and Pople, J. A. (1991) Gaussian-2 theory for molecular-energies of 1st-row and 2nd-row compounds, J. Chem. Phys., 94, 7221–7230.

[40] Curtiss, L. A., Redfern, P. C., Raghavachari, K., and Pople, J. A. (2002) Gaussian-3X (G3X) theory using coupled cluster and Brueckner energies, Chem. Phys. Lett., 359, 390–396.

[41] Uhlík, F., Slanina, Z., and Hinchliffe, A. (1993) An *ab initio* correlated study of structure, energetics and vibrations of $(O_2)_2$, J. Mol. Struct. (Theochem), 285, 273–276.

[42] Orlando, J. J., Tyndall, G. S., Nickerson, K. E., and Calvert, J. G. (1991) The temperature-dependence of collision-induced absorption by oxygen near $6\,\mu$m, J. Geophys. Res. D, 96, 20755–20760.

[43] Long, C. A. and Ewing, G. E. (1971) The infrared spectrum of bound state oxygen dimers, Chem. Phys. Lett., 9, 225–229.

[44] Long, C. A. and Ewing, G. E. (1973) Spectroscopic investigation of van der Waals molecules. I. The infrared and visible spectra of $(O_2)_2$, J. Chem. Phys., 58, 4824–4834.

[45] Slanina, Z. (1974) Dimerization of homonuclear biatomic molecules, Collect. Czech. Chem. Commun., 39, 228–235.

[46] Uhlík, F., Slanina, Z., and Hinchliffe, A. (1993) Gas-phase association of O_2: a computational thermodynamic study, Thermochim. Acta, 228, 9–14.

[47] Slanina, Z., Uhlík, F., De Almeida, W. B., and Hinchliffe, A. (1994) A computational thermodynamic evaluation of the altitude profiles of $(N_2)_2$, N_2–O_2, and $(O_2)_2$ in the Earth's atmosphere, Thermochim. Acta, 231, 55–60.

[48] Aquilanti, V., Ascenzi, D., Bartolomei, M., Cappelletti, D., Cavalli, S., Vitores, M. D., and Pirani, F. (1999) Molecular beam scattering of aligned oxygen molecules. The nature of the bond in the O_2–O_2 dimer, J. Am. Chem. Soc., 121, 10794–10802.

[49] Cacace, F., de Petris, G., and Troiani, A. (2001) Experimental detection of tetraoxygen, Angew. Chem., Int. Ed. Engl., 40, 4062–4062.

[50] Cacace, F., de Petris, G., and Troiani, A. (2002) Experimental detection of tetranitrogen, Science, 295, 480–481.

EQUILIBRIUM CONSTANTS FOR THE FORMATION OF WEAKLY BOUND DIMERS

A. A. Vigasin

Wave Research Center,

General Physics Institute, Russian Academy of Sciences,

Vavilova 38, Moscow, 119991, Russia

Abstract   Simplified expressions are proposed for the estimation of the equilibrium abundance of dimers. Physical background and accuracy of some popular relationships for the equilibrium constants are discussed. Classical calculations for truncated partition functions for true bound states are shown to yield reasonably accurate estimates of the equilibrium constant for diatomic dimers. Care should be taken, however, in extending quasi-diatomic relationships to polyatomic dimers.

Keywords: mass-action law / dimers / partition function / intermolecular potential / dimer content

1. Introduction

During the past decades extensive spectroscopic studies and quantum chemical calculations have provided a huge amount of data concerning individual molecular parameters of van der Waals dimers. Practical needs, e.g. atmospheric applications, often require, however, to convert these molecular parameters into the estimates of the self-associated or mixed dimer number density in the gas at equilibrium. Statistical physics allows for the prediction of these abundances making use of the law of mass-actions. The latter implies computations of the relevant partition functions which are based on the knowledge of the geometry, energetics, and vibrational spectra of initial and final products. In the classical approximation the calculation of the dimer partition function requires the knowledge of the potential energy surface (PES) characterizing the interaction between the pair of monomers.

111

C. Camy-Peyret and A.A. Vigasin (eds.),
Weakly Interacting Molecular Pairs: Unconventional Absorbers of Radiation in the Atmosphere, 111–124.
© 2003 *Kluwer Academic Publishers. Printed in the Netherlands.*

The advent of powerful computer tools, such as e.g. LEVEL 7.0 computer code [1], makes it possible to easily determine all the rovibrational energy levels for diatomic molecules or dimers. This means that the computations of the dimer rovibrational partition functions can be made accurately by direct summation over available quantum states. Nevertheless, simplified analytical expressions are worth considering in order to make clear the underlying physics and because they can be used in subsequent more sophisticated modeling of the collision states. An extension of simple analytical approaches to polyatomic dimers is a particularly interesting pathway since rigorous computations of high-lying levels is still unavailable for polyatomic dimers.

The so-called Rigid Rotor–Harmonic Oscillator (RRHO) approximation forms one of the popular basis for calculating equilibrium constants and is known to be reliably applicable in case of "normal" molecules. Care should be taken, however, when attempting to apply this commonly used formalism to weakly bound species. The reason for that is multifold:

- The extreme weakness of intermolecular bonds causes substantial population of very high-lying dimer states even at relatively low temperature. All true bound dimer states in the intermolecular potential are often equally populated at ambient temperature;

- Barriers to internal rotations are easy to overcome at ambient temperature, hence the anisotropy of interaction is often effectively averaged;

- Metastable states can be substantially populated. Among these the so-called shape resonance states are likely to be distinguished from Feshbach resonances, which are of a different nature;

- The zero-point energy may represent considerable fraction of the intermolecular well depth. This means that quantum corrections due to zero-point energy can not be neglected even though the overall problem may be treated in terms of classical physics.

The statistical physics description of metastable dimer states represents a serious problem. One of the possible solutions consists of direct counting the rovibrational dimer states, the energy of which exceeds that of the dissociation threshold. Such counting, however, is hard to apply because of severe difficulties in solving the Schrödinger equation for polyatomic pairs. Stogryn and Hirshfelder [2] were first to demonstrate how shape resonance states should be accounted for in calculating the equilibrium constant for a dimer consisting of structure-less monomers. Later on [3], their classical approach was extended to cover the case of polyatomic monomers. In what follows we concentrate on the calculation of the equilibrium constants for true bound dimeric states only. The goal of the present paper is to discuss both the physical nature of weakly bound dimers and the precautions one should use, when applying some popular expressions for the equilibrium constants recommended in the literature.

2. General outline

The validity of computations of dimer abundances is sensitive to various factors. Primarily one should examine the adequacy of the adopted intermolecular PES and the accuracy of the calculation of relevant partition functions. While considering the role of both factors Dardi and Dahler [4] concluded that classical integration using simplified model potentials (e. g. Lennard—Jones 6 : 12 potential) gives rise to notable errors in the calculated equilibrium constant for diatomic dimers. In contrast, good agreement is usually found between quantum and classical calculations based on identical potential well shapes.

In the forthcoming sections we pay due attention to simplified calculations of the equilibrium constant which allow rapid and reliable estimates using a minimum number of input parameters. Two approaches will be presented. One of these is based on the use of a truncated harmonic oscillator (THO) shape for the potential well which is appropriate for intermolecular stretch. This model allows for obtaining tractable analytical expressions. The latter are then compared with the popular relationship based on numerical calculations by Stogryn and Hirshfelder [2] for the Lennard—Jones potential. The proposed analytical form was suggested by Anderson et al. [5] and was employed then in a number of publications (see e. g. [6–8]) for atmospheric applications. In terms of molar fraction x_2 of diatomic dimers formed in a gas with a total pressure P, the relationship derived in [5] can be written in the form:

$$x_2 = b_0 \frac{P}{RT} \frac{2}{\chi} \exp\left(0.105 + \frac{0.364}{T^*} - 1.46 \ln T^*\right). \tag{1}$$

Here $b_0 = (2/3)\pi N_0 \sigma^3$, $R = kN_0$ is the gas constant, k and N_0 are the Boltzmann constant and the Avogadro's number, respectively, χ is the symmetry number ($\chi = 1$ for heteronuclear dimer and $\chi = 2$ for homonuclear dimer). The reduced temperature T^* stands for the ratio $T^* = kT/\epsilon$, ϵ and σ being the conventional parameters for the

Lennard—Jones (LJ) interatomic potential:

$$U(r) = 4\epsilon \left(\left(\frac{\sigma}{r}\right)^{12} - \left(\frac{\sigma}{r}\right)^{6} \right). \tag{2}$$

Formula (1) was derived in [5] by a least-squares fit of numerical data obtained by Stogryn and Hirschfelder [2] for true bound dimeric states. It has been shown in [9] that the specific temperature variation of x_2 in (1) is due to the limited number of true bound states in the interatomic potential well.

3. Diatomic dimers

Let us consider the following association equilibrium

$$A + B + M \rightleftharpoons D + M, \tag{3}$$

where a dimer D is formed from the monatomic monomers A and B in a bath of M particles. The equilibrium constant for the reaction (3)

$$K_p = \frac{P_D}{P_A \, P_B} \tag{4}$$

along with the Dalton law $P = P_D + P_A + P_B$ for the partial pressures allow for determination of the dimer molar fraction x_2. Presupposing weak self-association, i.e. when $A \equiv B$ one readily obtains: $x_2 \approx P K_p$.

Let us assume that intermolecular potential energy well[1] can be approximated by THO potential:

$$U(r) = -D_e + k_s \, (r - r_e)^2 \qquad |r - r_e| \leq \sqrt{D_e/k_s}. \tag{5}$$

Here k_s is the interparticle force constant, D_e stands for the potential well depth. With these assumptions the classical theory allows to obtain the following closed-form expression for the equilibrium constant for true bound dimer states [9]:

$$K_p = \Lambda^3 \frac{kT}{(h\omega)\,(hcB\chi)} \left[e^{\frac{D_e}{kT}} - \left(1 + \frac{D_e}{kT}\right) \right], \tag{6}$$

where $\Lambda = h/(2\pi\mu_c kT)^{1/2}$ stands for the thermal de Broglie wavelength corresponding to the reduced mass $\mu_c = m_A m_B/(m_A + m_B)$, k and h are the Boltzmann and Planck constants, respectively.

---

[1]This energy is supposed to characterize a non-rotating dimer.

Note that the term in square brackets in (6) was obtained by Jaffe [10] while examining the high-temperature rovibrational partition functions of conventional diatomics. Also Knuth [11] derived a formula similar to (6) by kinetic modeling of the dimer formation in noble gases. Assuming a truncated harmonic oscillator potential well shape Knuth obtained the following expression for the equilibrium constant for the rare gas dimers:

$$K_p = \frac{128\sqrt{\pi}}{45}\, n\, \sigma^3\, \frac{kT}{\epsilon}\left[e^{\frac{\epsilon}{kT}} - \left(1 + \frac{\epsilon}{kT}\right)\right]. \tag{7}$$

Here K_p is expressed in $\mathrm{cm}^3\,\mathrm{mole}^{-1}$, and n is the monomer number density. The numerical factor in this expression has been derived in [11] by adjusting the numerical constant to match that of Stogryn and Hirshfelder [2] (see also (1)) in the high-temperature limit.

Some more sophisticated and somewhat more exact closed-form approximations to the Stogryn and Hirshfelder's (SH) calculations have been suggested then by Meyer [12]. Meyer noted that the relative error of the Knuth's formula (7) with respect to the SH solution for the LJ potential reaches up to 100% at low temperature. In what follows we provide a brief comment on this deviation.

In contrast to (7), formula (6) is classically exact for diatomics and as such it could be the basis for further modifications. In particular, simple versions of (6) can be suggested which can be used irrespective of the true potential well shape. It is convenient for instance to introduce the vibrational frequency ω and the rotational constant B in cm^{-1}. The formula (6) holds then the form:

$$K_p^{\mathrm{THO}}[\mathrm{atm}^{-1}] = \frac{18.86}{\omega\, B\, \chi\, T^{0.5}\, \mu_c^{1.5}}\left[e^{\frac{\tilde{D}_e}{T}} - \left(1 + \frac{\tilde{D}_e}{T}\right)\right]. \tag{8}$$

Here μ_c is given in atomic mass units, the tilde over D_e means hereafter that this energy term is expressed in Kelvins, while ω and B are in cm^{-1}.

Alternative form of the equilibrium constant can be suggested if interatomic vibrational stretch frequency ω is expressed in terms of potential parameters. This can be done assuming e. g. the Lennard—Jones well shape (2). It is convenient to rewrite (2) in the following form:

$$U(r) = D_e\left(\left(\frac{r_e}{r}\right)^{12} - 2\left(\frac{r_e}{r}\right)^{6}\right). \tag{9}$$

Here r_e stands for the equilibrium interatomic separation. Taking the second derivative of (9) one obtains $k_s = 72D_e/r_e^2$. Consequently the vibrational stretching frequency has the form:

$$\omega[\mathrm{cm}^{-1}] = (2\pi c\, r_{\mathrm{e}})^{-1}\sqrt{72\, D_e/\mu_c}\,.$$

Also the rotational constant B can be written as:

$$B[\mathrm{cm}^{-1}] = h/(8\pi^2 c\,\mu_c\, r_e^2)\,.$$

Substitution of these into formula (6) results in:

$$K_p^{\mathrm{THO}}[\mathrm{atm}^{-1}] = \frac{N_0}{R}\,\frac{2\,\pi^{1.5}\, r_{\mathrm{e}}^3}{3}\,\frac{1}{\chi\,\sqrt{T\tilde{D}_e}}\left[e^{\frac{\tilde{D}_e}{T}} - \left(1 + \frac{\tilde{D}_e}{T}\right)\right]. \tag{10}$$

Here r_{e} is in cm, R is the gas constant $R = 82.053\ \mathrm{atm\,cm^3\,(K\,mole)^{-1}}$. Formula (10) can be rewritten with sufficient accuracy as:

$$K_p^{\mathrm{THO}}[\mathrm{atm}^{-1}] \approx \frac{0.027\, r_{\mathrm{e}}^3}{\chi\,\sqrt{T\,\tilde{D}_{\mathrm{e}}}}\left[e^{\frac{\tilde{D}_e}{T}} - \left(1 + \frac{\tilde{D}_e}{T}\right)\right], \tag{11}$$

where r_{e} is expressed in Å.

In the limit of high temperature the term in square brackets can be expanded:

$$e^{\frac{\tilde{D}_e}{T}} - \left(1 + \frac{\tilde{D}_e}{T}\right) \approx \frac{1}{2}\left(\frac{\tilde{D}_e}{T}\right)^2\left(1 + \frac{1}{3}\frac{\tilde{D}_e}{T} + \dots\right). \tag{12}$$

The first terms of the above expansion coincide with the expansion of

$$\frac{1}{2}\left(\frac{\tilde{D}_e}{T}\right)^2\exp\left(\frac{\tilde{D}_e}{3T}\right).$$

Consequently the following high-temperature approximation of formula (10) can be proposed:

$$K_p^{\mathrm{THO}} \approx \frac{\pi^{1.5}}{3\chi}\, r_{\mathrm{e}}^3\,\frac{N_0}{R}\,\frac{\tilde{D}_{\mathrm{e}}^{1.5}}{T^{2.5}}\exp\left(\frac{\tilde{D}_e}{3T}\right). \tag{13}$$

This formula may now be compared with what stems from the numerical fit (1). One can easily see that formula (1) can be rewritten in terms of the equilibrium constant as:

$$K_p^{\mathrm{SH}} = \frac{4\pi}{3\chi}\, e^{0.105}\,\sigma^3\,\frac{N_0}{R}\,\frac{\tilde{\epsilon}^{1.46}}{T^{2.46}}\exp\left(\frac{0.364\,\tilde{\epsilon}}{T}\right). \tag{14}$$

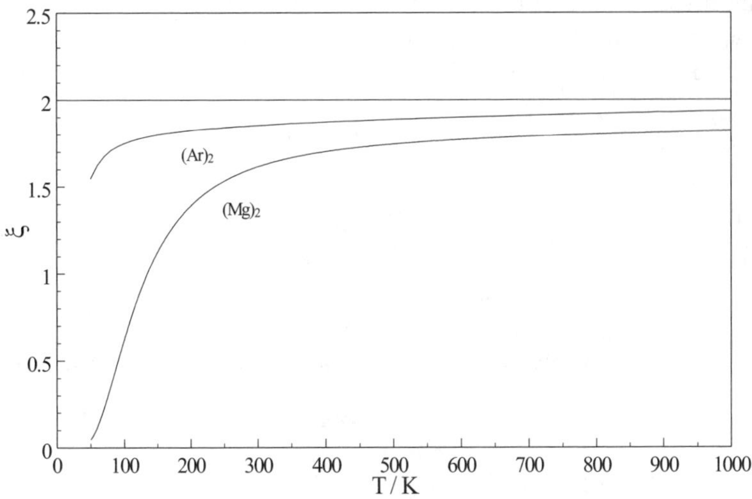

Figure 1. Temperature variations of the ratio ξ of SH (14) and THO (8) equilibrium constants for Ar and Mg dimers. Potential parameters here and below are adopted from [4]: $\tilde\epsilon = 120.45$ K, $\sigma = 3.42$ Å and $\tilde\epsilon = 620.13$ K, $\sigma = 3.23$ Å for Ar–Ar and Mg–Mg, respectively.

The meaning of $\tilde\epsilon$ is identical to that of $\tilde D_{\rm e}$, σ and $r_{\rm e}$ for the Lennard—Jones potential are related by $\sigma = r_{\rm e}(2)^{-1/6}$. It is seen that expressions (13) and (14) are very much alike: approximate formula (14) has 1.46 instead of the exact 1.5 for the power of $\tilde D_{\rm e}$ in (13). Also formula (14) has 2.46 instead of 2.5 for the exponent of T and $0.364\,\tilde\epsilon$ instead of $\tilde D_{\rm e}/3$ in the exponential. The numerical multiplier $2\sqrt{2}\exp(0.105)$ appears in (14) instead of the constant $\sqrt{\pi}$ in (13). Note that the ratio of the latter numerical factors is about 1.77. Figure 1 displays the ratio $\zeta - K_p^{\rm SH}/K_p^{\rm THO}$ of (14) and (8) formulas as a function of temperature. It can be seen that the factor ξ approaches asymptotically two as temperature rises.

Interestingly, this is by no means accidental. The higher the temperature, the more populated the high-lying levels of the intermolecular potential well. The limiting value for the vibrational partition function is simply the number of true bound vibrational levels. Therefore in the high-temperature limit the equilibrium constant is proportional to the number of bound states supported by the given intermolecular potential, that is $\lim_{T\to\infty} \xi = \beta = \text{const}$. In the next paragraph we will show that the number of levels in the LJ and THO potentials notably differs so that $\beta \approx 2$.

The number of vibrational levels in the LJ potential well for a non-rotating dimer was calculated by Stogryn and Hirshfelder in [2] using the WKB approximation. In terms of potential well depth ϵ and rotational constant B the limiting number of levels v^*_{LJ} can be represented as:

$$v^*_{\mathrm{LJ}} = 0.2385 \left(\frac{\epsilon}{h\,c\,B} \right)^{1/2} + \frac{1}{2}.$$

In the truncated harmonic oscillator well the number of vibrational levels is:

$$v^*_{\mathrm{THO}} = \frac{\epsilon}{h\,c\,\omega} + \frac{1}{2}.$$

Very roughly the ratio of both numbers can be represented as:

$$\beta = v^*_{\mathrm{LJ}}/v^*_{\mathrm{THO}} \approx 0.286\,\omega\,(\tilde{\epsilon}\,B)^{-1/2},$$

where ω and B are in cm^{-1}, $\tilde{\epsilon}$ is in K.

Substituting ω for the LJ potential into this formula results in $\beta \approx 2.86$ irrespective to the potential parameters.[2] This means that the number of the energy levels in the LJ potential well exceeds that in the THO potential. To obtain a more accurate estimate, the explicit calculations of Dardi and Dahler [4] can be used. Rationalizing the number of states obtained in [4] for the $(\mathrm{Ar})_2$ dimer, values of β from 1.83 to 2.03 have been found, the variations being due to the use of different shapes for the model potential.

Figures 2, 3 show the calculated equilibrium constants as a function of temperature for $(\mathrm{Ar})_2$ and $(\mathrm{Mg})_2$ dimers.

It is seen that formula (14) fits nicely the points corresponding to the exact classical and quantum calculations. Deviations are seen, however, at lower temperature (see e. g. Fig. 3) where the THO approximation seems to be more accurate. When multiplied by a factor of two the THO formula (8) turns out to become asymptotically as exact as SH formula (14). As mentioned above the formula (7) is inexact by a factor of about two at low temperature. The reason for that is due to the choice of the numerical multiplier in (7) to fit the SH solution. It can be speculated that the so-called high-temperature limit of the simplified expressions for the equilibrium constants (see approximations K_p^{SH} (14) or $2\,K_p^{\mathrm{THO}}$ (8)) becomes valid as the temperature exceeds roughly $0.5\,\epsilon/k$. This is the case in many situations, namely for major atmospheric applications, except for relatively strongly bound dimers

---

[2]This value was obtained for the non-rotating dimer. Allowance for rotational excitation will obviously cause β to decrease somewhat.

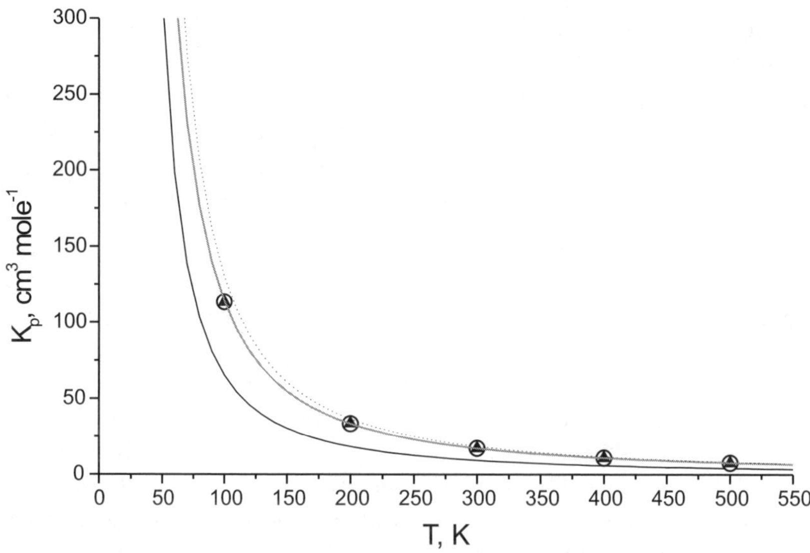

Figure 2. Calculated temperature variations of the equilibrium constant for $(Ar)_2$ formation. Heavier and lighter solid lines show K_{eq} values as calculated using relationships (14) and (8) for THO and SH potentials, respectively. Dotted line displays $2\,K_{eq}^{THO}$ (see text). Open circles and filled triangles refer, respectively, to quantum and classical calculations [4] assuming LJ potential.

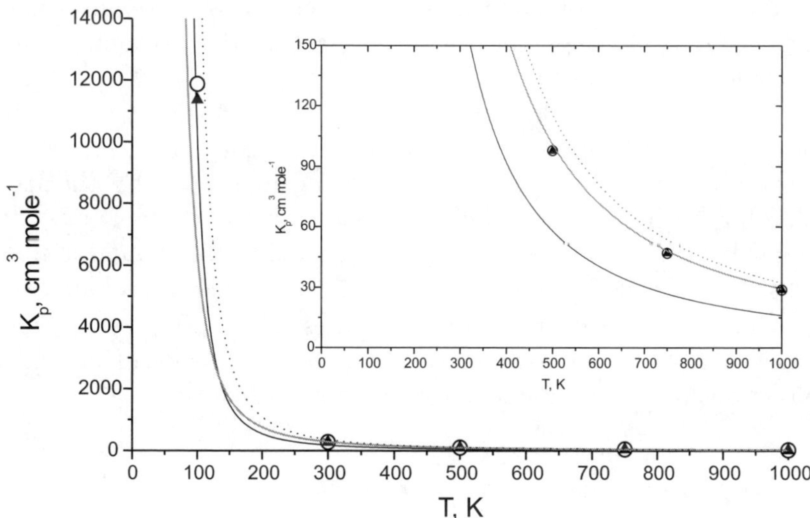

Figure 3. Same as in Fig. 2 for $(Mg)_2$. An enlarged portion of this graph is shown as an insert.

like e. g. the water dimers. In the other extreme when the temperature is roughly below $0.1\,\epsilon/k$, the classical computations fail due to the neglect of quantum effects.

The advantage of (6) against (14) consists of the possibility to extend this formula to polyatomic dimers. Note that almost equally accurate expressions for the equilibrium constant are given by relationships (6), (8), (10), and (11). All these are derived, however, for diatomic dimers only.

4. Quantum corrections

Care should be taken since the formulae derived in the previous section are valid within the limits of the classical approximation. The difference is ignored between the potential energy well depth D_e and the energy of dissociation D_0 which is due to the zero-point energy. This difference can notably influence the ability of a dimer to dissociate. To roughly account for this quantum correction one can substitute D_0 in place of D_e in the classical formulae for the equilibrium constant discussed above. Neglecting the zero-point energy is the main imperfection of the classical approach since the difference between D_e and D_0 is often comparable to dissociation energy itself.

Another, often less severe source of inaccuracy, consists in the use of a classical vibrational partition function instead of a quantum one. To make the necessary correction it is recommended to introduce a multiplier[3] $Q_v^q/Q_v^{cl} = (1 - \exp(-\hbar\omega/kT))^{-1}\,\hbar\omega/kT$ in the relevant formulae for the equilibrium constant.

In some cases the electronic excited states should be taken into account in the calculations of the equilibrium constant. The expression for the equilibrium constant has to be multiplied by the factor $Q_{el}^{dim}/(Q_{el}^{mon})^2$. The oxygen dimer provides a typical example of system with low-lying electronic states. It is known that the ground state for the oxygen molecule is a triplet state $^3\Sigma_g^-$. In a dimer the ground state is split into singlet, triplet and quintet states $^1A_g + {}^3B_{1u} + {}^5A_g$, whose energies increase in approximately equal steps of about 14 cm^{-1} [13]. The relevant partition functions can be represented as:

$$Q_{el}^{mon} = 3$$
$$Q_{el}^{dim} \approx 1 + 3\,e^{-(20/T)} + 5\,e^{-(40/T)}. \tag{15}$$

---

[3] If a dimer is characterized by several intermolecular vibrational degrees of freedom such multiplier should relate to each of them.

5. Polyatomic dimers

In a number of publications (see e. g. [6–8]) formula (1) is employed to estimate the fraction of dimers formed by polyatomic molecules. It is instructive to define the domain of applicability of this approach. Let us remind first that formula (1) (as well as its analogues (6), (8), (10), and (11)) was derived under the assumption of a spherically symmetrical interparticle potential which leads to only one intermolecular degree of freedom. The rovibrational energy of a dimer was also supposed not exceed the dissociation energy. This means that the shape resonances are totally disregarded. Given this assumption, the equilibrium constant for a polyatomic dimer can be represented using the dimer and monomer partition functions in the following form:

$$K_p = \frac{V \, Q_{\mathrm{dim}}^{\mathrm{RRHO}}}{k \, T \, Q_{\mathrm{mon}}^2} \left[e^{\frac{D_0}{kT}} - \left(1 + \frac{D_0}{kT} \right) \right]. \tag{16}$$

Here $Q_{\mathrm{dim}}^{\mathrm{RRHO}}$ means total dimer partition function calculated in the RRHO approximation i. e. without any restrictions on the energy of excitation. The term in the square brackets provides a reasonable correction to the RRHO rovibrational partition function in case of a diatomic dimer. In case of polyatomics this term allows for correcting the partition function for the stretching mode only. The calculation of the equilibrium constant using (16) is recommended for polyatomic dimers with certain precautions. The metastable states of the Feshbach type may be effectively accounted for in (16). In fact the use of a truncated harmonic oscillator model makes it possible to neglect the shape resonances caused by the end-over-end rotation of a dimer. In contrast, the excitation of librational intermolecular modes (which occur in the form of oscillations or free or hindered internal rotations) within a dimer can result in the formation of Feshbach resonance states in which the sum of the energy of internal modes in a dimer exceeds the dissociation energy. These resonance or metastable states are implicitly accounted for in formula (16) provided mode specific partition functions for internal movements are calculated presupposing no restrictions on the total excitation energy.

In the low-temperature extreme, when the energy of thermal excitation integrated over all the internal modes is lower than the dissociation energy of the dimer, the equilibrium constant relates only to true bound states. Assuming that the population of high-lying states is negligible and that all internal movements can be treated in terms of harmonic oscillations, the relevant partition functions can be summed up to infinity. Then the temperature dependence of the equilibrium constant will be governed by the factor:

$$T^{s-4}\left[e^{\frac{\tilde{D}_0}{T}} - \left(1 + \frac{\tilde{D}_0}{T}\right)\right], \tag{17}$$

where s is the number of intermolecular degrees of freedom.

In the opposite extreme i.e. when all the barriers to internal rotation are smaller than the average thermal excitation kT, librational movements transform into free internal rotations. It is easy to show that in this case the temperature dependence of the equilibrium constant reduces to:

$$T^{-0.5}\left[e^{\frac{\tilde{D}_0}{T}} - \left(1 + \frac{\tilde{D}_0}{T}\right)\right]. \tag{18}$$

This coincides with formulae (6), (8), (10), and (11) for diatomic dimers. It is worth mentioning that this same temperature dependence seems to hold independently on how complicated the monomers are (of which a dimer is formed).[4] Note, however, that in case of polyatomic dimers formula (18) relates to dimers combining true bound dimeric states with an ill-determined fraction of the metastable states. In fact the sum of the energy for free or hindered internal movements within a dimer can exceed the dissociation threshold. Whether this will happen or not at any selected temperature is determined by the relief of the multidimensional intermolecular PES for each concrete individual molecular system. Thus, the simplified formulae (14) and (1) as well as their analogues (6), (8), (10), and (11), generally account for some ill-defined fraction of polyatomic metastable dimers. Note that as the molecules involved in a pair or dimer become more and more complicated, the domain in phase space occupied by metastable states becomes larger and larger. A more elaborate discussion of metastable states and of their contribution to the partition function can be found in [3].

6. Conclusions

Practical needs to estimate the dimer abundance in equilibrium gases are covered through the use of simplified expressions suggested in the present paper. These expressions are based on the use of classical approximations along with the assumption that the potential energy

---

[4]All librational modes should, however, realize in the form of free internal rotations.

curve as a function of intermolecular separation can be represented by a truncated harmonic oscillator (THO) potential well.

A general formula for the equilibrium constant as calculated in THO approximation is given by (16). Useful relationships (6), (8), (10), and (11) are recommended for rapid estimates. The estimates based on the THO model are expected to be accurate in the high-temperature limit within a factor of two. The asymptotical high-temperature behavior of the THO formula is identical (accounting for the factor of two) to the temperature dependence of the popular numerical fit (1) based on the use of the Lennard—Jones potential. Care should be taken, however, in the use of the THO and SH formulae discussed above without considering quantum corrections to the zero-point energy and to the vibrational partition functions, which are particularly significant at low temperatures. In some cases the electronic partition function has also to be accounted for.

The THO model is easily extended to polyatomic dimers. However, it must be kept in mind that the use of diatomic approximations (such as THO or SH) for polyatomic dimers results in an implicit inclusion of unknown fraction of metastable states (Feshbach resonances) into the relevant partition functions and equilibrium constants. Thus, except of very low temperatures, the simplified estimates of the equilibrium abundance of true bound polyatomic dimers based either on relationships (16), (11), or (1) are only very approximate since metastable and bound states are treated on the same footing. Rigorous calculations require the use of a complete multidimensional PES which makes it possible to discriminate between metastable and true bound states in the phase space (see e. g. [3]).

Acknowledgments

The author thanks H.-D. Meyer for supplying a copy of his valuable thesis (Ref. [12]). Present work has become possible due to partial financial support from the Russian Basic Research Foundation and RFBR–CNRS Grants 02-05-64529 and 01-05-22002, respectively.

References

[1] Le Roy, R. J., LEVEL 7.0 A computer Program Solving the Radial Schrödinger Equation for Bound and Quasibound Levels, University of Waterloo Chemical Physics Report CP-642 (2000).

[2] Stogryn, D. E. and Hirshfelder, J. O. (1959) Contribution of bound, metastable and free molecules to the second virial coefficient and some properties of double molecules, J. Chem. Phys., 31, 1531–1545.

[3] Vigasin, A. A. (1991) Bound, metastable and free states of bimolecular complexes, Infrared Phys., 32, 461–470.

[4] Dardi, P. S. and Dahler, J. S. (1990) Equilibrium constants for the formation of van der Waals dimers: Calculations for Ar–Ar and Mg–Mg, J. Chem. Phys., 93, 3562–3572.

[5] Anderson, J. B., Andres, R. P., Fenn, J. B., and Maise, G. (1966) Studies of low density supersonic jets. In: J. H. de Leeuw, editor. *Rarefied Gas Dynamics*, Vol. 2, Academic Press, N.Y., pp. 106.

[6] Calo, J. M. and Narcisi, R. S. (1980) Van der Waals molecules — possible role in the atmosphere, Geophys. Research Lett., 7, 289–290.

[7] Calo, J. M. (1975) Dimer formation in supersonic water vapor molecular beams, J. Chem. Phys., 62, 4904–4910.

[8] Calo, J. M. and Brown, J. H. The calculation of equilibrium mole fractions of polar-polar, nonpolar-polar, and ion dimers, J. Chem. Phys., 61, 3931–3941.

[9] Vigasin, A. A. (1998) Mass-action law for highly excited dimers Chem. Phys. Lett., 290, 495–501.

[10] Jaffe, R. L. (1986) Rate constants for chemical reactions in high-temperature nonequilibrium air. In: J. N. Moss and C. D. Scott, editors. *Thermophysical Aspects of Re-entry Flows*, AIAA, New York.

[11] Knuth, E. L. (1977) Dimer-formation rate coefficients from measurements of terminal dimer concentartion in free-jet expansions, J. Chem. Phys., 66, 3515–3525.

[12] Meyer, H.-D., *Bildung von Dimeren in Edelgas-Düsenstrahlen*, MPI für Strömungsforschung, Report 5/1978, Göttingen, 1978.

[13] Bussery-Honvault, B. and Wormer, P. E. S. (1993) A van der Waals intermolecular potential for $(O_2)_2$, J. Chem. Phys., 99, 1230–1239.

COLLISION INDUCED FAR WINGS OF CO$_2$ AND H$_2$O BANDS IN IR SPECTRA

M. V. Tonkov,[*] N. N. Filippov

Institute of Physics, StPetersburg University,

Peterhof, StPetersburg, 198504, Russia

Abstract   The far wing shapes of CO$_2$ ν_3 band in infrared absorption spectra were calculated accounting for line mixing and finite collision duration effects. The result shows that the impact approximation can be used for the rough estimate of absorption intensity in the far wing region. This approach was used for H$_2$O absorption estimate in the 8–12 μm atmospheric window. According to our calculations the H$_2$O monomer absorption could be responsible for the part of absorption depending on the air pressure. However, the part of absorption depending on H$_2$O pressure squared can hardly be explained in the frame of a model presupposing collisions between monomer molecules.

Keywords: line mixing / band wings / 8–12 μm atmospheric window

1. Introduction

There is no unambiguous opinion in scientific literature on the IR radiation absorption mechanisms by water vapor in the region of the 8–12 μm atmospheric window. The most spread hypotheses relate this absorption to the wing of the rotational and vibrational-rotational bands or to the dimer formation in a gas phase. To choose between these two hypotheses we estimated the possible absorption produced by the far band wings of monomer molecules.

The calculation of a band wing shape is a rather difficult problem of spectroscopy since one should account for intermolecular interactions. We did not use for it the adiabatic approximation, which seems to be not suitable for vibration-rotation molecular spectra. Instead, we used

---

[*]Corresponding author. E-mail: `tonkov@niif.spb.su`; Fax: 7 (812) 428 72 40.

C. Camy-Peyret and A.A. Vigasin (eds.),

Weakly Interacting Molecular Pairs: Unconventional Absorbers of Radiation in the Atmosphere, 125–136.

the general shape theory, which account for line mixing and finite collision duration effects.

In this approach the band shape depends mainly on the rotational relaxation matrix. So, we have two main problems in the band shape calculation: the structure of relaxation matrix and the dependence of the matrix elements on frequency. We will show that two ways can give the reasonable results in the estimation of the absorption in the band wing: the neglect of the band structure in semiclassical calculation and, in same cases, the impact approximation. To examine these approaches we used the wing of the ν_3 CO_2 absorption band. After that we applied the impact approximation to the intensity estimation of H_2O band wings.

2. Band profile studies

The absorption coefficient $A(\omega)$ is given by the well-known expression:

$$A(\omega) = \frac{4\pi^2 (\omega + \omega_0)}{3\,\hbar\,c} \left[1 - \exp\left(-\frac{\hbar (\omega + \omega_0)}{k\,T} \right) \right] \Phi(\omega), \qquad (1)$$

where ω is the frequency detuning from the vibrational band origin ω_0 ($\omega_0 = 0$ for rotational band), $\Phi(\omega)$ is the spectral function, which contains the main information about the spectral shape.

The general formulation for overlapping line shapes prescribes to calculate the band shapes via the elements of rotational relaxation matrix $\boldsymbol{\Gamma}$ [1–3]. The elements of this matrix depend on the probability of transitions between the rotational states of molecules under collisions. The use of the line space formalism makes it possible to express the spectral function $\Phi(\omega)$ by the following formula [3]:

$$\Phi(\omega) = \frac{1}{\pi}\,\mathrm{Re} \left\langle\!\!\left\langle M \left| \frac{1}{i\,(\omega - \mathbf{L_0}) + \boldsymbol{\Gamma}(\omega)} \right| M \right\rangle\!\!\right\rangle, \qquad (2)$$

where $|M\rangle\!\rangle$ is a vector in the Liouville space. This vector can be written as

$$|M\rangle\!\rangle = \sum_m |M_m\rangle\!\rangle = \sum_m M_m\,|m\rangle\!\rangle,$$

where m represents the line index for the line m with $M_m = \sqrt{P_m}\,d_m$, $P_m = P_i$ is the population of initial state for the transition $m = i \to f$, $d_m = d_{if}$ is the reduced matrix element of the dipole moment, and $\mathbf{L_0}$ gives the diagonal matrix of unperturbed line frequency detuning ω_m from the band origin:

$$\mathbf{L_0}\,|m\rangle\!\rangle = \omega_m\,|m\rangle\!\rangle.$$

With this notation, the line strength is $S_m = M_m^2$ and $\sum_m S_m = S_\nu$ is the band strength. The matrix $\mathbf{\Gamma}(\omega)$ in Eq. (2) is the generalized rotational relaxation matrix in the symmetrized representation related to the commonly used nonsymmetrized matrix W by the relation

$$\Gamma_{m,m'} = (P_m/P'_m)^{1/2}\, W_{m,m'}. \tag{3}$$

The diagonal elements of $\mathbf{\Gamma}$ matrix determine the widths of lines (lineshifts are neglected in our work); the off-diagonal elements are responsible for non-additive effects (line mixing) when the lines overlap.

These relations should be supplemented by the sum rules:

$$\langle\!\langle m|\,\mathbf{\Gamma}(\omega)\,|M\rangle\!\rangle = 0; \qquad \langle\!\langle M|\,\mathbf{\Gamma}(\omega)\,|m'\rangle\!\rangle = 0$$

or

$$\sum_{m'} M_{m'}\,\Gamma(\omega)_{mm'} = 0; \qquad \sum_{m} M_m\,\Gamma(\omega)_{mm'} = 0, \tag{4}$$

which relate the off-diagonal elements of $\mathbf{\Gamma}$ matrix to the line widths.

3. Far wing approximation

Let us consider far band wings i.e.the region of frequency detuning ω which satisfies the relation

$$|\omega| \gg \sqrt{\tilde{\mu}_2}, \tag{5}$$

where $\tilde{\mu}_2$ is the reduced second central spectral moment of the band. In gases $\sqrt{\tilde{\mu}_2}$ is close to the width of rotational structure of a band. The spectral moments can be determined as

$$\mu_n = \int_{-\infty}^{\infty} \Phi(\omega)\,\omega^n\,d\omega. \tag{6}$$

The zeroth, first, and second moments do not depend on the intermolecular interactions. The reduced values are usually defined as $\tilde{\mu}_n = \mu_n/\mu_0$.

When the relation (5) is valid and $|\omega| \gg |\Gamma_{mm'}|$, the resolventa in Eq. (2) can be developed in power series in $(i\mathbf{L}_0 - \mathbf{\Gamma})/\omega$. This yields:

$$\Phi(\omega) = \pi^{-1}\,\mathrm{Re}\sum_n (i\,\omega)^{-n-1}\langle\!\langle M\,|\,(i\,\mathbf{L}_0 - \mathbf{\Gamma}(\omega))^n\,|M\rangle\!\rangle. \tag{7}$$

Due to sum rules (4) the terms containing $\boldsymbol{\Gamma}(\omega)\,|M\rangle\!\rangle$ and $\langle\!\langle M|\,\boldsymbol{\Gamma}(\omega)$ go to zero and we have for the terms up to ω^{-4} [3]:

$$\Phi(\omega) = \pi^{-1}\,\omega^{-4}\,\mathrm{Re}\langle\!\langle \dot{M}|\,\boldsymbol{\Gamma}(\omega)\,|\dot{M}\rangle\!\rangle. \tag{8}$$

If the dependence of relaxation operator on frequency can be neglected, the asymptotic behavior of the rotation-vibration band wing is proportional to ω^{-4} and not to ω^{-2}, as one could obtain for a sum of Lorentz lines. This result is in accordance with the Burstein and Temkin theory for classical rotors [4].

In the binary approximation for linear molecules with the moments of inertia I the Eq. (8) can be transformed into [3]:

$$\Phi(\omega) = \frac{n_B}{I^2\,\pi\,\omega^4}\,\mathrm{Re}\int\limits_0^\infty dt\,e^{-i\,\omega\,t}<G(0)\,G(t)>_{\mathrm{pair}}, \tag{9}$$

where n_B is the density of perturbing molecules and the vector G can be treated as a vector product of the dipole moment $\mathbf{d}$ and torque moment $\mathbf{N}$ arising in the collision.

For calculation of the wing shape (9) we used the classical limit in the translational and rotational motions. This equation can be greatly simplified in the case of non-adiabatic collisions approximation (translational motion is faster than rotational one), which seems to be reasonable for molecular collisions. In this case the dipole moment orientation does not change substantially during a collision and we can accept it to be constant. In this case we can approximate Eq. (9) by:

$$\Phi(\omega) = \frac{n_B\,\mu_2<N^2>}{\pi\,\omega^4\,(1+\exp(-\hbar\,\omega/k\,T))}\,\phi(\omega), \tag{10}$$

where

$$\phi(\omega) = Re\int\limits_0^\infty dt\,e^{-i\,\omega\,t}\,\frac{<N(0)\,N(t)>_{\mathrm{cl}}}{<N^2>}.$$

One can see that the torque shape expression $\phi(\omega)$ looks like that in the collision induced theory, but the dipole moment function is changed for the molecular torque. In classical limit the "integrated intensity" of the wing is proportional to $<N^2>$ and can be calculated via the pair distribution function $g(R,\Omega)$:

$$<N^2> = \mathrm{Re}\int\limits_0^\infty (N(R,\Omega))^2\,g(R,\Omega)\,R^2\,dR\,d\Omega. \tag{11}$$

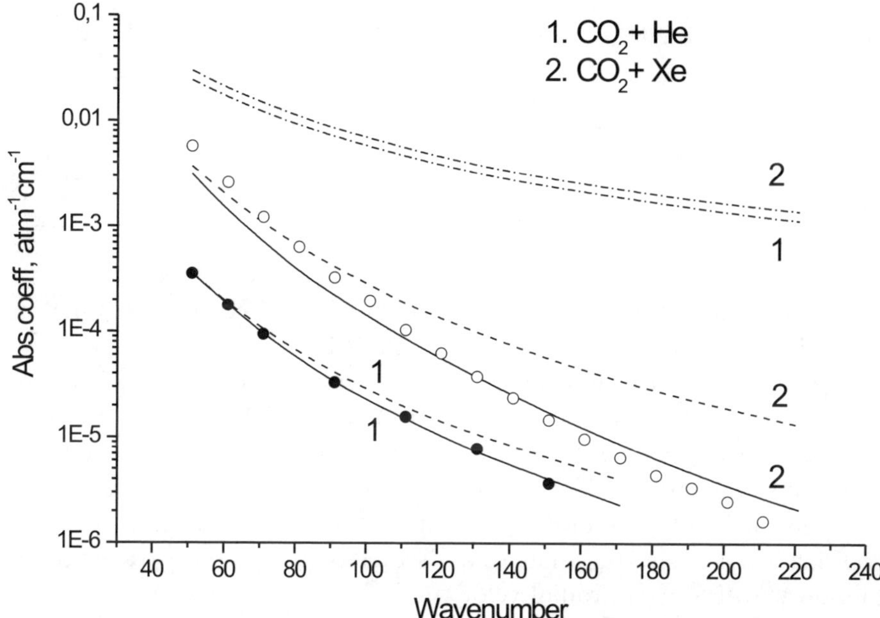

Figure 1. The absorption coefficient in the wing region of the ν_3 CO_2 band. The points are experimental data. The calculated shapes are: calculated using Eqs. (11)–(13), solid curves; impact approximation with $\phi(\omega) = 1$, dash curves; Lorentzian lines, dash-dot curves.

In case of very short collisions (impact approximation) $\phi(\omega) \approx \phi(0)$ $= 1$, and the wing shape is proportional to ω^{-4}.

This expression was used for the wing shape calculations of the IR absorption bands of CO and CO_2 molecules in rare gas mixture [5]. It was shown that for these cases the repulsive interactions are most important in rotational perturbations and the molecular trajectories were calculated in analytical form for the classical vibration-rotation motion. The obtained shapes of molecular torques were fit to a simplified expression

$$\phi(\omega) = C \exp\left(-\sqrt{(\tau_c\,\omega)^2 + b^2}\right), \qquad (12)$$

where C and $b = 0.87$ are the shape parameters and τ_c is the correlation time for molecular torques which can be treated as a collision duration. It is worth noting that this function describes very well the collision induced shapes. The calculated shapes are very close to the measured ones, the deviations can be explained by inaccuracy of the potentials used in our calculations (Fig. 1). The parameters $\nu_c = (2\pi\tau_c)^{-1}$ found

Table 1. The parameters ν_c for ν_3 CO_2 band wing, cm^{-1}.

Perturbing gas	$\nu_c^{\exp}$	ν_c^{calc}	ν_c^{tr}
He	160	174	116
Ar	80	93	49
Xe	50	79	39

from experimental data and those calculated from correlation function for molecular torques are given in the Table 1.

The temperature dependence of the wing intensity is determined by Eq. (11). The potential well depth for CO_2–CO_2 interaction, being estimated from experimental temperature wing dependence [5], has a reasonable value of 400 K.

One can see two interesting results in these calculations. First, the collision time is shorter than that predicted by the pure translational motion with the mean radial velocity

$$\tau_c^{\text{tr}} = r_0 \sqrt{\pi m / 2 k T}.$$

Even for repulsive forces, when $r_0 \approx 0.3\,\text{Å}$, ν_c^{tr} given in the Table 1 are essentially below than the experimental ones or those calculated with the more correct model. We suppose that the collision time decreases because rotational motion influences the effective collision velocity.

Second, the main reason of deviation of real shape from that calculated as a sum of Lorentz lines is line mixing, whereas finite collision duration plays minor role. This result allows us to use the impact approximation for the rough estimation of absorption intensity in the far wing region. In any case, this estimation should be the upper limit of the wing intensity.

4. Impact approximation

Let us examine the approach, which uses the impact approximation. This approach can be used for the cases of molecular collisions at least for the estimation of the upper limit of wing intensity for the molecules of the various types. For a band shape calculation in the impact approximation one can use the expression $(1, 2)$ in which rotational relaxation matrix does not depend on frequency.

Ab initio calculations of relaxation matrix $\mathbf{\Gamma}$ were carried out only for the simplest systems such as diatomic and linear molecules [6, 7].

In practice, for more complicated molecules the matrix elements are constructed using various empirical ways.

One of the used approaches is the strong collision model [8], which is based on the assumption that the relaxation time is equal to the duration between successive collisions for any rotational state of a molecule. The $\mathbf{\Gamma}^{SC}$-matrix is the corresponding relaxation matrix for this case. In the frame of this model the matrix elements $\Gamma_{mm'}$ can be expressed analytically:

$$\Gamma^{SC}_{mm'} = \tau_0^{-1}\left(\delta_{mm'} - M_m M_{m'}/S_\nu\right), \tag{13}$$

where τ_0 is the parameter of the model equal to the mean duration between collisions. With this model the matrix inversion in Eq. (2) can be performed analytically. The spectral function in this approach can be expressed as

$$\Phi(\omega) = \frac{1}{\pi}\,\mathrm{Re}\,\frac{S(\omega)}{1 - \tau_0^{-1}\,S(\omega)}, \tag{14}$$

where

$$S(\omega) = \sum_m \frac{S_m}{i\left(\omega - \omega_m\right) + \tau_0^{-1}}. \tag{15}$$

Since the diagonal matrix elements are equal to the line widths, the value of τ_0 can be found from the following equation

$$\sum_m (S_m/S_\nu)\,\Gamma^{\mathrm{theor}}_{mm} = \sum_m (S_m/S_\nu)\gamma_m, \tag{16}$$

using the experimental line widths γ_m.

However, the model has two main shortcomings: first, it cannot explain the difference in the action of various perturbing gases, and there is no difference for the matrix elements corresponding to the lines of the same branch and of different branches. In contrast, as it was shown in Refs. [9,14], the interaction between lines within the same branch and among the lines from different branches usually differs significantly.

The coupling of lines of different branches is weakened to some extent in ABC model [10], where the relaxation matrix is a linear combination of $\mathbf{\Gamma}^{SC}$ and $\mathbf{\Gamma}^{iso}$:

$$\mathbf{\Gamma}^{ABC} = C_{\mathrm{branch}}\,\mathbf{\Gamma}^{SC} + (1 - C_{\mathrm{branch}})\,\mathbf{\Gamma}^{iso}, \tag{17}$$

where C_{branch} defines the interbranch interaction. The matrix $\mathbf{\Gamma}^{SC}$ is the relaxation matrix for the strong collision model. The $\mathbf{\Gamma}^{iso}$ matrix corresponds to the isolated branch model, where the coupling of the lines in different branches is forbidden (the elements responsible for the

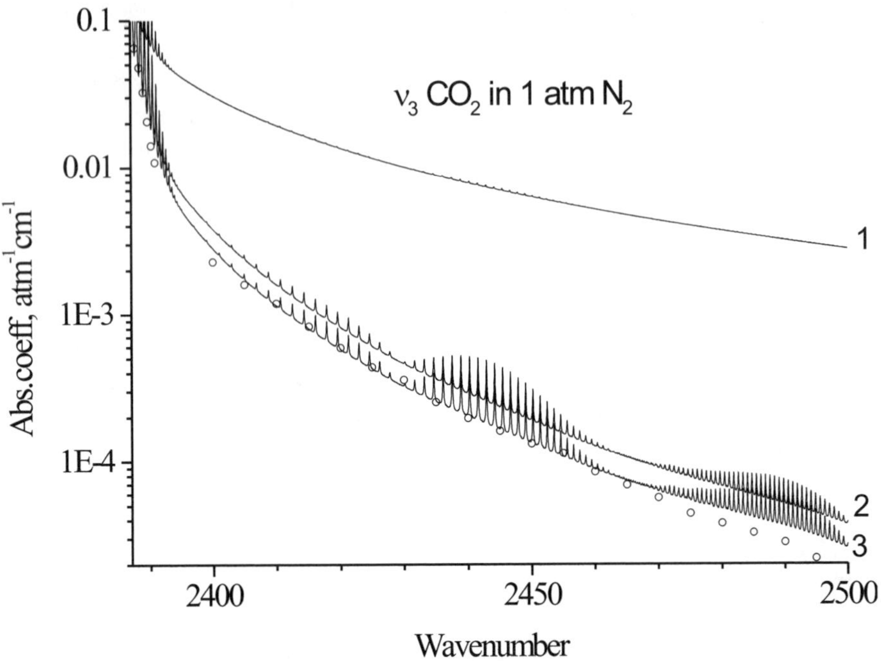

Figure 2.   The absorption coefficient in the wing region of the ν_3 CO$_2$ band. The points are experimental data. The calculated shapes are: Lorentzian lines (1), the strong model (2), ABC model with $C_{branch} = 0.6$ (3).

interaction of different branch lines are taken to be zero). The C_{branch} parameter defines the degree of rotational branch coupling and can be varied from 0 to 1. It should be found from experimental data in the spectral region, where the line mixing effect is the most pronounced. In the framework of this model it is also possible to invert the matrix in Eq. (2) analytically. The ABC relaxation matrix differs from the "real" matrix, however, when the spectral shape is produced by many overlapped lines, the use of this model gives good results for many perturbing gases including the case of real atmosphere [10].

We adjusted the calculated shapes to the measured ones varying the contribution of different matrices to a final relaxation matrix by means of non-linear optimization procedure for CO$_2$–N$_2$ case. The experimental data were received in our laboratory. The measurements were made in the wing region only (between lines). The line parameters for shape calculation were taken from HITRAN database [11]. The variation of the C_{branch} coefficient with pressure was in the limits of experimental inaccuracy. In case of CO_2–N_2 mixture (Fig. 2) the obtained matrix is

$$\mathbf{\Gamma}^{\mathrm{ABC}} = 0.6\,\mathbf{\Gamma}^{\mathrm{SC}} + 0.4\,\mathbf{\Gamma}^{\mathrm{iso}},$$

and the calculated shape is very close to the measured one. The deviation in the high frequency part of the wing can be explained by the effect of the finite duration of collisions.

5. H$_2$O band wings

Since the results of our calculations for the CO_2 band wing in the impact approximation were acceptable, we decided to apply this approach to the H_2O band wing calculations. We estimated the H_2O absorption in the 8–$12\,\mu$m spectral region in impact approximation since the translational and rotational velocities for H_2O molecule are greater than those for CO_2 molecule. The estimated ν_c^{tr} parameter for radial translational motion is about $70\,\mathrm{cm}^{-1}$ for H_2O collisions, so the real value should be more than $100\,\mathrm{cm}^{-1}$. In any case our calculation should provide the upper limit of the real absorption coefficient.

We have accepted the broadening effect of air to be identical to that of nitrogen in our calculation. The line parameters in shape calculation were taken from the HITRAN database [12]. The self-broadening coefficients of the ν_2 band were used in the rotational spectrum for H_2O–H_2O collisions since both bands belong to the B-type. The experimental data for the water continuum absorption were taken from the Arefiev's analysis [13].

The comparison of the calculated spectrum with the experimental one for H_2O–N_2 collisions is given in Fig. 3a. We used for comparison the quadratic part of the measured absorption only since it should be related to band wing absorption. The spectral shape predicted by the strong collision model is above the measured values. This is in accordance with the fact that the strong collision model gives the upper limit of the real absorption coefficient. The ABC model provides good agreement of the measured and the calculated values. It shows that the monomer absorption could play an important role in the observed absorption.

The results for H_2O–H_2O are opposite (Fig. 3b), i.e. the experimental points are above the calculated shape even for strong collision model. So, to explain this absorption one should consider additional mechanisms.

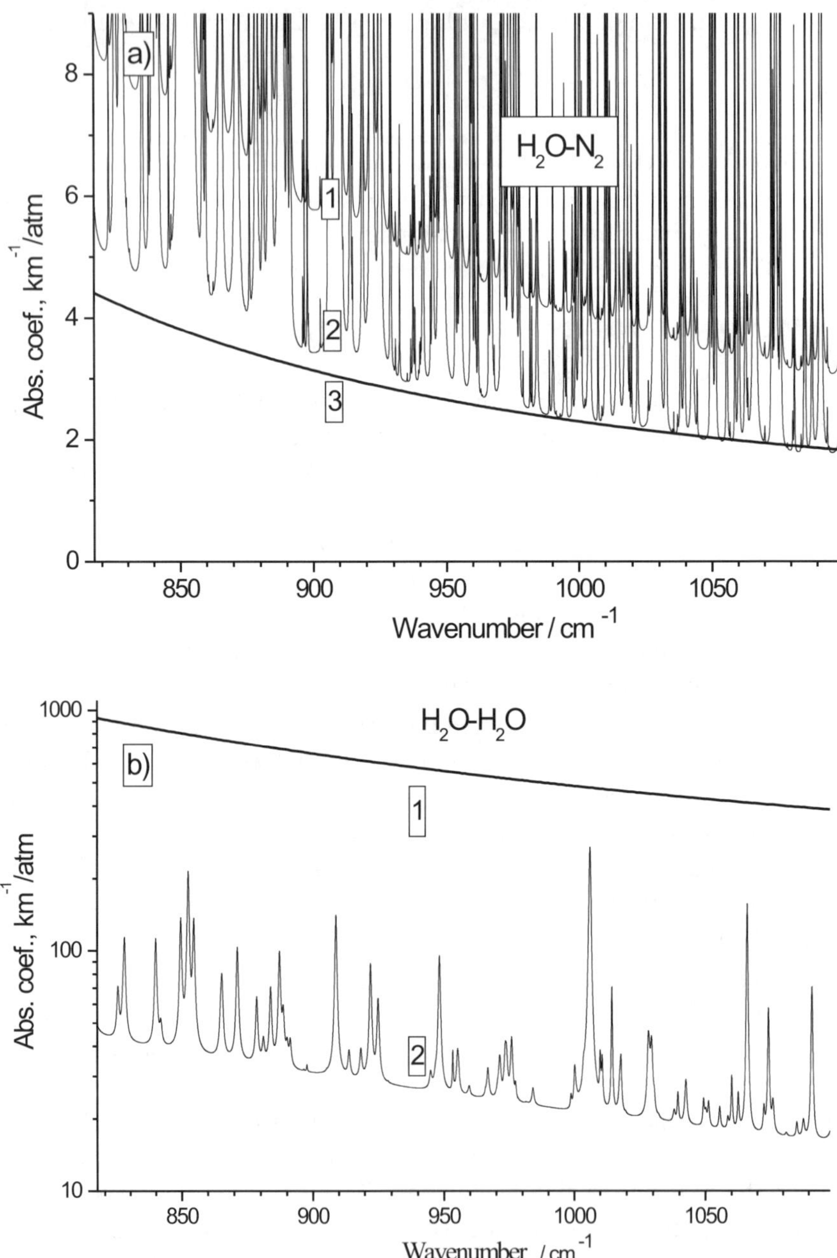

Figure 3.   *a*) The absorption of H_2O vapor due to collisions with N_2 molecules. Data calculated with the strong collisions and ABC at $C_{\mathrm{branch}} = 0.6$ models $(1, 2)$; experimental data (3). *b*) The absorption of H_2O vapor due to collisions with H_2O molecules: experimental data (1); data calculated with the strong collisions approximation (2).

6. Conclusions

The simple relaxation model based on the strong and weak collision approximation gives the possibility to estimate the intensity of IR absorption in the region of the far wing of vibration-rotation bands. Such a model can be also used for empirical calculations of the spectral profiles in the wing region.

The use of this approach for the interpretation of the continuum IR absorption by water vapor in the 8–$12\,\mu$m region shows that H_2O monomers could be responsible for that part of absorption which vary linearly with the air pressure. However, the part of absorption depending on H_2O partial pressure squared, which is more important in real atmosphere, could not be explained in the frame of collisions of monomer molecules.

Acknowledgments

This work was supported in part by the Russian Foundation for Basic Research (Project No. 02-03-32752).

References

[1] Lévy, A., Lacome, N., and Chackerian, Ch. (1992) Collisional line mixing. In: K. N. Rao and A. Weber, editors. *Spectroscopy of the Earth's Atmosphere and Interstellar Medium*, NY, Academic Press, pp. 261–337.

[2] Filippov, N. N. and Tonkov, M. V. (1996) Line mixing in the infrared spectra of simple gases at moderate and high densities, Spect. Acta A., 52, 901–918.

[3] Filippov, N. N. and Tonkov, M. V. (1998) Kinetic theory of band shapes in molecular spectra of gases: Application to band wings, J. Chem. Phys., 108, 3608–3619.

[4] Burstein, A. I. and Temkin, S. I. (1994) *Spectroscopy of Molecular Rotation in Gases and Liquids*, University Press, Cambridge.

[5] Tonkov, M. V. and Filippov, N. N. (1991) Molecular torque dynamics in binary collisions and band wing shapes of CO and CO_2, Khim. Fiz., 10, 922–929 (in Russian).

[6] Boissoles, J., Boulet, C., Robert, D., and Green, S. (1987) IOS and ECS line coupling calculation for the CO–He system: Influence on the vibration-rotation band shapes. J. Chem. Phys., 87, 3436–3446.

[7] Boissoles, J., Boulet, C., Bonamy, L., and Robert, D. (1989) Calculation of absorption in the microwindows of the 4.3 mcm CO_2 band from an ECS scaling analysis, JQSRT, 42, 509–520.

[8] Bulanin, M. O., Dokuchaev, A. B., Tonkov, M. V., and Filippov, N. N. (1984) Influence of line interference on the vibration-rotation band shape, JQSRT, 31, 521–543.

[9] Filippov, N. N. and Tonkov, M. V. (1993) Semiclassical analysis of the line mixing in the infrared bands of CO and CO_2, JQSRT, 50, 111–125.

[10] Tonkov, M. V., Filippov, N. N., Timofeev, Yu. M., and Polyakov, A. V. (1996) A simple model of line mixing effect for atmospheric application: theoretical background and comparison with experimental profiles, JQSRT, 56, 783–795.

[11] Rothman, L. S., Rinsland, C. P., Goldman, A., Massie, S. T., Edwards, D. P., Flaud, J.-M., Perrin, A., Camy-Peyret, C., Dana, V., Mandin, J.-Y., Schroeder, J., McCann, A., Gamache, R. R., Wattson, R. B., Yoshino, K., Chance, K. V., Jucks, K. W., Brown, L. R., Nemtchinov, V., and Varanasi, P. (1998) The HITRAN molecular spectroscopic database and HAWKS (HITRAN atmospheric workstation): 1996 edition, JQSRT, 60, 665–710.

[12] Arefiev, V. N. (1990) Molecular Absorption in Gases and Attenuation of Laser Radiation in the Atmosphere. Thesis for a doctoral grade. Obninsk (in Russian).

[13] Tonkov, M. V. and Filippov, N. N. (1983) Influence of molecular interactions on the form of the vibrational-rotational bands in the spectra of gases: properties of the spectral function, Opt. Spectrosc. (USSR), 54, 475–478.

[14] Bulanin, M. O., Tonkov, M. V., and Filippov, N. N. (1984) Study of collision-induced rotational perturbations in gases via the band shape of infrared bands, Can. J. Phys., 62, 1306–1314.

FAR-WING LINE SHAPES:
APPLICATION TO THE WATER CONTINUUM

R. H. Tipping

Department of Physics and Astronomy,
University of Alabama, Tuscaloosa AL 3548-0324, U.S.A.

Q. Ma

Department of Applied Physics, Columbia University, and
Institute for Space Studies, Goddard Space Flight Center,
2880 Broadway, New York, NY 10025,U.S.A.

Abstract   A far-wing line shape theory that satisfies the detailed balance principle has been developed invoking only the binary-collision and quasistatic approximations. This first-principles theory has been applied to calculate the far-wing line shapes and the corresponding absorption for H_2O–H_2O and H_2O–N_2, which for historical reasons, are called the self- and foreign-broadened water continua. Using sophisticated interaction potentials that give good agreement with other transport data, and the coordinate representation, in which the required traces become multi-dimensional integrals over the angular coordinates necessary to specify the positions before and after the collision (11-dimensional for the self and 9-dimensional for the foreign continuum, respectively), we can obtain converged results using modest computational resources. Results obtained are in very good agreement with existing laboratory data, and comparisons with empirical continua are discussed.

Keywords: far-wing line shapes / self and foreign water continua

1. Introduction

In order to model the radiative properties, deduce the abundance of the constituents and other physical properties of planetary atmospheres from measured spectra, one has to know the spectroscopy of the dominant species present. To this end, a number of databases containing laboratory information about the frequencies, intensities, and Lorentzian widths for atoms and molecules have been compiled [1, 2].

137

C. Camy-Peyret and A.A. Vigasin (eds.),
Weakly Interacting Molecular Pairs: Unconventional Absorbers of Radiation in the Atmosphere, 137–145.
© 2003 *Kluwer Academic Publishers. Printed in the Netherlands.*

However, these data are often insufficient to model accurately planetary atmospheres for a number of reasons: the temperatures are either much lower or higher than the range of laboratory data used to compile the databases; collisions alter both the line shapes, especially in the far wings, and the intensities of the allowed lines (line-mixing [3, 4]); new molecular bands having different selection rules are induced (collision-induced absorption [5–7]); and molecular dimers, both metastable and bound [8], are formed and alter the absorption profiles. Therefore, experimental data have to be supplemented by theoretical calculations.

Most planetary atmospheres have spectral "windows" both in the visible, which allow incoming sunlight to penetrate the atmosphere, and in the infrared, which allow outgoing longer wavelength radiation to escape, thus maintaining the energy balance. These windows are framed by the high- and low-frequency wings of the dominant molecular absorption bands, and do not contain many strong allowed absorption lines. However, there are weak, continuous absorptions. In the Earth's atmosphere both in the far infrared and infrared regions there are two main sources for the continuous absorption of radiation over a range of frequencies: the self and foreign water continua arising from the far wings of the allowed pure rotational and vibration-rotational dipole transitions, and collision-induced absorption (CIA) arising from transient dipole moments induced during binary collisions. Both these absorptions scale quadratically with the number density, ρ^2 for a pure gas, or as the product of the ρ's for a mixture. In the case of zenith measurements, the water continua dominate and empirical models are included in most atmospheric radiative transfer programs. However, for low humidity conditions, such as long-path limb measurements through the stratosphere, the collision-induced translation-rotational and fundamental absorptions of N_2–N_2, N_2–O_2 and O_2–O_2 can be important [9, 10].

During the past few years, we have developed a first-principles far-wing line shape theory making only the binary collision and quasistatic approximations, both of which are valid for the low-density gases of interest [11–15]. The theory explicitly satisfies the detailed balance requirement [16]. In earlier work, we utilized for the basis set the product of the two unperturbed molecular wave functions to describe the internal degrees of freedom; the translational motion was treated classically and the average over the separation r between the centers of mass of the two molecules was done analytically using conservation of energy during the collisions. In this representation, the anisotropic part of the interaction potential is off diagonal and had to be diagonalized in order to obtain the eigenvalues and eigenvectors needed to carry out the trace operation for calculating the line shape. For simple systems such as CO_2–Ar, this presented no difficulties, but for more complicated systems, such as H_2O–H_2O, the size to the matrix increases very fast as more rotational states are included. As a result, the matrix had to be truncated, which resulted in convergence problems and a loss of accuracy. In recent work we have avoided this problem by adopting the coordinate representation where the internal state $|\xi\rangle$ is expressed as

$$|\xi\rangle = |\delta(\Omega_a - \Omega_{a\xi})\,\delta(\Omega_b - \Omega_{b\xi})\rangle, \tag{1}$$

where the subscript a and b denote the absorber and bath molecules, respectively. For linear molecules, Ω_ξ represents the two angles θ_ξ and ϕ_ξ needed to specify the orientation of the axis with respect to a space-fixed coordinate system, while for an asymmetric top molecule it represents the three Euler angles α_ξ, β_ξ and γ_ξ. The advantages of this representation are that the anisotropic interaction potential, $V(r, \Omega_a, \Omega_b)$, is diagonal and one can include as many state as desired in the calculations. The disadvantages are that the density operator is off diagonal and that the sums over internal quantum numbers become multi-dimensional integrals. The diagonalization of the density operator is straightforward, and although it is time consuming, it only has to be done once for a fixed temperature T. The dimensionality depends on the types of molecules: for two linear molecules, for one linear and one asymmetric top, or for two asymmetric tops, it is 7, 9, and 11, respectively. For the latter two cases, one has to use a Monte Carlo method, and even then, in order to obtain converged results, a large amount of CPU time is required. This, of course, depends on the complexity of the interaction potential model assumed.

For the H_2O–H_2O case, we tested several of the interaction potentials available in the literature, but none gave good results for the far-wing absorption. In particular, the far-wing line shapes depend very sensitively on the short-range anisotropic gradients. Consequently, we adopted the following model potential:

$$V(r, \Omega_a, \Omega_b) = \sum_{i \in a} \sum_{j \in b} \frac{q_i\, q_j}{r_{ij}} + \sum_{k \in a} \sum_{l \in b} V_{kl} \exp\left(-\frac{r_{kl}}{\rho_{kl}}\right) - \frac{B}{r^6}, \tag{2}$$

where q_i, V_{kl}, and B are adjustable parameters. We assume that for each H_2O, there are two point charges $+q$ located on the H atoms and one charge $-2q$ at a position along the symmetry axis at a distance δ from the O atom. We require that these charges give the correct permanent dipole moment and good values for the quadrupole moment components. We assume that there are three repulsive force centers: two located on the two H atoms and one on the O atom. In addition to yielding good agreement with laboratory data for the self-continuum at $T = 296$ K, we also require this potential to reproduce the measured differential scattering cross-sections and T-dependent second virial coefficient. Once we obtain the parameters, we use this potential to calculate the far-wing absorption at other temperatures for comparison with experiment.

2. Results

In this section, we will describe qualitatively the line shapes obtained for H_2O–H_2O and H_2O–N_2, and compare the results with experimental laboratory data [17, 18] and with the empirical CKD 2.4 results [19]. In our calculations, we use the band-average shape, in which we assume that all the spectral lines have the same shape; this is not necessary, but it greatly reduces the computational requirements. In the case of H_2O–H_2O, the shapes are superlorentzian near the line center (where the quasistatic theory is not valid) to approximately $400\,\mathrm{cm}^{-1}$ displacement from the center. For higher displacements, the line shapes fall off exponentially as one can prove theoretically from the analyticity of the dipole-moment correlation function. When the frequency detuning is considered explicitly, the shapes obtained are asymmetric, where the high-frequency side falls off more slowly with frequency displacement than the low-frequency wing [15]. The shapes show a large, negative temperature dependence, and in the important spectral region between 600–$1250\,\mathrm{cm}^{-1}$, they decrease by almost an order of magnitude in the temperature range 220–330 K that is relevant to the Earth's atmosphere. For explicit comparisons, please see Refs. [13–15] and references therein to earlier work. Suffice it to say, that very good agreement over several orders of magnitude for the absorption coefficient is obtained with the data of Burch *et al.* [17, 18]. In addition, excellent agreement with the recent cavity ringdown measurements of Cormier *et al.* [20] for three frequencies between 931–$969\,\mathrm{cm}^{-1}$ is also obtained.

In comparison with the empirical CKD 2.4 results in the frequency interval between 600–$1250\,\mathrm{cm}^{-1}$, the theoretical calculations predict more absorption; the difference is very small near $600\,\mathrm{cm}^{-1}$, but this difference generally increases with both increasing frequency and temperature. The agreement is not surprising because the original CKD model was based primarily on the laboratory data of Burch *et al.* For the theoretical calculations, we obtained the adjustable parameters in the interaction potential, Eq. (2), by requiring good agreement with the data of Burch *et al.* at 296 K, in addition to good agreement with the differential scattering cross section, and the temperature dependence of second virial coefficient. The major difference between the empirical model and the theoretical calculations is in the temperature-dependence. Because the laboratory data were obtained at $T = 296\,\mathrm{K}$ and higher, to obtain empirical results for lower T, it is necessary to extrapolate. This is done by the *ad hoc* algorithm

$$\mathrm{Self}(T) = \mathrm{Self}(T_0) \exp\left(-G(\omega)\left(T - T_0\right)\right), \tag{3}$$

where $T_0 = 296\,\mathrm{K}$, and the frequency-dependent factor $G(\omega)$ is given by

$$G(\omega) = \ln\left(\mathrm{Self}(260)/\mathrm{Self}(296)\right)/\left(296 - 260\right). \tag{4}$$

This factor introduces significant differences for the T-dependence within the bands and in the window regions. For the theoretical calculations, the temperature dependence is incorporated correctly in the theory, thus leading to a more physical determination of the T-dependence.

For the case of H_2O–N_2, the theoretical line shapes obtained [14] are different from those for the self-continuum [15]. They are much less superlorentzian in magnitude and over a smaller range of displacements from the line center to approximately $100\,\mathrm{cm}^{-1}$, and they exhibit smaller, though non-negligible, negative T-dependence. Again, the agreement between the theoretical results and the experimental data of Burch *et al.* [17, 18] is very good. For the theoretical results and the experimental data of Burch *et al.* [17, 18] is very good. For the foreign-continuum in the spectral range between 600–1250 cm^{-1}, differences between the theoretical absorption coefficient and the CKD 2.4 results are very much greater than for the self continuum. While they agree well at $600\,\mathrm{cm}^{-1}$ and $1250\,\mathrm{cm}^{-1}$, the theoretical results predict more absorption, the difference reaching more than an order of magnitude around $1000\,\mathrm{cm}^{-1}$. Mlawer *et al.* [21] present in their Fig. 1, comparisons between the CKD 0 and CKD 2.2 models, together with the experimental data of Burch *et al.*For some reason, they do not plot the experimental data for the spectral region between $600\,\mathrm{cm}^{-1}$ and $1000\,\mathrm{cm}^{-1}$.

Comparisons between the theoretical results and experimental data in other spectral regions have been presented previously and are not discussed in detail here, except for the one experimental data point at $9466\,\mathrm{cm}^{-1}$ [22]. The CKD 2.4 results predict that the self continuum is slightly larger than the foreign continuum for the experimental conditions; the corresponding total absorption is larger than the measurement, but within the rather large experimental uncertainties. This empirical result has not changed from the initial version CKD 0. In contrast, the theoretical value obtained is 5.2×10^{-10} whereas the experimental result is 6×10^{-10}. However, for the theoretical results, the self continuum is approximately four times that of the foreign continuum. Given the large T-dependence of the self continuum, one can conclude that for atmospheric applications, the CKD and theoretical results will differ significantly in this spectral region. The reason that this comparison is important is that the experimental data at $9466\,\mathrm{cm}^{-1}$ was not used to constrain either the empirical model or the theoretical calculations, and thus provides an independent check on the validity of the results.

We would like to make some other comments concerning the definition of the continuum, and why one would expect that the theoretical results are lower than the empirical results in those spectral regions containing large contributions from local lines; i.e. within vibration-rotational bands. In the theoretical calculations, we exclude all contributions from local lines within $\pm 25\,\mathrm{cm}^{-1}$; in the CKD definition, the "pedestal" contribution is included [19]. Furthermore, in the CKD model, the local contributions are calculated assuming Lorentzian line shapes to $\pm 25\,\mathrm{cm}^{-1}$. As mentioned above, both the self and foreign line shapes are superlorentzian, and thus would result in larger local absorption. This effect is hard to quantify, because the "true" line shape is not known over its entire profile.

We would like to discuss one final topic: the importance of collision-induced absorption as a possible source of the water continuum. Mlawer *et al.* [21], have proposed a revised formulation of the foreign continuum by including a low-frequency component centered at the ν_2 band origin that they attribute to collision-induced absorption. Their motivation was to improve the empirical continuum so as to be in better agreement with the experimental measurements of Tobin *et al.* [23]. While the agreement can be improved by introducing additional adjustable parameters, they provide no theoretical justification for the magnitude of the collision-induced absorption inferred. In fact, theoretical calculations [24] show that the dominant dipole-induced dipole collision-induced mechanism is negligible for frequencies below $2000\,\mathrm{cm}^{-1}$. However, this is not the case for frequencies between 2500–$2600\,\mathrm{cm}^{-1}$, because of the collision-induced double transitions, in which the N_2 molecule makes a fundamental Raman transition, while the H_2O molecule makes a pure rotational dipole transition. In his analysis of experimental data in this spectral region, Burch [18] has subtracted out the collision-induced N_2–N_2 contribution, but not the collision-induced H_2O–N_2 contribution in determining the foreign continuum. In fact, this was not possible because both the collision-induced absorption and the foreign continuum have the same density dependence and both are broad, featureless absorptions. One can only separate them by calculating the collision-induced absorption theoretically and subtracting this from the experimental measurement. This would affect the numerical values obtained for the foreign continuum from laboratory measurements. For atmospheric applications, both mechanisms will absorb and should be taken into account in this spectral region. In conclusion, over the past few years, we have developed a first-principles line shape theory applicable to any type of colliding molecular pairs. Only the binary collision and quasistatic approximations are necessary, and both of these are well satisfied for the systems of interest. The accuracy of the theoretical results for the line shapes and the corresponding absorption coefficients depends only on the accuracy of the interaction potential. Conversely, the far-wing line shapes and the corresponding absorption coefficients can provide strong constraints for the accuracy of these potential models. Finally, although in this paper, we have focused on the H_2O–H_2O and H_2O–N_2 systems, one can easily carry out calculations for other systems such as H_2O–O_2, CO_2–CO_2, H_2O–CO_2, etc., that are important in other planetary atmospheres.

Acknowledgments

This work was supported in part by the Department of Energy Interagency Agreement under the Atmospheric Radiation Measurement Program, and by NASA under Grant NAG5-8269. One of the authors (QM) would like to thank the National Energy Research Supercomputer (Livermore, CA) for computer time and facilities provided.

References

[1] Rothman, L. S., Gamache, R. R., Tipping, R. H., Rinsland, C. P., Smith, M. A. H., Benner, D. C., Devi, V. M., Flaud, J.-M., Camy-Peyret, C., Perrin, A., Goldman, A., Massie, S. T., and Brown, L. R. (1992) The HITRAN Molecular Database: Editions of 1991 and 1992, J. Quant. Spec. Rad. Transfer, 48, 469–507.

[2] Jacquinet-Husson, N., Aire, E., Ballard, J., Barbe, A., Bjoraker, G., Bonnet, B., Brown, L. R., Camy-Peyret, C., Champion, J. P., Chedin, A., Chursin, A., Clerbaux, C., Duxbury, G., Flaud, J. M., Fourrie, N., Fayt, A., Granier, G., Gamache, R., Goldman, A., Golovko, Vl., Guelachvili, G., Hartmann, J. M., Hilico, J. C., Hillman, J., Leferve, G., Lellouch, E., Mikhailenko, S. N., Naumenko, O. V., Nemtchinov, V., Newnham, D. A., Nikitin, A., Orphal, J., Perrin, A., Reuter, D. C., Rinsland, C. P., Rosenmann, L., Rothman, L. S., Scott, N. A., Selby, J., Sinitsa, L. N., Sirota, J. M., Smith, A. M., Smith, K. M., Tyuterev, Vl. G., Tipping, R. H., Urban, S., Varanasi, P., and Weber, M. (1999) The 1997 Spectroscopic GEISA Databank, J. Quant. Spec. Rad. Transfer, 62, 205–254.

[3] Levy, A., Lacome, N., and Chackerian, Jr. C. (1992) Collisional line mixing. In: K. Narahari Rao and A. Weber, editors. *Spectroscopy of the Earth's Atmosphere and Interstellar Medium*, Academic Press, San Diego, pp. 261–337.

[4] Ozanne, L., Ma, Q., Nguyen-Van-Thanh, Brodbeck, C., Bouanich, J. P., Hartmann, J. M., Boulet, C., and Tipping, R. H. (1997) Line-mixing, finite duration of collision, vibrational shift, and non-linear density effects in the ν_3 and $3\nu_3$ bands of CO_2 perturbed by Ar up to 1000 bars, J. Quant. Spec. Rad. Transfer, 58, 261–277.

[5] Frommhold, L (1993) *Collision-Induced Absorption in Gases*, Cambridge University Press, Cambridge.

[6] Tipping, R. H., Brown, A., Ma, Q., Hartmann, J.-M., Boulet, C., and Lievin, J. (2001) Collision-induced absorption in the ν_2 fundamental band of CH_4: I. Determination of the quadrupole transition moment, J. Chem. Phys., 115, 8852–8857.

[7] Brown, A. and Tipping, R. H. (2001) Theoretical study of collision-induced double transitions in CO_2–X_2 (X_2–N_2, and O_2) pairs, J. Mol. Spectrosc., 205, 319–322.

[8] Moreau, G., Boissoles, J., Le Doucen, R., Boulet, C., Tipping, R. H., and Ma, Q. (2001) Metastable dimer contributions to the collision-induced fundamental absorption spectra of N_2 and O_2 pairs, J. Quant. Spec. Rad. Transfer, 70, 99–113.

[9] Moreau, G., Boissoles, J., Boulet, C., Tipping, R. H., and Ma, Q. (2000) Theoretical study of the collision-induced fundamental absorption spectra of O_2–O_2 pairs for temperatures between 193 K and 273 K, J. Quant. Spec. Rad. Transfer, 64, 87–107.

[10] Moreau, G., Boissoles, J., Le Doucen, R., Boulet, C., Tipping, R. H., and Ma, Q. (2001) Experimental and theoretical study of the collision-induced fundamental absorption spectra of N_2–O_2 and O_2–N_2 pairs, J. Quant. Spec. Rad. Transfer, 69, 245–256.

[11] Ma, Q., Tipping, R. H., and Boulet, C. (1996) The frequency detuning and band-average approximations in a far-wing line shape theory satisfying detailed balance, J. Chem. Phys., 104, 9678–9688.

[12] Ma, Q. and Tipping, R. H. (1998) The distribution of density matrices over potential-energy surfaces: Application to the calculation of the far-wing line shapes for CO_2, J. Chem. Phys., 108, 3386–3399.

[13] Ma, Q. and Tipping, R. H. (1999) The averaged density matrix in the coordinate representation: Application to the calculation of the far-wing line shapes for H_2O, J. Chem. Phys., 111, 5909–5921.

[14] Ma, Q. and Tipping, R. H. (2000) The Density matrix of H_2O–N_2 in the coordinate representation: A Monte Carlo calculation of the far-wing line shape, J. Chem. Phys., 112, 574–584.

[15] Ma, Q. and Tipping, R. H. (2002) The Frequency detuning correction and the asymmetry of line shapes: The far-wings of H_2O–H_2O, J. Chem. Phys., 116, 4102–4115.

[16] Ma, Q., Tipping, R. H., Hartmann, J. M., and Boulet, C. (1995) Detailed balance in far-wing line shape theories: comparisons between different formalisms, J. Chem. Phys., 102, 3009–3010.

[17] Burch, D. E. and Gryvnak, D. A. (1979) Method of calculating the H_2O transmission between 333 and 633 cm^{-1}, AFGL-TR-79-0054.

[18] Burch, D. E. and Alt, R. L. (1984) Continuum absorption by H_2O in the 700–1200 cm^{-1} and 2400–2800 cm^{-1} windows, AFGL-TR-84-0128.

[19] Clough, S. A., Kneizys, F. X., and Davies, R. W. (1989) Line shape and the water vapor continuum, Atmos. Res., 23, 229–241.

[20] Cormier, J. G., Ciurylo, R., and Drummond, J. R. (2002) Cavity Ringdown Spectroscopy measurements of the infrared water vapor continuum, J. Chem. Phys., 116, 1030–1034.

[21] Mlawer, E. J., Clough, S. A., Brown, P. D., and Tobin, D. C. (1998) Collision-induced effects and the water vapor continuum, Eighth ARM Science Team Meeting Proceedings, Tucson, March 23–27.

[22] Fulghum, S. F. and Tilleman, M. M. (1991) Interferometric calorimeter for the measurement of water vapor absorption, J. Opt. Soc. Am. **B**, 8, 2401–2413.

[23] Tobin, D. C., Strow, L. L., Lafferty, W. J., and Olson, W. B. (1996) Experimental investigation of the self- and N_2-broadened continuum within the ν_2 band of water vapor, Appl. Opt., 35, 4724–4734.

[24] Brown, A. and Tipping, R. H., Collision-induced absorption in dipolar molecule-homonuclear diatomic pairs, this volume, 93–99.

II

LABORATORY STUDIES

COLLISION-INDUCED ABSORPTION IN THE CO$_2$ FERMI TRIAD FOR TEMPERATURES FROM 211 K TO 296 K

Y. I. Baranov

Institute of Experimental Meteorology,
Lenina 82, Obninsk, Kalouzhskaya obl., 249020, Russia

G. T. Fraser, W. J. Lafferty

NIST, 100 Bureau Drive,
Gaithersburg, 20899-8440 MD, U.S.A.

A. A. Vigasin

Wave Research Center,
General Physics Institute, Russian Academy of Sciences,
Vavilova 38, Moscow, 119991, Russia

Abstract   Absorption spectra of pure CO_2 have been recorded in the vicinity of the $2675\,\mathrm{cm}^{-1}$ Fermi triad for temperatures between 211 K and 296 K. The $2\nu_1$, $\nu_1 + 2\nu_2$, $2\nu_2$ collision-induced components have been extracted from the measured spectra, including for the low frequency band at $2547\,\mathrm{cm}^{-1}$, which is strongly masked by the ν_3 wing absorption. Dimeric features are clearly seen on top of the structureless profiles. Integrated intensities of the Fermi-triad components are determined as a function of temperature.

Keywords: carbon dioxide / Fermi triad / collision-induced spectra / dimers

1. Introduction

Spectroscopic signatures of the carbon dioxide dimer in collision-induced absorption (CIA) spectra were detected for the first time by Mannyk and Stryland [1] in the $\nu_1, 2\nu_2$ Fermi doublet region. The spectral resolution and numerical analysis were, however, insufficient

149

C. Camy-Peyret and A.A. Vigasin (eds.),
Weakly Interacting Molecular Pairs: Unconventional Absorbers of Radiation in the Atmosphere, 149–158.
© 2003 *Kluwer Academic Publishers. Printed in the Netherlands.*

to allow for accurate modeling of the band profiles. This same region has been recently reinvestigated using Fourier-transform spectroscopy, which has allowed detection of dimeric features in the spectra taken at various temperatures [2,3]. In the region of the Fermi doublet, pure CIA absorption features are strongly overlapped by electric-dipole allowed bands belonging to asymmetrical isotopic forms of CO_2 with a natural abundance of approximately 0.4%. These features are subtraction from the collision-induced absorption profile using line parameters from the HITRAN database [4]. The overtone region, which is the subject of the present paper, provides a better separation of the collision-induced absorption features from the isotopic bands than the Fermi dyad region, allowing a more detailed investigation of the dimer features.

In the present investigation we have recorded new FTIR spectra of CO_2 in the region of the Fermi triad $2(\nu_1, 2\nu_2)$ at temperatures ranging from 211 K to 296 K. The pure CIA profiles of these bands were obtained after removal of the contribution from isotopically substituted molecules and the ν_3-band wing. The CIA profiles were analyzed to determine integrated intensities, which were compared with results in the literature. From the central, most intense component of the triad, the dimeric contributions were determined, yielding the integrated intensity and the fraction of dimer states as a function of temperature.

2. Experimental

An initial investigation of the CIA in the region of CO_2 $2(\nu_1, 2\nu_2)$ overtone was undertaken in Obninsk at $0.5\,cm^{-1}$ resolution using a Perkin-Elmer 1720, Fourier-transform, infrared spectrometer coupled to a 1 m baselength, 30.09 m optical pathlength, coolable, multipass White cell. Later, a series of spectra in the same region was recorded at NIST at room temperature and at pressures from 101 kPa to 808 kPa (1 atm to 8 atm). The spectra were recorded at $0.5\,cm^{-1}$ resolution with a BOMEM FTIR spectrometer coupled to a 2 m baselength, 84 m optical pathlength, White cell. A low-noise InSb detector was used in the experiments at NIST, allowing high quality spectra with high signal-to-noise ratio (see Fig. 1).

Figure 2 presents the Obninsk' CIA spectrum obtained in the region of the Fermi triad at low temperature along with the spectrum taken at NIST near room temperature. It is seen that the spectral bandshapes are practically identical. Note that these spectra have had the interfering lines of the isotopic species removed using the HITRAN database [4]. That structural features atop all the CIA bands arise from the $(CO_2)_2$ dimer.

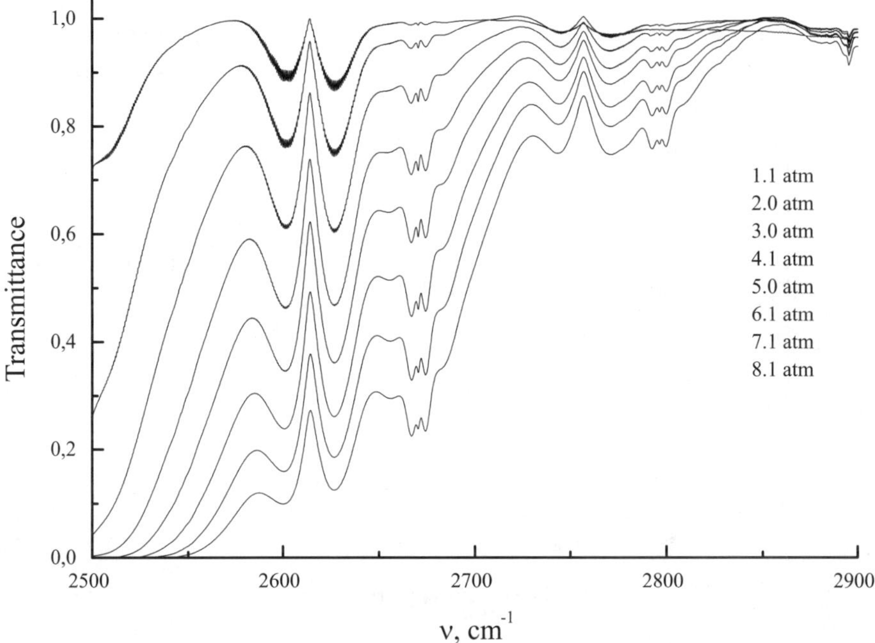

ν, cm^{-1}

Figure 1. Transmittance spectra recorded at NIST. Spectral curves are for CO_2 pressures from $101\,\mathrm{kPa}$ to $808\,\mathrm{kPa}$ in $101\,\mathrm{kPa}$ (1 to 8 atm in 1 atm) steps (from top to bottom).

3. Data processing and results

3.1. Retrieval of CIA profiles

Overlapping of the dipole forbidden and allowed isotopic bands is less pronounced in the region of the Fermi triad than in the region of the Fermi dyad. The main difficulty in treating the spectra arises from the very strong wing of the ν_3 CO_2 band which varies rapidly in intensity in the region of the low frequency component of the Fermi triad. In Figure 1 it appears there are no traces of CIA band in the vicinity of $2500\,\mathrm{cm}^{-1}$. Following numerical removal of the isotope contamination, the contribution of the base ν_3 profile was also removed from the spectral absorption coefficient.

Experimental and theoretical investigations of the far wing profiles of carbon dioxide bands have used both exponential $\exp(-\delta\nu/\tau)$ and power $(\delta\nu)^{-n}$ functions for fitting absorption coefficients in the ν_3 wing region. The power law model was adopted in the present work because it provided a better reproduction of the wing structure. The

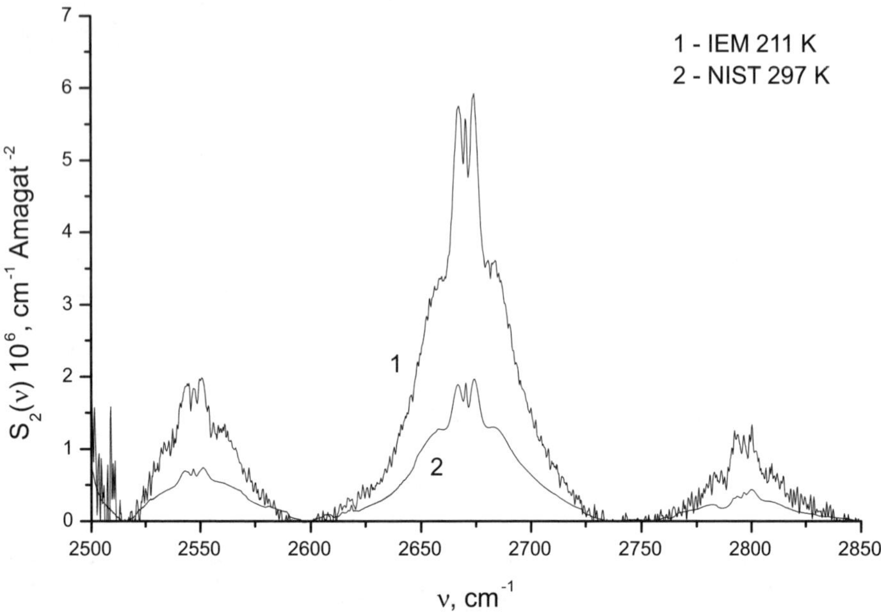

Figure 2.   CIA spectra of the Fermi triad. Low temperature spectrum was recorded in Obninsk (Institute of Experimental Meteorology), room temperature spectrum is from NIST, Gaithersburg.

next step in the removal of the ν_3 wing contribution consisted of establishing the spectral intervals for fitting the model parameters. Initial trial fits demonstrated that the the choice of spectral intervals in which absorption is expected to belong to the ν_3 wing is limited. It is clearly seen from Fig. 2, that only four narrow intervals centered at $2515\,\mathrm{cm}^{-1}$, $2595\,\mathrm{cm}^{-1}$, $2740\,\mathrm{cm}^{-1}$ and $2865\,\mathrm{cm}^{-1}$ may be considered as free from collision-induced contributions to the absorption coefficient and, therefore, these intervals can be used to estimate the wing model parameters. After fitting the absorption within these intervals with the power law function (see Fig. 3) we succeeded in subtracting the ν_3 wing contribution from the recorded spectra to obtain pure CIA associated with the three $2(\nu_1, 2\nu_2)$ Fermi coupled bands. Note that the parameter n is bracketed between 5 and 6. Figures 2 and 3 demonstrate that this removal once accomplished makes it possible to position the center of the low-frequency CIA band at $2547\,\mathrm{cm}^{-1}$.

3.2. Integrated CIA intensity in the Fermi triad region

Integration of the CIA profile may cause problems because of imperfections in removal of the permitted isotopically substituted CO_2

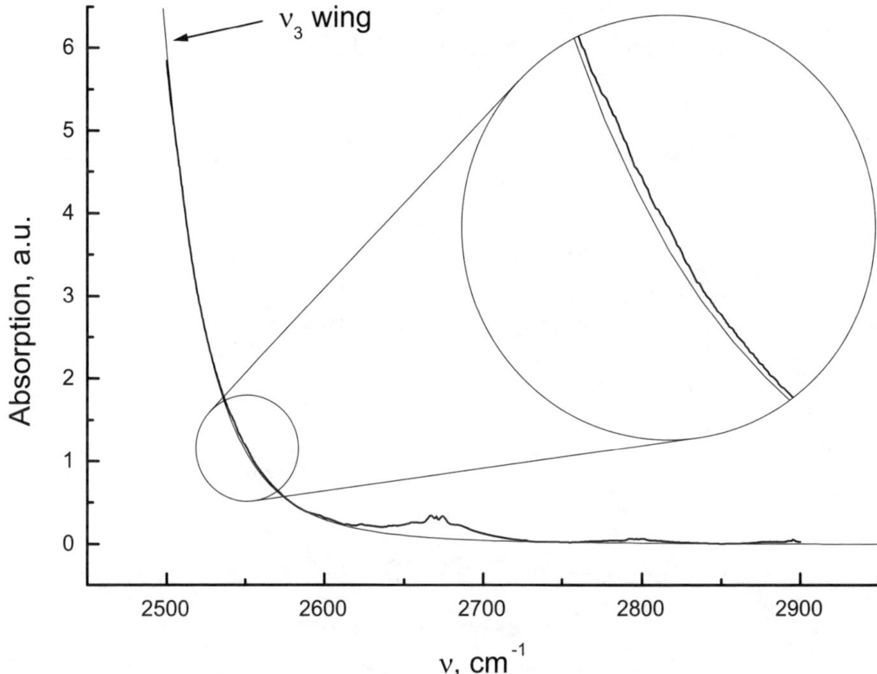

Figure 3.   Retrieval of the low frequency component of the Fermi triad in the vicinity of the ν_3 wing.

contribution. Insufficient knowledge of the line broadening and line intensity parameters results in notable inaccuracy in the wings of the CIA bands. Therefore, in most cases the integration was extended over a frequency range roughly $\pm 70\,\mathrm{cm}^{-1}$ from the band center, ensuring that the absorption is negligible near the ends of the integration limits. The measured values of the integrated CIA intensities in the overtone range are collected in Table 1.

Temperature variations of the CIA Fermi doublet $2(\nu_1, 2\nu_2)$ were studied in our previous work [3]. The available CIA intensity vs. temperature data for both the Fermi doublet and the Fermi triplet regions are displayed in Fig. 4.

We note that the data by Thomas and Linevsky [5] exhibit no temperature dependence. This result is certainly due to a weak temperature dependence rather than due to the complete lack of a temperature variation of the intensity. Interestingly, the results by Thomas and Linevsky [5] for the $2670\,\mathrm{cm}^{-1}$ band are an order of magnitude lower than other available measurements.

Table 1. Integrated intensities of the Fermi triad components at various temperatures.

Temperature, K	Intensity, 10^5 cm^{-2}			Ref.
	$2547\,\mathrm{cm}^{-1}$	$2670\,\mathrm{cm}^{-1}$	$2796\,\mathrm{cm}^{-1}$	this work
211	6.3 ± 0.2	20.1 ± 0.5	3.4 ± 0.5	
235	4.3 ± 0.1	15.7 ± 0.5	2.6 ± 0.3	
264		12.0 ± 0.5		
297	3.0 ± 0.3	10.0 ± 0.5	1.5 ± 0.3	
		$2671\,\mathrm{cm}^{-1}$	$2797\,\mathrm{cm}^{-1}$	[6]
273		17.5	3.0	
296		16.2	2.7	
336		14.1	2.2	
359		13.3	2.0	
		$2669.2\ \mathrm{cm}^{-1}$		[5]
294, 329, 367		2.42		

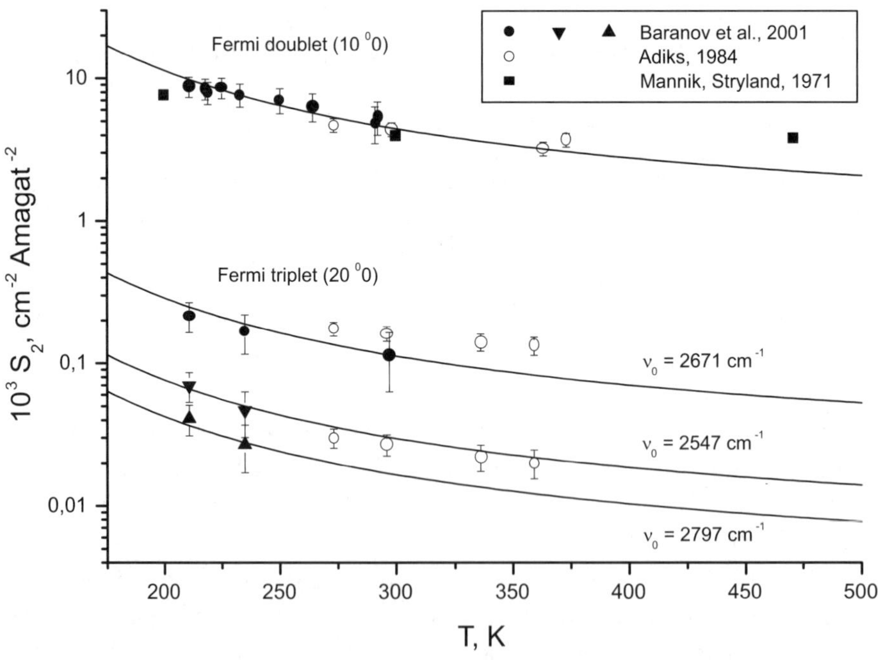

Figure 4. Temperature variations of the integrated intensity in the Fermi doublet and triplet regions.

The high density measurements from Adiks [6] are in good agreement with our results (see Table 1). Note that all data from [1, 5, 6] refer to the measurements at elevated pressures up to several dozens of atmospheres. It looks like in the high density spectra the density-squared contribution from the wings of the allowed bands plays a role. This can result in somewhat overestimated intensities of the recorded CIA bands. On the other hand, the accuracy of our measurements taken at significantly lower densities is not as high due to the weakness of the induced bands in the region of the overtone transitions.

Very roughly, one expects that the temperature variations of all observed bands obey a simple exponential law: $\exp(D_0/T)$, where $D_0 = 246$ K. This latter value is characteristic of the angularly averaged Lennard—Jones intermolecular potential (see e. g. [7]). In the limited temperature range appropriate to our data the temperature dependence of the central peak intensity in the Fermi triplet can be approximated by

$$\log S(2670\,\mathrm{cm}^{-1}) = -0.02293 - 0.00312\,T.$$

3.3. Intensity of dimeric absorption

We will now discuss the dimeric contribution to the central, most intense peak in the overtone region. To obtain a true dimer profile from the measured CIA spectra requires removal of the base absorption due to unbound pairs. Although, we do not know the exact profile of this pedestal in the CIA bands, we estimate that it roughly has a Lorentzian shape. This suggestion is based mainly on our previous experience in the region of the Fermi doublet where the wings of the CIA bands are well approximated by Lorentzian profiles. In all cases, the structureless pedestal for the central triad peak was simulated by the following profile:

$$\frac{C}{202 + (\nu - 2671.4)^2}.$$

Only coefficients C are different for a set of spectra under consideration. A typical base profile is shown in Fig. 5 along with the retrieved dimer constituent.

Also shown is the result of a line-mixing fit of this profile undertaken with the use of a strong-collision model, the details of which can be found elsewhere [8, 9]. It is seen that the calculated profile is almost symmetrical. The key parameter of the line-mixing model, the relaxation time, τ, is on the order of 3×10^{-13} s, in agreement with analogous parameters for other investigated systems (see [10]).

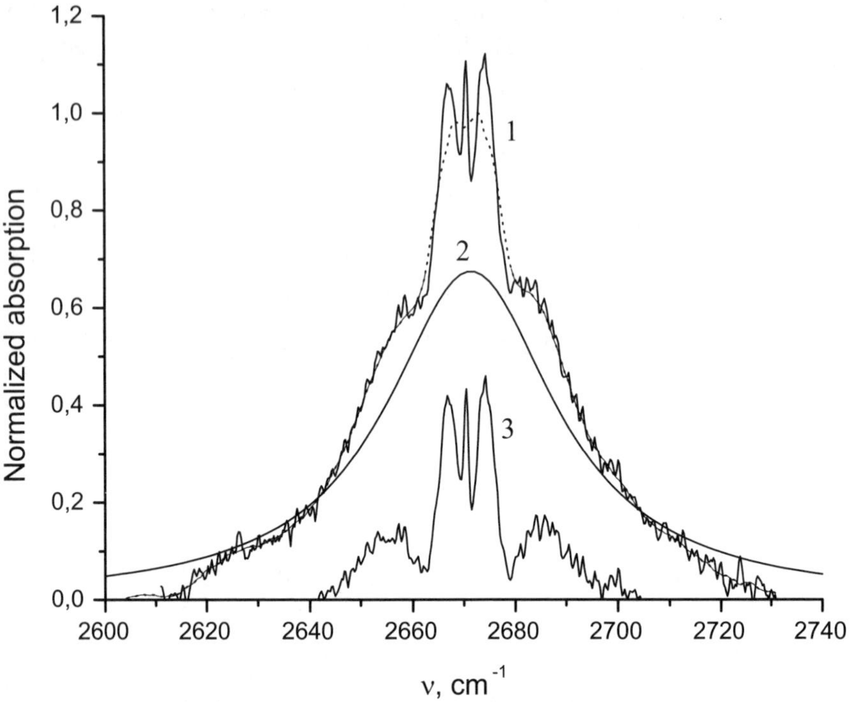

Figure 5.   Typical base profile: (1) from experimental CIA spectrum (dash line shows smoothed contour), (2) from line-mixing fit to a smoothed profile followed by adjustment the base profile intensity, (3) dimer profile.

Table 2 contains dimer intensities normalized to the total CIA intensity for the $2670\,\mathrm{cm}^{-1}$ band. Relative intensities shown in Table 2 are indicative of the temperature variations of the fraction of true dimeric states. Figure 6 shows how these data fit into our previous analysis of the CIA in the Fermi doublet region. Also shown is the result of our statistical mechanics calculations using a sophisticated intermolecular potential energy surface suggested by Bouanich [7]. It is seen that the data obtained in the course of graphical decomposition of CIA profiles are in good agreement with our calculations. This confirms both our statistical mechanical model and the procedure adopted for decomposition of the spectral profiles.

4. Conclusions

All three components of the Fermi triad have been detected in the CIA spectra of compressed carbon dioxide from 211 K to 296 K. The low frequency component is hidden by the intense wing of the allowed ν_3

Table 2. Temperature variations of the dimer contribution to the central CIA band in the region of the Fermi triad.

Temperature, K	Relative intensity $S_{\mathrm{dim}}/S_{\mathrm{tot}}$
211.	0.21925
235.	0.210831
297.16	0.19978

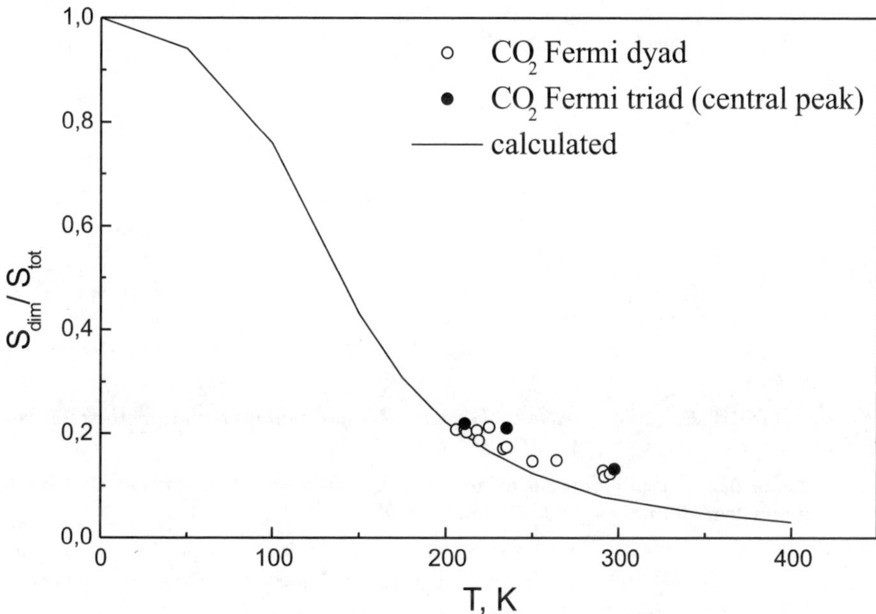

Figure 6.   Temperature variations of the normalized dimer intensity (from Ref. 3).

CO_2 band. Nevertheless, all the Fermi coupled bands are retrieved and are shown to be identical in their shapes; the central peak being notably more intense than the low and high frequency components. Dimeric features manifest themselves very clearly in all the recorded spectra even near room temperature and 101 kPa (1 atm) pressure of carbon dioxide.

Acknowledgments

The authors thank NASA Upper Atmosphere Research Project for partial support of this work. A. A. V. and Y. I. B. acknowledge support from the Russian Foundation for Basic Researches through Grant 02-05-64529.

References

[1] Mannyk, L. and Stryland, J. C. (1972) The ν_1 band of carbon dioxide in pressure-induced absorption. II. Density and temperature dependence of the intensity; Critical phenomena, Can. J. Phys., 50, 1355–1362.

[2] Baranov, Y. I. and Vigasin, A. A. (1999) Collision-induced absorption by CO_2 in the region of ν_1, $2\nu_2$, J. Molec. Spectrosc., 193, 319–325.

[3] Vigasin, A. A., Baranov, Y. I., and Chlenova, G. V. (2002) Temperature variations of the interaction induced absorption of CO_2 in the $\nu_1, 2\nu_2$ region: FTIR measurements and dimer contribution, J. Molec. Spectrosc., 213, 51–56.

[4] Rothman, L. S., Rinsland, C. P., Goldman, A., Massie, S. T., Edwards, D. P., Flaud, J.-M., Perrin, A., Camy-Peyret, C., Dana, V., Mandin, J.-Y., Schroeder, J., Mccann, A., Gamache, R. R., Wattson, R. B., Yoshino, K., Chance, K. V., Jucks, K. W., Brown, L. R., Nemtchinov, V., and Varanasi, P. (1998) The HITRAN molecular spectroscopic database and HAWKS (HITRAN Atmospheric Workstation): 1996 Edition, JQSRT, 60, 665–710.

[5] Thomas, M. E. and Linevsky, M. J. (1989) Integrated intensities of N_2, CO_2, and SF_6 vibrational bands from 1800 to $5000\,\mathrm{cm}^{-1}$ as a function of density and temperature, JQSRT, 42, 465–476.

[6] Adiks, T. G. (1982) Experimental Study of the CO_2 IR Absorption Spectra as Applied to the Windows of Transparency of Venusian Atmosphere. Ph. D. Thesis, Institute of Atmospheric Physics, USSR Academy of Sciences, Moscow (in Russian).

[7] Bouanich, J.-P. (1992) Site-site Lennard—Jones potential parameters for N_2, O_2, H_2, CO, and CO_2, JQSRT, 47, 243–250.

[8] Vigasin, A. A. (1996) On the nature of collision-induced absorption in gaseous homonuclear diatomics, JQSRT, 56, 409–422.

[9] Vigasin, A. A. (2000) Collision-induced absorption in the region of the O_2 fundamental: Bandshapes and dimeric features, J. Molec. Spectrosc., 202, 59–66.

[10] Vigasin, A. A. (2003) Bimolecular absorption in atmospheric gases, this volume, pp. 23–47.

LABORATORY STUDIES OF
OXYGEN CONTINUUM ABSORPTION

Y. I. Baranov,* G. T. Fraser, W. J. Lafferty

Optical Technology Division,
National Institute of Standards and Technology,
Gaithersburg, MD 20899-8441, U.S.A.

B. Maté

Instituto de Estructura de la Materia,
Consejo Superior de Investigaciones Cientificas,
Serrano 123, Madrid 28006, Spain

A. A. Vigasin

Wave Research Center,
General Physics Institute, Russian Academy of Sciences,
Vavilova 38, Moscow 119991, Russia

Abstract   Laboratory studies of the mid and near infrared collision-induced absorption bands of O_2 are reviewed, with an emphasis on the 6.4 μm and 1.27 μm bands and on Ar, O_2, N_2, and CO_2 collision partners. These absorption bands are important in the radiative balance of the atmosphere and contribute to the background absorption in remote-sensing radiometers. Also discussed is the ripple structure observed on the 6.4 μm band, which is absent from the 1.27 μm band. This structure has been attributed to dimers or to line mixing involving the weakly allowed electric-quadrupole absorptions. Finally, the strong enhancement of the collision-induced absorption by CO_2 collisions is shown.

Keywords: collision-induced absorption / oxygen fundamental / magnetic-dipolar absorption / line-mixing

---

*Permanent address: Institute of Experimental Meteorology, Lenina 82, Obninsk 249020, Kalouzhskaya obl., Russia

159

C. Camy-Peyret and A.A. Vigasin (eds.),
Weakly Interacting Molecular Pairs: Unconventional Absorbers of Radiation in the Atmosphere, 159–168.
© 2003 *Kluwer Academic Publishers. Printed in the Netherlands.*

Molecular oxygen continuum or collision-induced absorption plays an important role in the atmosphere. Oxygen continuum absorption from the near infrared through the visible is responsible for 1% of the approximately $80\,\mathrm{W\,m^{-2}}$ globally averaged, total atmospheric solar absorption [1]. The $6.4\,\mu$m mid-infrared continuum absorption contributes to the background extinction in satellite filter radiometers used to sense NO_2, water vapor, and sulfate aerosol levels in the atmosphere. Satellite instruments affected include CLAES (Cryogenic Limb Array Etalon Spectrometer), where approximately 33% to 40% of the extinction in Channel 3 is due to oxygen continuum absorption [2], HIRDLS (High Resolution Dynamics Limb Sounder) [3], ATMOS (Atmospheric Trace Molecule Spectroscopy Experiment) [4], and ISAMS (Improved Stratospheric and Mesospheric Sounder) [5]. To ensure accurate atmospheric retrievals of NO_2, water vapor, and sulfate aerosol concentrations, uncertainties of less than 5% in the $6.4\,\mu$m continuum absorption cross sections are desired [6–7].

In principle continuum absorption studies also furnish information about intermolecular pair potentials. Such information is difficult to extract from the absorption profiles due to their relatively low information content, primarily because the profiles are weighted by a room-temperature thermal distribution of collision pairs. High-resolution spectroscopy of bound molecular complexes or dimers has been demonstrated to provide a very detailed and quantitative picture of intermolecular interactions. Complexes of atmospheric interest studied include the homogeneous dimers of water [8] and of CO_2 [9] and the heterogeneous dimers H_2O–CO_2 [10] and H_2O–N_2 [11].

In the present paper we review our recent investigations [12–14] of the continuum absorption in O_2, induced by collisions with O_2, N_2, Ar, and CO_2. We concentrate on the continuum absorption in the vicinity of the weak $1.27\,\mu$m, $v = 0 \leftarrow 0$ component of the O_2 $a^1\Delta_g \leftarrow X^3\Sigma_g^-$ magnetic-dipole-allowed and $6.4\,\mu$m, $v = 1 \leftarrow 0$ electric-quadrupole-allowed discrete bands. On top of both of these bands a broad continuum is induced by collisions, which grows in intensity for pure O_2 with the pressure squared, in contrast to the discrete structure which grows linearly with pressure.

The spectra were investigated at $0.5\,\mathrm{cm^{-1}}$ resolution using a Fourier-transform spectrometer coupled to a $2\,$m base-length, $84\,$m optical path, White cell. Sample temperatures ranged from $205\,$K to $296\,$K and samples densities extended up to nearly $10\,$amagat, or 10 times the density of an ideal gas at standard temperature and pressure (STP). Non-ideal gas corrections were necessary to derive an accurate value for the gas density. In the near infrared, a tungsten lamp source, CaF_2 beam splitter, and an InSb detector were used, while in the mid-infrared a globar source, KBr beam splitter, and HgCdTe detector were used.

Figure 1 shows the collision-induced absorption spectrum of pure O_2 in the vicinity of the electric-dipole-forbidden, $6.4\,\mu$m, $v = 1-0$ vibrational fundamental band of O_2. Figure 2 shows a similar spectrum in the vicinity of the electric-dipole-forbidden, $1.27\,\mu$m, $v = 0 \leftarrow 0$, $a^1\Delta_g \leftarrow X^3\Sigma_g^-$ electronic band of O_2. The difference in shape of the two collision-induced profiles reflects the underlying structure of the weak discrete Σ–Σ electric quadrupole and Δ–Σ magnetic dipole bands, which have the same rotational-state selection rules as the collision-induced bands. This difference is illustrated more clearly in Figures 3 and 4 where theoretical stick spectra of the discrete bands at 296 K are shown together with simulated spectra of the collision-induced absorption profiles generated by imposing a $40\,\mathrm{cm}^{-1}$ Lorentzian full-width-half-maximum (FWHM) lineshape profile on the individual discrete transitions. The origin of the broad linewidth is the transient nature of the transition dipole induced during the collision, which is responsible for the intensity of the collision-induced bands. This simple model neglects the ensemble of translational states simultaneously excited with the rotational and vibrational states of the O_2 molecules, which could be included by using a more sophisticated Boltzmann-modified Lorentzian function with a fourth power tail, as discussed by McKellar [15] for N_2 collision-induced absorption.

The mid and near infrared collision-induced absorption bands also exhibit a number of other differences. Most striking, is the ripple structure observed on the $6.4\,\mu$m band which is absent from the $1.27\,\mu$m band. This structure has been attributed to torsional subbands associated with bound and metastably bound oxygen-containing dimers [15, 16] or to line mixing of the electric-quadrupole-induced lines within a collision-induced band [17, 18]. Interestingly, the valleys of the ripple pattern are coincident with the predicted discrete electric quadrupole lines, as shown in Figure 5.

Figure 5 also shows that the position and amplitude of the maxima and minima for the ripple structure are independent of gas collision partner (O_2, N_2, or Ar) inducing the absorption. This observation makes it difficult to attribute the ripple structure to dimers. Spectra of O_2 containing complexes are expected to vary — both in frequency and amplitude of the torsional subbands — with binding partner due to the differences in the strengths and corresponding anisotropies of the interactions in O_2–Ar, O_2–N_2, and O_2–O_2. An additional concern with the dimer model is the strength of the ripples as they progress from the band center. For dimer bands, the intensity for these features comes

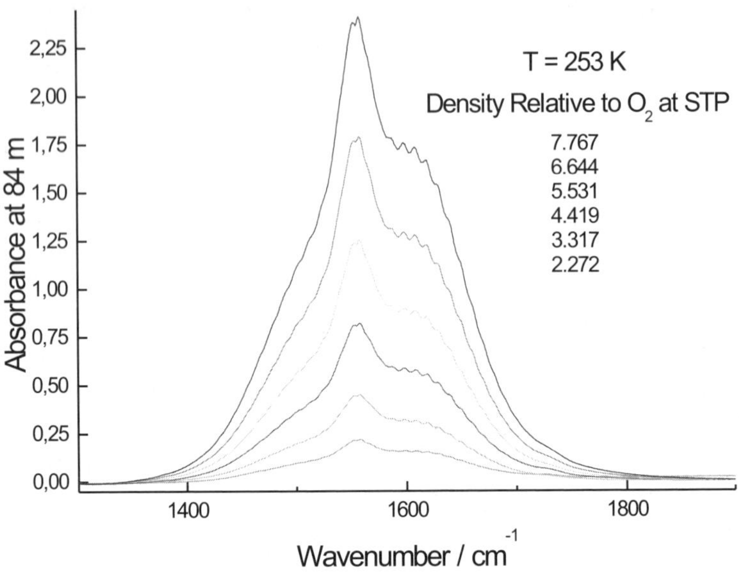

Figure 1.   Continuum absorption profiles at $253\,\mathrm{K}$ for the $6.4\,\mu\mathrm{m}$, $v = 1-0$ vibrational fundamental of O_2 in pure O_2 for densities from 2.27 to 7.76 times that of an ideal gas at STP. The pathlength is $84\,\mathrm{m}$.

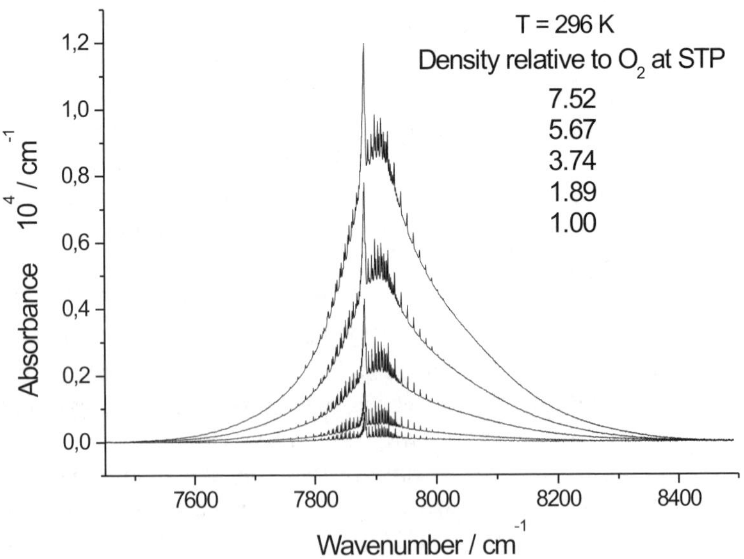

Figure 2.   Continuum absorption profiles at $296\,\mathrm{K}$ for the $1.27\,\mu\mathrm{m}$, $v = 0 \leftarrow 0$, $a^1\Delta_g \leftarrow X^3\Sigma_g^-$ electronic band of O_2 in pure O_2 for densities from 1.00 to 7.52 times that of an ideal gas at STP. Spectra were recorded at a pathlength of $84\,\mathrm{m}$.

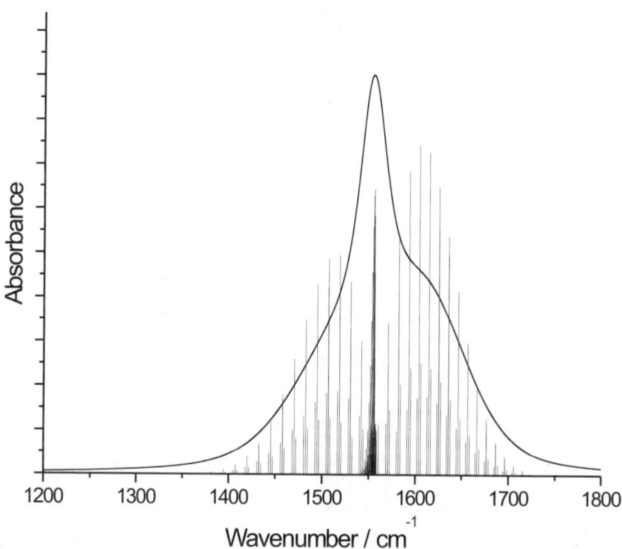

Figure 3. Continuum absorption profile for the 6.4 μm band of O_2 generated by taking the predicted underlying discrete electric quadrupolar $v = 1-0$ band of pure O_2 and superimposing a $40\,\mathrm{cm}^{-1}$ full-width-at-half-maximum Lorentzian lineshape on each of the lines.

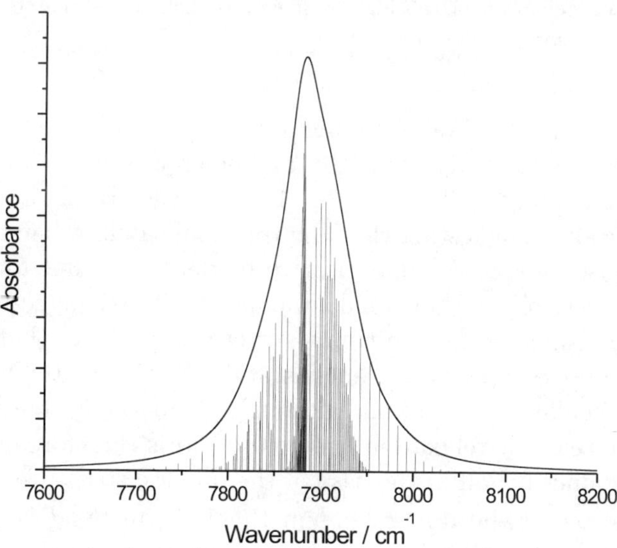

Figure 4. Continuum absorption profile for the 1.27 μm band of O_2 generated by taking the predicted underlying discrete structure of the $v = 0 \leftarrow 0$, $a^1\Delta_g \leftarrow X^3\Sigma_g^-$ electronic band of pure O_2 and superimposing a $40\,\mathrm{cm}^{-1}$ full-width-at-half-maximum Lorentzian lineshape on each of the lines.

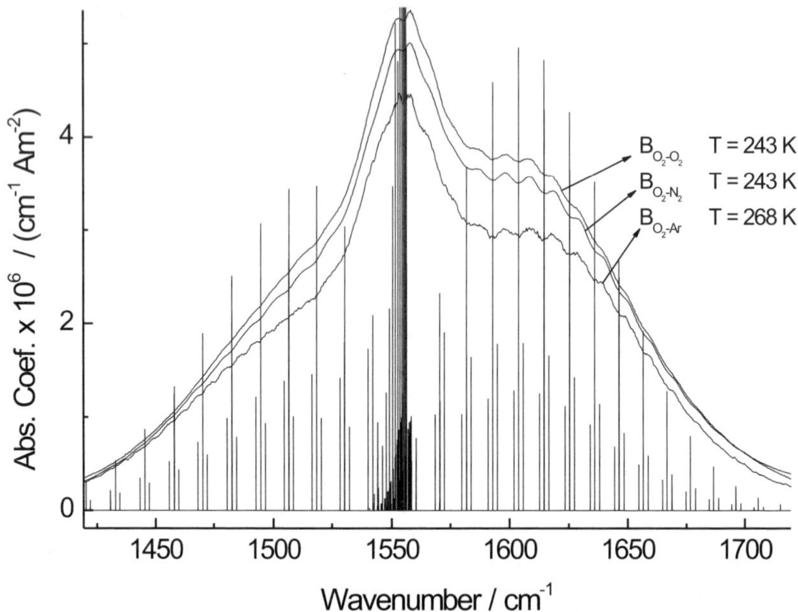

Figure 5. The collision-induced absorption coefficient for the 6.4 μm band for O_2, N_2, and Ar collision partners. Note the lack of variation of the ripple structure with collision partner. The stick spectrum is the predicted $v = 1-0$ electric quadrupole band of O_2. Examination of the high-frequency branch shows that the rotational lines fall in the valleys of the ripple structure.

solely from the van der Waals interaction, as the quadrupole component of the transition intensity from the monomer is small. Moreover, at high excitation energies the strength of the van der Waals interaction is effectively weaker, suggesting that the torsional satellite band strengths should likewise be low. No quantitative model to account for the large intensity in these alleged torsional subbands has been suggested to support the dimer picture for the ripple structure. We note that line mixing among induced absorption lines is also sensitive to the strength of the intermolecular perturbation, although the effect is due to the decrease of the relevant relaxation time only, and is therefore expected to be less pronounced than in the case of the dimeric structure.

The absence of the dimer lines in the 1.27 μm band is surprising. One possible explanation based on the appearance of the underlying monomer band is that the torsional structure of the 1.27 μm band is more compressed, masking any ripple structure. Another possibility is that the dimer structure is not resolved due to rapid electronic predissociation of the complex. The lack of ripple structure in the 1.27 μm

Table 1. Integrated collision-induced absorption coefficients.

Band, μm	$S_{O_2-O_2}$, $\mathrm{cm}^{-2}(\mathrm{molec/cm}^3)^{-2}$	$S_{O_2-N_2}$, $\mathrm{cm}^{-2}(\mathrm{molec/cm}^3)^{-2}$	$S_{O_2-CO_2}$, $\mathrm{cm}^{-2}(\mathrm{molec/cm}^3)^{-2}$
1.27	$4.847(22) \times 10^{-43}$	$0.941(50) \times 10^{-43}$	$2.95(40) \times 10^{-43}$
6.4	$9.658(91) \times 10^{-43}$	$9.86(30) \times 10^{-43}$	$39.8(40) \times 10^{-43}$

band can be interpreted in terms of the line mixing model by suggesting more rapid rotational relaxation in the electronic band than in the vibrational fundamental. This increase in rotational relaxation rate can be caused by electronic predissociation of the metastable dimer, resulting in destruction of the ripple structure found in the case of the O_2 vibrational fundamental. The successful modeling of the absence of ripple structure in the $1.27\,\mu$m band using the line-mixing model developed for the $6.4\,\mu$m band would further verify the role of line mixing in creating the ripple structure and is presently being pursued by one of the authors.

The different nature of the $6.4\,\mu$m and $1.27\,\mu$m bands is also seen from examination of the collision-induced absorption coefficients obtained from the band analysis. The integrated band intensities, S, are related to the integrated binary collision coefficients, $S_{O_2-O_2}$ and S_{O_2-X}, and the pressure-broadening coefficient, S_{O_2}, of the underlying discrete band by

$$S = \frac{1}{L} \int \ln\left(\frac{I_0(\nu)}{I(\nu)}\right) d\nu = S_{O_2-O_2}\,\rho_{O_2}{}^2 + S_{O_2-X}\,\rho_{O_2}\,\rho_X + S_{O_2}\,\rho_{O_2}, \quad (1)$$

where ρ_{O_2} and ρ_X are the densities of O_2 and X, with X = N_2, CO_2, or Ar, L is the optical pathlength, and $I_0(\nu)$ and $I(\nu)$ are the light intensities before and after transmission through the sample.

Values for the integrated collision-induced absorption coefficients, S, are tabulated in Table 1 for the two bands. As seen in the table, for the $1.27\,\mu$m band the absorption coefficient changes dramatically between O_2 and N_2, whereas for the $6.4\,\mu$m band it is effectively unchanged. The large difference between the O_2 and N_2 $1.27\,\mu$m coefficients is most likely the consequence of a preference for a magnetic-dipolar collision partner to induce magnetic dipolar absorption at $1.27\,\mu$m, whereas an electric quadrupolar probe is preferred for inducing electric quadrupolar absorption at $6.4\,\mu$m. Further evidence for the different natures of the $6.4\,\mu$m and $1.27\,\mu$m collision-induced absorption bands is seen from examination of the $S_{O_2-CO_2}$ coefficient, which is nearly a factor of 13 greater for the $6.4\,\mu$m band than for the $1.27\,\mu$m band, presumably due to the large electric quadrupole moment of CO_2. The magnitudes of the quadrupole moments alone do not explain, however, the observed differences in the integrated absorption coefficients. The square of the quadrupole moments for $O_2 : N_2 : CO_2$ are roughly in the proportion $1 : 12 : 115$, whereas the integrated intensities for the $6.4\,\mu$m band induced by these collision partners scale as $1 : 1 : 4$. The differences in the effective but unknown separation, R, during the collision also play a role due to the approximate R^8 dependence of the integrated collision-induced absorption intensity.

As shown in Table 1, CO_2 has a significantly greater ability than N_2 to induce continuum absorption in O_2, presumably due to the larger electrostatic quadrupole moment of CO_2 relative to N_2. Since CO_2 is a trace component in the atmosphere this effect plays a negligible role in the atmospheric continuum absorption by O_2. Water vapor, however, is not a trace atmospheric constituent, and its partial pressure is largest at low altitudes where the continuum absorption dominates. Water vapor like CO_2 has a large electrostatic moment, in this case an electric dipole moment, capable of driving continuum absorption in other gases. To assess the magnitude of this effect we note that the integrated continuum absorption for the O_2 band will increase from 2.6% to 13% for air at $101\,$kPa (1 atm) for a H_2O mole fraction of 0.013% (i. e. 10 Torr partial pressure), when $S_{O_2-H_2O} = S_{O_2-CO_2}$ to $S_{O_2-H_2O} = 3\,S_{O_2-CO_2}$. This effect is greater than a target goal of 5% suggested by some for accounting for the $6.4\,\mu$m O_2 continuum absorption. For pure O_2 samples we have successfully achieved this target goal of 5%, as shown in Figure 6.

Clearly, future efforts need to be directed at further understanding the role of H_2O vapor in inducing continuum absorption in gases, particularly if H_2O vapor is found to be more efficient at inducing of continuum absorption than CO_2.

Acknowledgments

We would like to thank the NASA Upper Atmosphere Research Program for partial support of this research. Y.I.B. and A.A.V. acknowledge partial support from the Russian Foundation for Basic Researches through Grant 02-05-64529.

References

[1] Pfeilsticker, K., Erle, F., and Platt, U. (1997) Absorption of solar radiation by atmospheric O_4, J. Atmos. Sci., 54, 933–939.

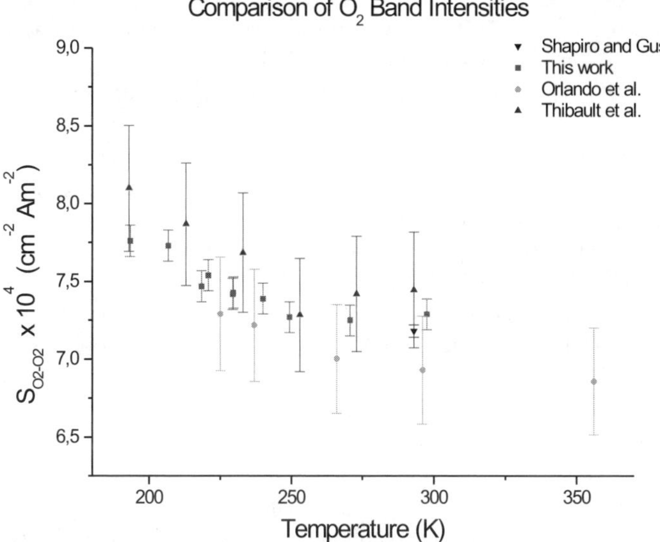

Figure 6. $S_{O_2-O_2}$ binary collision induced absorption coefficient for the $6.4\,\mu$m vibrational fundamental of CO_2 as a function of temperature. The figure includes additional measurements not presented in our initial study of this system [13]. Our measurements are compared with the results of Shapiro and Gush [19], Orlando *et al.* [20], and Thibault *et al.* [21]. Expanded uncertainties are given with a $k=2$ coverage factor (2 standard deviations).

[2] Massie, S. T., Gille, J. C., Edwards, D. P., Bailey, P. L., Lyjak, L. V., Craig, C. A., Cavanaugh, C. P., Mergenthaler, J. L., Roche, A. E., Kumer, J. B., Lambert, A., Grainger, R. G., Rodgers, C. D., Taylor, F. W., Russell III, J. M., Park, J. H., Deshler, T., Hervig, M. E., Fishbein, E. F., Waters, J. W., and Lahoz, W. A. (1996) Validation studies using multiwavelength Cryogenic Limb Array Etalon Spectrometer (CLAES) observations of stratospheric aerosol, J. Geophys. Res., 101, 9757–9773.

[3] Edwards, D. P., Gille, J. C., Bailey, P. L., and Barnett, J. J. (1995) Selection of sounding channels for the High Resolution Dynamics Limb Sounder, Appl. Optics, 34, 7006–7018.

[4] Rinsland, C. P., Yue, G. K., Gunson, M. R., Zander, R., and Abrams, M. C. (1994) Mid-infrared extinction by sulfate aerosols from the Mt. Pinatubo eruption, J. Quant. Spectrosc. Radiat. Transfer, 52, 241–252.

[5] Goss-Custard, M., Remedios, J. J., Lambert, A., Taylor, F. W., Rogers, C. D., Lopez-Puertos, M., Zaragoza, G., Gunson, M. R., Suttie, M. R., Harries, J. E., and Russell III, J. M. (1996) Measurements of water vapor distributions by the improved stratospheric and mesospheric sounder: Retrieval and validation, J. Geophys. Res. Atmos., 101, 9907–9928.

[6] Rinsland, C. P., Smith, M. A. H., Seals, Jr., R. K., Goldman, A., Murcray, F. J., Murcray, D. G., Larsen, J. C., and Rarig, P. L. (1982) Stratospheric measurements of collision-induced absorption by molecular oxygen, J. Geophys. Res., 87, 3119–3122.

[7] Rinsland, C. P., Zander, R., Namkung, J. S., Farmer, C. B., and Norton, R. H. (1989) Stratospheric infrared continuum absorptions observed by the ATMOS instrument, J. Geophys. Research, 94, 16303–16322.

[8] Dyke, T. R., Mack, K. M., and Muenter, J. S. (1977) Structure of water dimer from Molecular-Beam Electric Resonance Spectroscopy, J. Chem. Phys., 66, 498–510.

[9] Jucks, K. W., Huang, Z. S., Miller, R. E., Fraser, G. T., Pine, A. S., and Lafferty, W. J. (1988) Structure and vibrational dynamics of the CO_2 dimer from the sub-Doppler infrared spectrum of the 2.7 μm Fermi dyad, J. Chem. Phys., 88, 2185–2195.

[10] Peterson, K. I. and Klemperer, W. (1984) Structure and internal rotation of H_2O–CO_2, HDO–CO_2, and D_2O–CO_2 van der Waals complexes, J. Chem. Phys., 80, 2439–2445.

[11] Leung, H. O., Marshall, M. D., Suenram, R. D., and Lovas, F. J. (1989) Microwave spectrum and molecular structure of the N_2–H_2O complex, J. Chem. Phys., 90, 700–712.

[12] Maté, B., Lugez, C. L., Fraser, G. T., and Lafferty, W. J. (1999) Absolute intensities for the O_2 1.27 μm continuum absorption, J. Geophys. Research, 104, 30585–30590.

[13] Maté, B., Lugez, C. L., Solodov, A. M., Fraser, G. T., and Lafferty, W. J. (2000) Investigation of the collision-induced absorption by O_2 near 6.4 μm in pure O_2 and O_2/N_2 mixtures, J. Geophys. Research, 105, 22225–22230.

[14] Fraser, G. T. and Lafferty, W. J. (2001) 1.27 μm O_2 continuum absorption in O_2/CO_2 mixtures, J. Geophys. Research, 106, 31749–31753.

[15] McKellar, A. R. W. (1988) Infrared spectra of the $(N_2)_2$ and N_2–Ar van der Waals molecules, J. Chem. Phys., 88, 4190–4196.

[16] Moreau, G., Boissoles, J., Le Doucen, R., Boulet, C., Tipping, R. H., and Ma, Q. (2001) Metastable dimer contributions to the collision-induced fundamental absorption spectra of N_2 and O_2 pairs, J. Quant. Spectrosc. Radiat. Transfer, 70, 99–113.

[17] Vigasin, A. A. (1996) On the nature of collision-induced absorption in gaseous homonuclear diatomics, J. Quant. Spectrosc. Radiat. Transfer, 56, 409–422.

[18] Vigasin, A. A. (2000) Collision-induced absorption in the region of the O_2 fundamental: Bandshapes and dimeric features, J. Mol. Spectrosc., 202, 59–66.

[19] Shapiro, M. M. and Gush, H. P. (1966) The collision-induced fundamental and first overtone bands of oxygen and nitrogen, Can J. Phys., 44, 949–963.

[20] Orlando, J. J., Tyndall, G. S., Nickerson, K. E., and Calvert, J. G. (1991) Temperature dependence of collision-induced absorption by oxygen near 6 μm, J. Geophys. Research, 96, 20755–20760.

[21] Thibault, F., Menoux, V., Le Doucen, R., Rosenmann, L., Hartmann, J.-M., and Boulet, C. (1997) Infrared collision-induced absorption by O_2 near 6.4 μm for atmospheric applications: Measurements and empirical modeling, Appl. Opt., 36, 562–567.

MOLECULAR BEAM SCATTERING EXPERIMENTS ON SPECIES OF ATMOSPHERIC RELEVANCE: POTENTIAL ENERGY SURFACES FOR CLUSTERS AND QUANTUM MECHANICAL PREDICTION OF SPECTRAL FEATURES

V. Aquilanti, M. Bartolomei, D. Cappelletti, E. Carmona-Novillo,
E. Cornicchi, M. Moix-Teixidor, M. Sabidó,* F. Pirani

*INFM and Università di Perugia I-06123,
Perugia, Italy*

Abstract   Accurate intermolecular potential energy surfaces for the major components of the atmosphere, leading to the characterization of the O_2–O_2, N_2–N_2 and N_2–O_2 dimers, have been obtained from the analysis of scattering experiments from our laboratory, also exploiting where available second virial coefficient data. A harmonic expansion functional form describes the geometries of the dimers and accounts for the relative contributions to the intermolecular interaction from components of different nature. For O_2–O_2, singlet, triplet and quintet surfaces are obtained accounting for the role of spin-spin coupling. The new surfaces allow the full characterization of structure and internal dynamics of the clusters, whose bound states and eigenfunctions are obtained by exact quantum mechanics. Besides the information on the nature of the bond, these results can be of use in modelling the role of dimers in air and the calculated rotovibrational levels provide a guidance for the analysis of spectra, thus establishing the ground for atmospheric monitoring. The same approach is currently being extended to simple hydrocarbons and water molecules interacting with rare gas atoms or simple molecules.

Keywords: scattering / interaction potentials / atmospheric relevance

---

*Present address: Universidad de Barcelona, Spain

C. Camy-Peyret and A.A. Vigasin (eds.),
Weakly Interacting Molecular Pairs: Unconventional Absorbers of Radiation in the Atmosphere, 169–182.
© 2003 *Kluwer Academic Publishers. Printed in the Netherlands.*

1. Introduction

In recent years experimental and theoretical interest has been focused on the study of weakly bound complexes, and we consider here both those regarding the major components present in the atmosphere and other species relevant for its modeling. High resolution spectra for van der Waals clusters can give insight into the nature of intermolecular forces and of internal dynamics, but results for the case of non polar molecules are limited. For them, the interactions involve dispersion forces (induced multipole-induced multipole) which usually are weaker than those arising from electrostatic effects (permanent multipole-permanent multipole), and their complexes rarely have transition moments active in the medium infra-red or microwave ranges. As a consequence, experimental information on the equilibrium geometry and on the bond energy for diatom-diatom complexes is scarce and a full characterization of the interaction is often lacking. Therefore, new experiments are necessary and a simultaneous analysis of different experimental data is recommended.

We report here our work on dimers of the major components of the atmosphere, a preliminary account having been given [3], other papers being in press on N_2-N_2 [2] and O_2-O_2 [4]. Further examples will be anticipated on recent experiments with molecular beams of simple hydrocarbons or water molecules scattered by rare gas targets.

2. Description of the interaction

We give some background information on the interactions and their representations. From *spectroscopic* studies available for $(N_2)_2$, the infrared spectrum [6], suggested that the equilibrium conformation is **T**-shaped with the center-of-mass of the monomers approximately separated by 3.7–4.2 Å. The N_2-N_2 dimer has been studied by *ab initio* methods which have provided several alternatives for the interaction potentials [7–14], yielding different results for the equilibrium geometry and for the bond energy of the ground state. The latest *ab initio* methods yielded as most stable geometries a **T**-shaped and a **Z** one (see Fig. 1) with bond energy in the range 70–80 cm^{-1} [10, 11], a frequency of the stretching van der Waals mode of 22 cm^{-1} and internal rotation barriers with a maximum at 30 cm^{-1}, in fair agreement with experimental results of Long *et al.* [6]. Spectroscopic and theoretical information on the N_2-O_2 system is very scarce, see e. g. [15].

Another experimental source of information is the *second virial coefficient*. Low temperature data depends on the full anisotropic potential energy surface while at high temperature only the spherical part of the interaction is of relevance. Data are available for O_2–O_2 in a wide temperature range and accurate results have been recently derived for the N_2–N_2 system from new acoustic measurements [16]. For N_2–O_2 measurements are lacking, however data can be indirectely derived as average of those for O_2–O_2 and N_2–N_2, being the correction due to the "excess" of the second virial coefficient for mixtures negligible, because the nature of the interaction in N_2–N_2, O_2–O_2 and N_2–O_2 is similar [17, 18].

The measurement of quantum mechanical interference effects in the *scattering cross-sections* provides data which, together with accurate second virial coefficients, yields information on the intermolecular potential and on the dimer structure. We have demonstrated the use of this technique for the O_2–O_2 case [1]. In the velocity dependence of the integral cross-section, the "glory" oscillations, overimposed to a smooth average component, are a probe of the depth of the potential well and of its position, while the absolute value of the cross-section depends on the long range attraction. In our laboratory, experiments have been performed by using "hot" molecules ($T_{\rm rot} = 10^3$ K) as projectile and "cold" molecules, kept at liquid air temperature ($T_{\rm rot} = 10^2$ K), as target. At low velocity the measured cross-sections mainly probe the spherical component of the interaction, since the collisional time is larger than the period of a molecular rotations—the critical time needed to induce an angular averaging of the interaction. As the collisional velocity increases the anisotropy enters into play, since the collisions tends to assume a "sudden" character. The novel technique developed in our group, first reported in "Nature" in 1994 [19], for cooling oxygen to the lowest vibro-rotational state and for aligning the rotational angular momentum, allows the control of the relative orientation of the colliding molecules and provides more direct information on the interaction anisotropy from the measurement of changes in the smooth component, in the amplitude and in the extrema position in the "glory" pattern, as a function of the projectile molecular alignment degree.

Along these lines, a simultaneous analysis of our experimental cross-sections and of available second virial coefficients has been carried out to obtain a reliable interaction for dimers of the major components of the atmosphere. Previous results on the N_2–N_2 system [7] have been reanalyzed, and a characterization of the N_2–O_2 complex is also presented. These results indicate that most of the bonding in the dimers comes from van der Waals (repulsion + dispersion) and electrostatic (permanent quadrupole–permanent quadrupole) forces. (Chemical, or spin-spin contributions are not negligible for $(O_2)_2$, which is an open-shell–open-shell system [1].)

We find useful to represent the interaction potential [1, 9, 20] as a harmonic expansion, separating radial and angular dependencies

$$V(R, \theta_a, \theta_b, \phi) = 4\pi \sum_{l_a, l_b, l} V_{l_a l_b l}(R)\, \mathcal{Y}_{l_a l_b l}(\theta_a, \theta_b, \phi),$$

where $\mathcal{Y}_{l_a l_b l}(\theta_a, \theta_b, \phi)$ are bipolar spherical harmonics (angles are as in Fig. 1). The radial coefficients $V_{l_a l_b l}(R)$ include different types of contributions to the interaction potential (electrostatic, dispersion, repulsion due to overlap, induction, spin-spin coupling, ...). As for $(O_2)_2$ [1], we look for a compact expansion, truncating the series to a small number of physically motivated terms. The number of terms used are *six* for the N_2–O_2 system corresponding to the number of configurations of the dimer (see Fig. 1: for N_2–N_2 and O_2–O_2 this number of terms is reduced at *five* and *four*, see below).

Different appropriate expansions will apply for hydrocarbons or water interactions, accounting for their symmetry.

The radial coefficients are derived from the analysis of experimental data assisted by empirical correlation rules [21]. The V_{000} term has the meaning of a typical isotropic van der Waals interaction, arising from a short-range repulsion, associated to the spherical "size" of both partners, and a long-range dispersion attraction, depending on the spherical polarizability of one of the monomers. The V_{202} and V_{022} coefficients are associated to the orientational anisotropy of each diatom when the other one is considered as a spherical partner: this is the anisotropy expected in the diatom-"pseudoatom" limits. For N_2–N_2 and O_2–O_2 these terms are equal but not in the case of N_2-O_2 where two different limits are possible. Specifically, in the analysis of the scattering of "hot" O_2 (10^3 K) by "cold" N_2 (10^2 K), O_2 acts as a "pseudoatom," and therefore cross-sections at high velocity provide information on the potential in this limiting case. V_{220} and V_{222} are coefficients which introduce corrections to the repulsion arising from the mutual orientation of both molecules. V_{224} is an electrostatic component exclusively depending on the quadrupole-quadrupole interaction, which can be accurately estimated. (This term is negligible in the O_2–O_2 system.) Induction contributions due to permanent quadrupole-induced multipole interactions, which should be included in the V_{202}, V_{220} and V_{222} coefficients, play only a minor role especially in the intermediate region of the intermolecular distance R, which are those of interest here. The analysis of the total cross-section measured as a function of the collision velocity

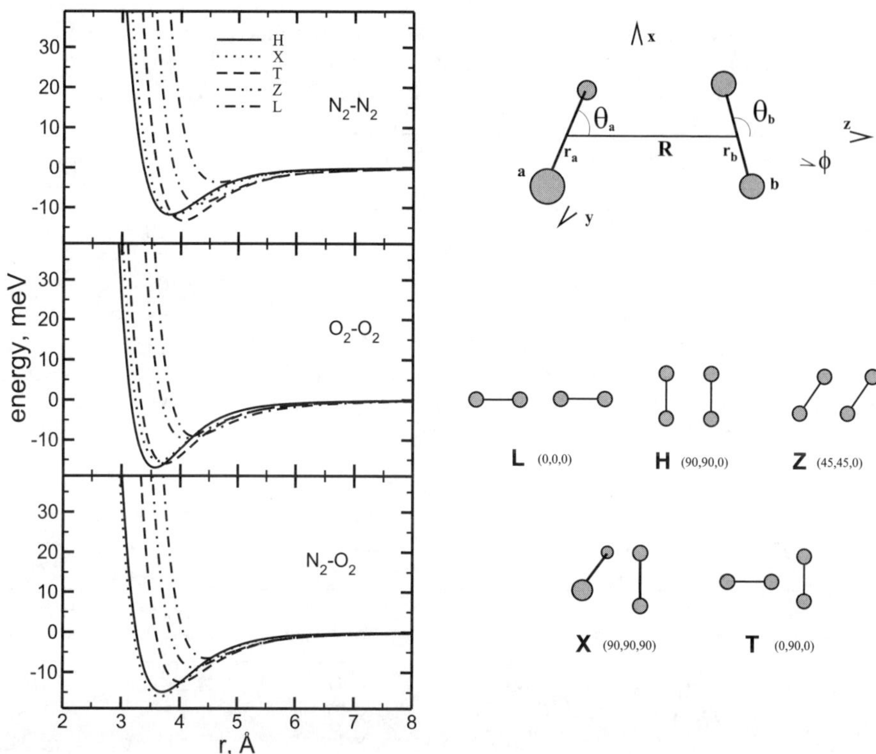

Figure 1. Characteristic configurations (angular coordinates are $(\theta_a, \theta_b, \phi)$, where ϕ is $\phi_a - \phi_b$) and corresponding cuts of potential energy surfaces $(1\,\text{Å} = 10^{-8}\,\text{cm})$. The O$_2$–O$_2$ curves are for the singlet surface of Ref. [1]; for the N$_2$–O$_2$ the two **T** curves are indistinguishable on this drawing.

provides direct information on the V_{000} component on the low velocity range, and yields information on the strength of the terms V_{202} and V_{220} from the "glory" quenching observed at high velocity. The second virial coefficient is affected by all the radial terms in the low temperature range while in the high temperature limit mainly depends on the spherical component.

The intermolecular potential for the limiting configurations of the dimers are shown in Fig. 1, while relevant features of the interactions are compared in Table 1. For the O$_2$–O$_2$ dimer the additional contribution due to the spin-spin coupling, for which information had been obtained from the analysis of the "glory" quenching in the low velocity range [1], had been included in the expansion and yielded the splittings among singlet, triplet and quintet surfaces. Properties of the singlet

Table 1. Equilibrium distances R_m (in Å) and binding energies E (in cm^{-1}; $1\,\text{cm}^{-1} = 0.1240\,\text{meV}$) for selected geometries (for $(O_2)_2$, in the singlet state [1]). Uncertainties are estimated as $\pm 6\,\text{cm}^{-1}$ for E and $\pm 0.07\,\text{Å}$ for R_m.

| | $V_{N_2-N_2}$ | | $V_{O_2-O_2}$ | | $V_{N_2-O_2}$ | |
	E	R_m	E	R_m	E	R_m
H	95	3.81	137	3.56	120	3.70
X	97	3.87	123	3.63	130	3.66
T	107	4.03	129	3.74	101	4.01
					98	3.99
Z	74	4.30	77	4.03	71	4.24
L	30	4.65	73	4.26	53	4.47
V_{000}	79	4.11	107	3.81	88	4.02

Table 2. Selected vibrational frequencies (harmonic: calculated from second derivatives; exact: close-coupling calculations). All values are in cm^{-1}.

		ω_R	$\omega_{(\phi_b - \phi_a)}$	Zero point energy	Binding energy
N_2-N_2	exact[a]	30.78	2.47	27.28	79.87
	harmonic[a]	28.70	12.77	35.73	71.40
	exact[b]	33.2	8.1	47.1	74.94
	harmonic[b]	39.2	13.9	48.7	73.3
O_2-O_2	exact[a]	29.26	13.45	22.59	120.26
	harmonic[a]	48.28	11.32	65.97	71.15
	exact[c]	23.59	6.92	79.82	73.18
	harmonic[c]	43	18	78	75
N_2-O_2	exact[a]	28.8	9.7	12.1	117.6
	harmonic[a]	46.56	8.05	44.68	85.1

[a]Present work;   [b]ref. [7];   [c]ref. [23].

surfaces only are given here for comparison, although results have also been obtained for the other ones.

Interestingly, the three dimers differ both for their geometry and for the nature of the bond. For N_2-N_2 (no electronic spin) the basic feature which determines the equilibrium geometry is the quadrupole-quadrupole interaction, which favors the **T**-configuration (see Fig. 1). In the case of oxygen the equilibrium geometry obtained for the ground singlet state is the **H**-configuration (see Fig. 1). A role is played by the equilibrium configuration from the spin-spin interaction, in spite that its contribution to the bond is approximately only 15% [1]. This leads

to the **H** geometry where the two O_2 molecules are parallel, and also
the bond forces are stronger than in the nitrogen case because of the
strength of the spin-spin interaction. Finally, a **X**-configuration (see
Fig. 1) is found to be the most stable for N_2–O_2: here no role is played
by spin interaction and the quadrupole-quadrupole interaction is not
strong enough to stabilize the **T**-configuration. These potential energy
surfaces are such that their handling and the physical interpretation of
their terms make them "realistic" in the sense that they reproduce mi-
cro and macroscopic quantities experimentally available. Spectroscopic
information, when available, has not been fully analyzed (see e. g. [26]),
and more spectra may be taken in the future, whose analysis will require theoretical guidance.

3. Cluster dynamics: bound states and spectral features

A particular effort has been therefore addressed to the study of the
dynamics within the dimer. Calculations of the bound rovibrational
states of the dimers have been performed for $J \leq 6$ by solving the sec-
ular problem over the exact Hamiltonian. We have calculated the rotovibrational levels for the new surfaces for the dimers N_2–N_2, N_2–O_2
and for O_2–O_2 (for all three surfaces, singlet, triplet, quintet) [1]. A
summary of results and their discussion follows. Full account of all
available data has been given in [2, 4].

Our potentials enable us to computing rovibrational energy levels of
the dimers treating each monomer O_2, N_2 as a rigid rotor, decoupling
intermonomer (van der Waals) and intramonomer vibrations (by almost
two orders of magnitude larger).

The theory to solve the Schrödinger equation for this Hamiltonian
is reported in Ref. [20]. The full quantum mechanical calculations of
bound states are carried out using the program BOUND [22], where
the intermolecular distance R is treated as a scattering coordinate, and
a basis set expansion is used for the remaining degrees of freedom. The
coupled equations are then solved using the standard techniques of scattering theory, but with bound state boundary conditions. This method
is found to be particularly appropriate for van der Waals complexes,
where there is wide-amplitude vibrational motion along the intermolecular coordinate [23]. Previously, the N_2–N_2 and O_2–O_2 dimers had been
studied in the body-fixed formulation [8, 24], which has some advantages when one neglects Coriolis couplings, an approximation that we
do not make but assess below.

It is found that some of the levels which are degenerate without Coriolis coupling are actually split. Because of Coriolis mixing, there are energy levels (with $J > 0$ but different K) which perturb each other leading to a stabilization of the order of $1\,\mathrm{cm}^{-1}$ in some cases. The splitting is of the order of $0.03\,\mathrm{cm}^{-1}$ for the $J = 1$ case in some levels, but for higher J can be more than 20 times larger. Computing time for the exact close coupling calculation is only twice than for the helicity decoupling approximation for $J = 1$, but becomes quite demanding as J increases. For example, for $J = 6$ it is 40 times larger.

We calculated also other spectral aspects, vibrational frequencies, rotational constants, *etc.* The results can be compared with previous works [8, 24] where the potential surfaces used are very different: we found that although some features are similar, zero point and dissociation energies differ (Table 2 should be compared with corresponding ones in Refs. [8, 24]).

Surprisingly we have seen that although the used potential energy surfaces can have different topography, comparable results are obtained for spectroscopic observables, in spite of the fact that geometry and values of well depths and positions differ significantly, what differs is their ability to reproduce experimental data (integral cross-sections and second virial coefficient).

We also calculated harmonic frequencies through the second derivative of the potential around the equilibrium geometry of the dimers also given in Table 2. From these values we obtain the energy levels of each vibrational level (harmonic), and the values calculated are also very different from the exact results, because the interaction forces are very weak and the potential anisotropy is important.

We can conclude that the range and strength in the bonds of the three dimers N_2–N_2, O_2–O_2 and N_2–O_2 present characteristics more of those of clusters than of weakly bound molecules and that the interaction between the monomers is very anharmonic. Implications for the interpretation of recent laboratory [26] and atmospheric [27] spectroscopic observations, as well as of current measurements of high pressure behavior of oxygen [28] are amenable to be discussed in this framework. This work provides the ground for the interpretation of complicated band features in rotational spectral, as exemplified by the case of oxygen. Still appears to be valid the statement [25] that spectral analysis alone is not sufficient to extract information on structure and bonding, and the combined use of scattering and gaseous properties information is therefore confirmed to be crucial.

4. Scattering experiments: simple hydrocarbons and water

Extensive scattering experiments have been recently performed in our laboratory on hydrocarbons–rare gas and water–rare gas systems. In particular the integral cross sections have been measured for the C_6H_6–He, Ne, Ar [29], C_2H_2–Ne, Ar, Kr [30, 31], C_2H_4–Ne, Ar, Kr [30, 31], C_2H_6–Ne, Ar, Kr [30, 31], C_3H_8–Ar [30] and H_2O–He, Ne, Ar, Kr, Xe systems.

In the case of benzene containing dimers, molecular beam scattering experiments were performed at sufficiently low collision energy (in the thermal range) and angular resolution high enough to permit for the first time the observation of the "glory" interference effect in rare gas atom–benzene molecule collisions. The analysis of the energy dependence of the total cross sections allowed us to characterize the range, strength and anisotropy of the interaction for the C_6H_6–He, –Ne and –Ar dimers [29]. These experimental data represent a crucial test of the topography of the involved potential energy surfaces which had been proposed in a recent paper [21] and given in a suitable analytical representation [32]. They are also of relevance for the modelling of the recently discovered phenomenon of alignment of benzene in supersonic seeded flows [33].

Details of the experiments using beams of C_2H_2, C_2H_4 and C_2H_6 will be reported elsewhere while we present here the results obtained for the C_3H_8–Ar dimer: integral cross section data for the propane–Ar systems are reported in Fig. 2 while the second virial coefficients data are shown in Fig. 3.

The potential energy surface employed in the dynamical calculations has been obtained by a recent semi-empirical method [21] and fitted by a flexible functional form.

Calculations of the integral cross sections have been performed within a semiclassical dynamical scheme making use of the IOS (Infinite Order Sudden) approximation. The second virial coefficient has been calculated including radial and angular quantum corrections [34].

To give a flavor of the accuracy of our technique in the case of water molecules, we report here the test of the most recent available *ab initio* potential for the H_2O–He dimer [35] on the measured integral cross section data. Since in our experimental conditions water molecules in the beam are kept at $500\,K$, the molecules rotate so fast to fully average the interaction during the collision and we calculated the cross sections with the interaction potential properly averaged over all orientations in space.

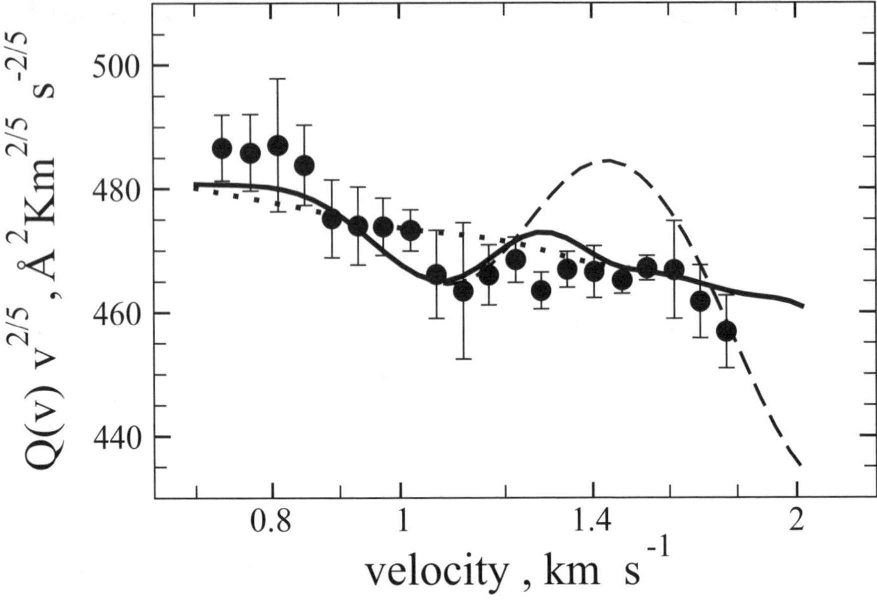

Figure 2.   Absolute integral cross section $Q(v)$ for the Ar–C_3H_8 as a function of the beam velocity v and plotted as $Q\,v^{2/5}$ to emphasize the glory structure. The dash line is calculated using only the isotropic component of the interaction. The dotted line using only the anisotropic component. The full line, has been calculated using the isotropic component for velocities lower than 1.0 Km/s, the sudden infinite-order approximation (IOS) for velocities between 1.5 and 2.5 Km/s, and a splining of the results of the two dynamical models for velocities between 1.0 and 1.5 Km/s.

Results are reported in Fig. 4, and show that the calculations underestimate the cross sections. This is due to the fact that the *ab initio* calculated potential underestimates the interaction (in the distance range between 3.5 and 7.0 Å) of about 10%. Further work includes the whole series of noble gases and several small molecules.

FORTRAN subroutines of the PES used in this work are available at the web site `www.tech.ing.unipg.it/PES/` and any question or comment should be addressed to David Cappelletti.[1]

Acknowledgments

This work is supported by the Italian CNR, MURST, INFM and ENEA, and by European Union contracts. E. Carmona-Novillo acknowledges the award of a Marie-Curie Fellowship.

---

[1] `prometeo@dyn.unipg.it`

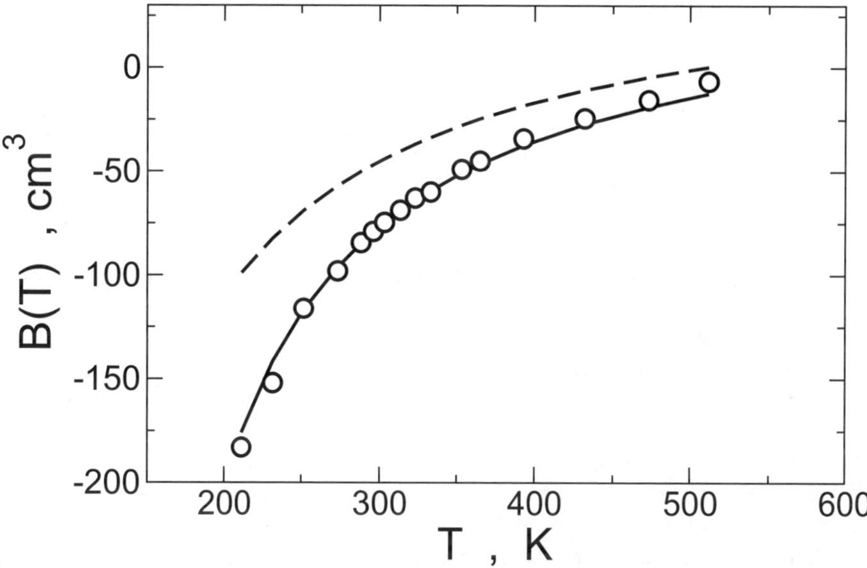

Figure 3. The second virial coefficient $B(T)$ for the C_3H_8–Ar dimer as a function of the temperature. The dash and full lines are calculations respectively with the spherical only and a full anisotropic potential energy surface.

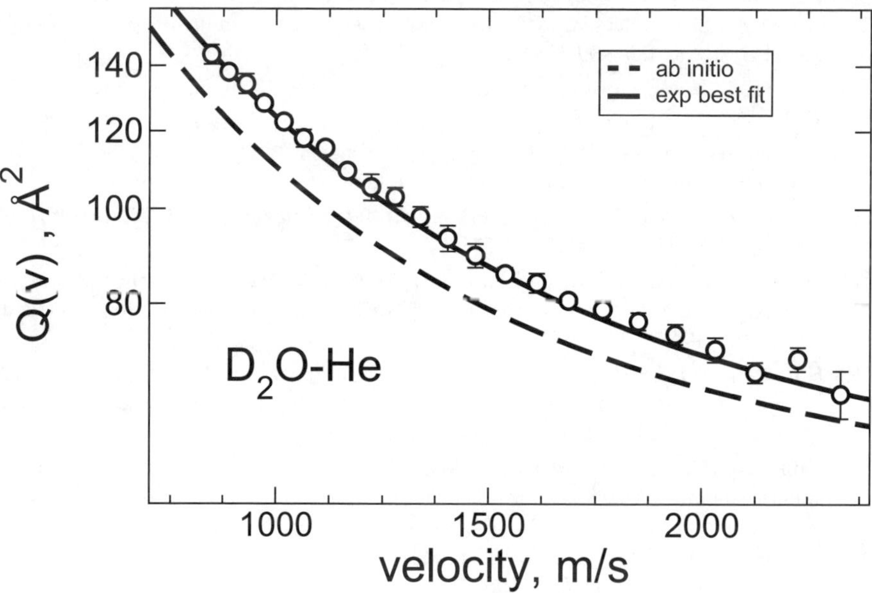

Figure 4. Total integral cross sections as a function of the beam velocity for the D_2O–He system.

References

[1] Aquilanti, V., Ascenzi, D., Bartolomei, M., Cappelletti, D., Cavalli, S., De Castro Vitores, M., and Pirani, F. (1999) Quantum interference scattering of aligned molecules: Bonding in O_4 and role of spin coupling, Phys. Rev. Lett., 82, 69–72; Molecular Beam scattering of aligned oxygen molecules. The nature of the bond in the O_2–O_2 dimer, J. Am. Chem. Soc., 121, 10794–10802.

[2] Aquilanti, V., Bartolomei, M., Cappelletti, D., Carmona–Novillo, E., and Pirani, F. (2002) The N_2–N_2 system: An experimental potential energy surface and calculated rotovibrational levels of the molecular nitrogen dimer, J. Chem. Phys., 117, 615–627.

[3] Aquilanti, V., Bartolomei, M., Cappelletti, D., Carmona–Novillo, E., and Pirani, F. (2001) Dimers of the major components of the atmosphere: Realistic potential energy surfaces and quantum mechanical prediction of spectral features, Phys. Chem. Chem. Phys., 3, 3891–3894.

[4] Aquilanti, V., Carmona–Novillo, E., and Pirani, F. (2002) Quantum mechanics of molecular oxygen clusters: rotovibrational dimer dynamics from realistic potential energy surface, Phys. Chem. Chem. Phys., 4, 4970–4978.

[5] Aquilanti, V., Bartolomei, M., Carmona–Novillo, E., and Pirani, F. (2003) The asymmetric dimer N_2–O_2: Characterization of the potential energy surface and quantum mechanical calculation of rotovibrational levels, J. Chem. Phys., 118, 2223–2234.

[6] Long, C. A., Henderson, G., and Ewing, G. E. (1973) The infrared spectrum of the $(N_2)_2$ van der Waals molecule, Chem. Phys., 2, 485–489.

[7] Cappelletti, D., Vecchiocattivi, F., Pirani, F., Heck, E. L., and Dickinson, A. S. (1998) An intermolecular potential for nitrogen from a multi-property analysis, Mol. Phys., 93, 485–499.

[8] Tennyson, J. and van der Avoird, A. (1982) Quantum dynamics of the van der Waals molecule $(N_2)_2$: An *ab initio* treatment, J. Chem. Phys., 77, 5664–5681.

[9] van der Avoird, A., Wormer, P. E. S., and Jansen, A. P. J. (1986) An improved intermolecular potential for nitrogen, J. Chem. Phys., 84, 1629–1635.

[10] Couronne, O. and Ellinger, Y. (1999) An *ab initio* and DFT study of $(N_2)_2$ dimers, Chem. Phys. Lett., 306, 71–77.

[11] Wada, A., Kanamori, H., and Iwata, S. (1998) *Ab initio* MO studies of van der Waals molecule $(N_2)_2$: Potential energy surface and internal motion, J. Chem. Phys., 109, 9434–9438.

[12] Hamdani, A. H., Shen, A., Dong, Y., Gao, H., and Ma, Z. (2000) Theoretical and experimental research on excimer like $(N_2)_2$ dimer: potential energy curves and spectra, Chem. Phys. Lett., 325, 610–618.

[13] Uhlík, F., Slanina, Z., and Hinchliffe, A. (1993) Computational studies of atmospheric chemistry species. Part VI. An *ab initio* correlated study of structure, energetics and vibrations of $(N_2)_2$, J. Mol. Struct. (Theochem), 282, 271–275.

[14] Stallcop, J. R. and Partridge, H. (1997) The N_2–N_2 potential energy surface, Chem. Phys. Lett., 281, 212–220.

[15] Slanina, Z., Uhlík, F., and De Almeida, W. B. (1994) A computational thermodynamic evaluation of the altitude profiles of $(N_2)_2$, N_2–O_2, and $(O_2)_2$ in the Earth's atmosphere, Thermochimica Acta, 231, 55–60.

[16] Ewing, G. E. and Trusler, J. P. M. (1992) Second acoustic virial coefficients of nitrogen between 80 and 373 K, Physica A, 184, 415–436.

[17] Brewer, A. J. and Vaughn, J. W. (1969) Measurement and correlation of some interaction second virial coefficients from -125° to 50°C, J. Chem. Phys., 50, 2960–2968.

[18] Hall, K. R. and Iglesias-Silva, G. A. (1994) Cross second virial coefficients for the system $N_2 + O_2$ and $H_2O + O_2$, J. Chem. Eng. Data, 39, 873–875.

[19] Aquilanti, V., Ascenzi, D., Cappelletti, D., and Pirani, F. (1994) Velocity dependence of collisional alignment of oxygen molecules in gaseous expansions, Nature (London), 371, 399–402.

[20] Green, S. (1975) Rotational excitation in H_2–H_2 collisions: Close-coupling calculations, J. Chem. Phys., 62, 2271–2277.

[21] Pirani, F., Cappelletti, D., and Liuti, G. (2001) Range, strength and anisotropy of intermolecular forces in atom-molecule systems: an atom-bond pairwise additivity approach, Chem. Phys. Lett., 350 286–296.

[22] Hutson, J. M. BOUND, Computer Code, Version 5 (1993), distributed by Collaborative Computational Project no. 6 of the Science and Engeneering Research Council (UK).

[23] Hutson, J. M. (1994) Coupled channel methods for solving the bound-state Schrödinger equation, Comp. Phys. Comm., 84, 1–18.

[24] Bussery-Honvault, B. and Veyret, V. (1999) Quantum mechanical study of the vibrational-rotational structure of $[O_2(^3\Sigma_g^-)]_2$, Phys. Chem. Chem. Phys., 1, 3387–3393.

[25] Long, C. A. and Ewing G. E. (1971) The infrared spectrum of bound state oxygen dimers in the gas phase, Chem. Phys. Lett., 9, 225–229; (1973) Spectroscopic investigation of van der Waals molecules. I. The infrared and visible spectra of $(O_2)_2$, J. Chem. Phys., 58, 4824–4834; (1975) Structure and properties of van der Waals molecules, Acc. Chem. Res., 8, 185–192.

[26] Campargue, A., Biennier, L., Kachanov, A., Jost, R., Bussery-Honvault, B., Veyret, V., Churassy, S., and Bacis, R. (1998) Rotationally resolved absorption spectrum of the O_2 dimer in the visible range, Chem. Phys. Lett., 288, 734–742.

Biennier, L., Romanini, D., Kachanov, A., Campargue, A., Bussery-Honvault, B., and Bacis, R. (2000) Structure and rovibrational analysis of the $[O_2(^1\Delta_g)_{v=0}]_2 \rightarrow [O_2(^3\Sigma_g^-)_{v=0}]_2$ transition of the O_2 dimer, J. Chem. Phys., 14, 6309–6321.

[27] Pfeilsticker, K., Bösch, H., Camy-Peyret, C., Fitzenberger, R., Harder, H., and Osterkamp, H. (2001) First atmospheric profile measurements of UV/visible O_4 absorption band intensities: Implications for the spectroscopy and the formation enthalpy of the O_2–O_2 dimer, Geophys. Res. Lett., 28, 4595–4598.

[28] Gorelli, F. A., Ulivi, L., Santoro, M., and Bini, R. (1999) The epsilon phase of solid oxygen: evidence of an O_4 molecule lattice, Phys. Rev. Lett., 83, 4093–4096.

[29] Cappelletti, D., Bartolomei, M., Pirani, F., and Aquilanti, V. (2002) Molecular beam scattering experiments on benzene-rare gas systems: probing the potential energy surfaces for the C_6H_6–He, –Ne and –Ar dimers, J. Phys. Chem., 106, 10764–10772.

[30] Sabidó, M. (2001) Master Degree Thesis, Universitá di Perugia & Universitat de Barcelona.

[31] Bartolomei, M. (2002) Ph. D. Thesis, Universitá di Perugia.

[32] Pirani, F., Porrini, M., Cavalli, S., Bartolomei, M., and Cappelletti, D. (2003) Potential energy surfaces for the benzene–rare gas systems, Chem. Phys. Lett., 367, 405–413.

[33] Pirani, F., Cappelletti, D., Bartolomei, M., Aquilanti, V., Scotoni, M., Vescovi, M., Ascenzi, D., and Bassi, D. (2001) Orientation of benzene in supersonic expansions, probed by IR-laser absorption and by molecular beam scattering, Phys. Rev. Lett, 86, 5035–5038.

[34] Pack, R. T. (1983) First quantum corrections to second virial coefficients for anisotropic interactions: Simple, corrected formula, J. Chem. Phys., 78, 7217–7222.

[35] Hodges, M. P., Wheatley, R. J., and Harvey, A. H. (2002) Intermolecular potential and second virial coefficient for the water–helium complex, J. Chem. Phys., 116, 1397–1405.

COLLISION-INDUCED ABSORPTION
OF GASEOUS OXYGEN
IN THE HERZBERG CONTINUUM

M. B. Kiseleva, G. Ya. Zelikina, M. V. Buturlimova, A. P. Burtsev

Physics Research Institute,

Saint-Petersburg State University, Department of Molecular Spectroscopy,

Ulyanovskaya 1, Stary Petergof, 198504, Saint-Petersburg, Russia

Abstract   The values of the binary absorption coefficient for gaseous molecular oxygen in the region of the Herzberg photodissociation continuum (200–230 nm) have been measured in the temperature range 110–295 K. The induced absorption of oxygen in the 200–220 nm spectral region was found to result from two bands with quite different temperature dependence of their intensity: for the Herzberg III band the values of the binary absorption coefficient increase by $\sim 50\%$ as the temperature decreases from 295 to 110 K, while for the other band no distinct temperature dependence of the binary absorption coefficient was observed. A transition dipole moment function including both short-range term (overlap of the electron shells of interacting molecules) and long-range term (interaction of the transient quadrupole moments) is proposed to describe the temperature dependence of the binary absorption coefficients in the Herzberg III band.

Keywords: oxygen / collision-induced absorption / Herzberg continuum / binary absorption coefficients / low temperature measurements / transition dipole moment function

1. Introduction

Induced absorption of molecular oxygen in the spectral region 200–280 nm is mainly determined by the absorption of collision complexes O_2–O_2 in the Herzberg III band corresponding to the $(^3\Sigma_g^-, {}^3\Sigma_g^-) \to (^3\Sigma_g^-, {}^3\Delta_u)$ transition. For a free oxygen molecule this electronic transition is forbidden in the electric dipole approximation. However, intermolecular interactions in a collision complex result in mixing of the wavefunctions of an O_2–O_2 pair and hence in the appearance of a non-zero

183

C. Camy-Peyret and A.A. Vigasin (eds.),
Weakly Interacting Molecular Pairs: Unconventional Absorbers of Radiation in the Atmosphere, 183–192.
© 2003 *Kluwer Academic Publishers. Printed in the Netherlands.*

induced transition dipole moment. It has been shown in terms of perturbation theory [1] that this induced transition concerned borrows its intensity from the longer wavelength allowed dipole transition in O_2 $X^3\Sigma_g^- \to B^3\Sigma_u^-$ (Schumann-Runge band). The transition dipole moment function $M(r)$ depends on the distance between two oxygen molecules in a collision complex. The form of the function $M(r)$, which can be derived from the experimental data on the temperature dependence of the induced absorption intensity, is of considerable importance, providing information about the mechanism of formation of the complex under study.

2. Experiment

Absorption spectra of gaseous oxygen were recorded using a single-beam spectrophotometer with a monochromator built on a Seya-Namioka design. To obtain spectra at low temperatures we used a cryostat with a cell having an optical path length 2.6 cm. The temperature of a sample could be varied within the range 90–295 K and determined with an uncertainty less than 1 K.

The observed absorption of oxygen in the spectral region 200–280 nm could be represented by the following expression:

$$D = \log(I_0/I) = \mu_1 \, \rho \, L + \mu_{11} \, \rho^2 \, L = D_1 + D_2, \tag{1}$$

where D is the optical density; I_0 and I are the intensities of the incident and transmitted beams, respectively; L is the optical path length in the cell (cm); ρ is the density of oxygen (Amagat); μ_1 is the absorption coefficient for a free O_2 molecule (Amagat^{-1}cm^{-1}); μ_{11} is the binary absorption coefficient (BAC) for a collision pair O_2–O_2 (Amagat^{-2}cm^{-1}).

The values of the BAC of oxygen at low temperatures were determined in a limited spectral region (200–230 nm), corresponding to the maximum of BAC values. Unfortunately, for $\lambda > 230$ nm reliable values of the optical density at low temperatures could not be obtained using a 2.6 cm long cell. Under the experimental conditions used in this work absorption of free oxygen molecules D_1 did not exceed 2–4%. To remove this contribution from the observed absorption of oxygen reliable data on μ_1 from [2] were used. The density of gaseous oxygen in the cell under a given pair of temperature and pressure was determined using compressibility data. The values of the BAC in the temperature range 120–295 K were obtained from spectra of oxygen with a density $\rho \approx 25$ Amagat. These experimental conditions allowed us to derive reliable values of the optical density and at the same time to use the approximation of binary interactions in the following data analysis. At lower temperatures experiments were carried out with smaller densities of oxygen, as we were limited by the density of saturated vapour. Thus, for temperatures within the range 110–115 K oxygen densities 14–19 Amagat were used. According to our estimates, the resultant uncertainty in the values of the BAC, determined from a single spectrum, varied from 3% for $\lambda = 200$ nm (maximum μ_{11} values) and $T = 295$ K up to 10% for $\lambda = 230$ nm (minimum μ_{11} values) and $T = 110$ K.

3. Analysis of the experimental data

Thorough investigations of the induced absorption of oxygen in mixtures with foreign gases X in the wavelength region 190–242 nm were carried out earlier in our laboratory [3–5]. All in all spectra of 20 different O_2–X mixtures were studied and the values of the BAC for O_2–X collision pairs (μ_{12}) at room temperature were determined. The curves $\log \mu_{12}(\lambda)$ for different mixtures were found to be similar to each other, demonstrating a "plateau" in the region 197–215 nm. The form of these experimental curves corresponds to the form of the Herzberg III band of oxygen predicted theoretically [6]. At the same time the curve $\log \mu_{11}(\lambda)$ for pure oxygen was found to follow that for mixtures only for $\lambda > 215$ nm. For shorter wavelengths the dependence of μ_{11} on λ is much stronger, revealing the presence of an unknown induced absorption band (hereafter called the Z-band). It should be noted that a similar increase of the BAC values of pure oxygen in the wavelength region 210–180 nm was also observed in [7], but without further discussion.

In the present work spectra of gaseous oxygen in the spectral region 200–230 nm were recorded in the temperature range 110–295 K. Values of the BAC for different temperatures, obtained from several experiments, as well as the fitting curves $\mu_{11}(T)$ are presented in Fig. 1 for several wavelengths. The root mean square deviation of the points from the corresponding curves does not exceed 2%. Table 1 contains the average values of the BAC of oxygen for different temperatures determined from the curves $\mu_{11}(T)$.

From the data of Table 1 the curves $\mu_{11}(\lambda)$ were plotted for each temperature. The temperature dependence of the BAC values for the Herzberg III band ($\mu_{11}^{H}(T)$) was obtained from the data in the region 220–230 nm, where the induced absorption of oxygen is entirely determined by this band. Assuming that the form of the Herzberg III band does not depend on temperature and taking the values of the BAC at $\lambda = 225$ and 230 nm as reference points, we have separated the bands in the wavelength region 200–220 nm (see Fig. 2). Thus, the temperature

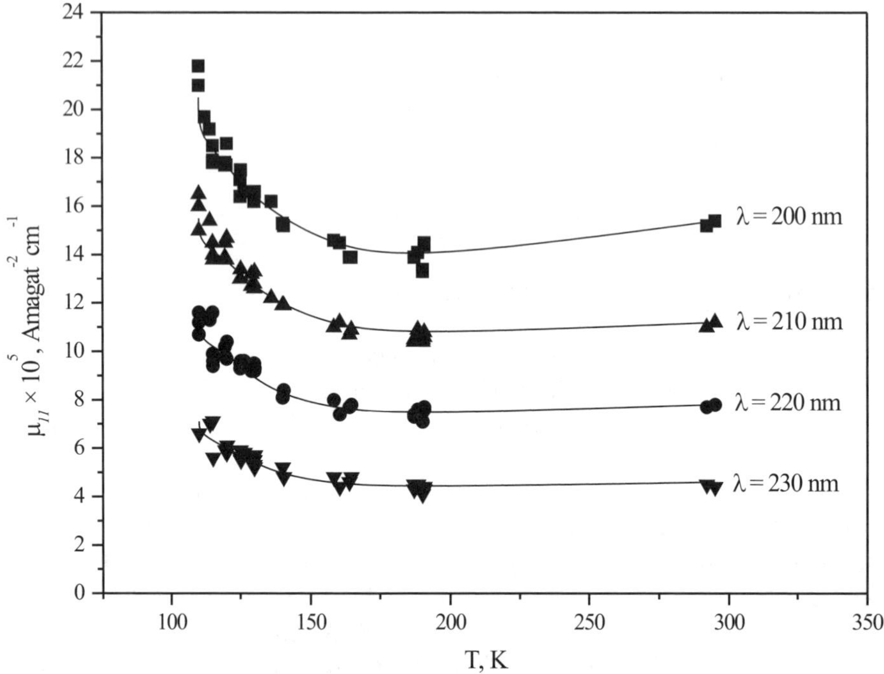

Figure 1. Temperature dependence of the binary absorption coefficient of oxygen in the spectral region 200–230 nm.

Table 1. Values of the binary absorption coefficient of oxygen $\mu_{11} \times 10^5$ (Amagat^{-2}cm^{-1}) for the spectral region 200–230 nm in the temperature range from 110 to 295 K.

T, K	λ, nm						
	200	205	210	215	220	225	230
295	15.4	12.8	11.2	9.3	7.8	6.0	4.6
190	13.8	12.3	10.7	9.0	7.4	5.8	4.4
160	14.3	12.6	11.0	9.5	7.6	6.0	4.5
140	15.5	13.6	11.9	10.3	8.2	6.6	4.9
130	16.4	14.5	12.7	10.9	8.9	7.2	5.4
125	16.9	15.0	13.2	11.4	9.4	7.5	5.7
120	17.8	15.6	13.8	11.8	9.9	8.0	6.0
115	18.9	16.4	14.5	12.4	10.5	8.6	6.6
110	20.5	17.4	15.5	13.2	11.3	(9.3)	(7.1)

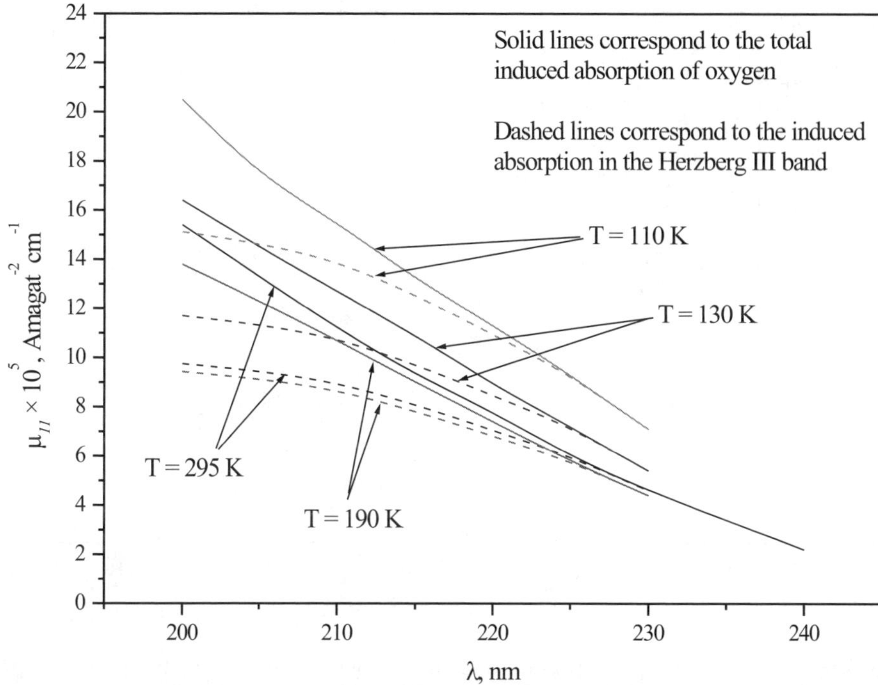

Figure 2. Separation of the two induced absorption bands of oxygen—Herzberg III band and Z-band—in the spectral region 200–220 nm.

Table 2. Values of the coefficient $\mu_{11}^{*H(Z)} \equiv \mu_{11}^{H(Z)}(T)/\mu_{11}^{H(Z)}(295)$ for the Herzberg III band and Z-band of oxygen in the temperature range from 110 to 295 K.

T, K	295	190	160	140	130	125	120	115	110
Herzberg III band, $\mu_{11}^{H}(T)$	1.0	0.97	1.0	1.08	1.18	1.24	1.31	1.43	(1.54)
Z-band, $\mu_{11}^{*Z}(T)$	1.0	0.93	0.87	1.0	0.99	0.98	0.98	0.85	(0.9)

dependence of the BAC values for the Z-band ($\mu_{11}^{Z}(T)$) was obtained using the values of μ_{11}^{Z} in the region 200–210 nm.

For discussing the induced absorption bands we have used the quantities μ_{11}^{*H} and μ_{11}^{*Z} obtained by scaling the BAC values at temperature T with respect to that at room temperature: $\mu_{11}^{*H(Z)} \equiv \mu_{11}^{H(Z)}(T)/\mu_{11}^{H(Z)}(295)$. The values of $\mu_{11}^{*H}(T)$ and $\mu_{11}^{*Z}(T)$ for different temperatures, obtained by averaging the data for $\lambda = 225$ and 230 nm and for $\lambda = 200$, 205 and 210 nm, respectively, are given in Table 2.

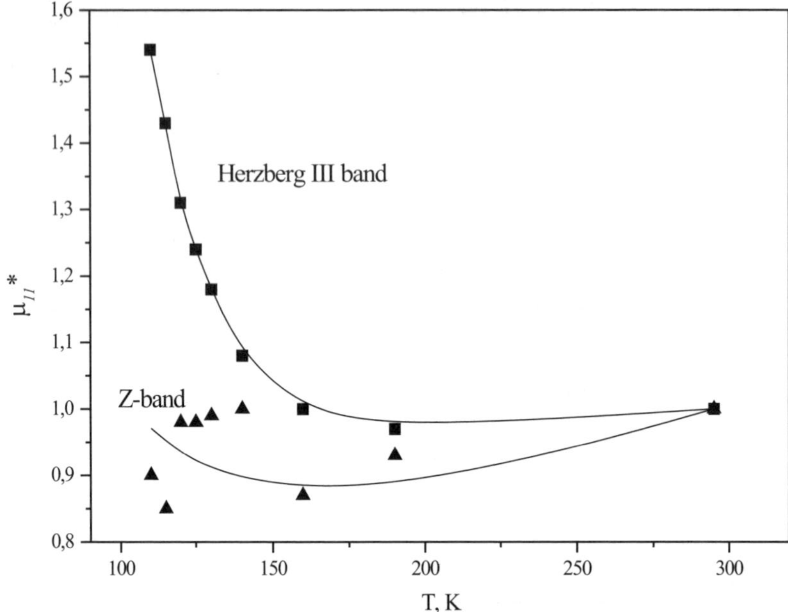

Figure 3.   Temperature dependence of the coefficient μ_{11}^{*} for two induced absorption bands of oxygen—Herzberg III band and Z-band.

The temperature dependence of $\mu_{11}^{*H}(T)$ and $\mu_{11}^{*Z}(T)$ is represented in Fig. 3. The scatter of data for the Z-band is much larger than for the Herzberg III band, since the μ_{11}^{Z} values, which were determined after separation of the two bands, have larger uncertainties. One can see that temperature dependence of the BAC values for the two induced absorption bands of oxygen are essentially different: while for the Herzberg III band the values of the BAC increase by $\sim 55\%$ as the temperature decreases from 295 to 110 K, for the Z-band the values of the BAC do not exhibit any distinct temperature dependence.

4. Discussion

The integrated intensity $A(T)$ of an induced electronic transition is known to be related to the transition dipole moment function $M(r)$ by the expression:

$$A(T) = C \int_{0}^{\infty} |M(r)|^2 \, g(r,T) \, r^2 \, dr, \qquad (2)$$

where C is a normalization factor; $g(r,T)$ is the pair distribution function; r is the intermolecular distance; T is the temperature. For a chosen function $M(r)$ the value of the normalization factor can be determined from the most accurate data on A, for example from those, obtained at room temperature. In the binary interactions approximation $g(r,T)$ can be represented in a simple form:

$$g(r,T) = \exp\left(-V(r)/kT\right), \tag{3}$$

where $V(r)$ is the intermolecular potential and k is the Boltzmann constant.

It should be noted that for the Herzberg III induced absorption band and for the Z-band it is rather difficult to obtain experimental data on the integrated intensity because of the superposition of several absorption bands in the wavelength region $\lambda < 210\,\mathrm{nm}$. According to the Franck-Condon principle, the form of an electron absorption band is mainly determined by the relative position of the potential curves for the two electronic states participating in the transition. Thus, assuming that the form of the induced absorption bands under study should not depend on temperature, while analyzing temperature dependence of their intensity we substituted the integrated intensity $A(T)$ by the coefficient $\mu_{11}^{*}(T)$ obtained in the limited spectral region. We have used equations (2) and (3) to calculate theoretical curve $\mu_{11}^{\nabla}(T)$ for a chosen parametric form of $M(r)$. The values of the parameters of a chosen $M(r)$ function for the Herzberg III band and Z-band were determined by a least-squares fit, comparing the curves $\mu_{11}^{\nabla}(T)$ and $\mu_{11}^{*H}(T)$ ($\mu_{11}^{*Z}(T)$). All calculations were carried out with the 22-parameter O_2–O_2 potential [8].

As a first step an exponential function $M(r)$ was considered:

$$M(r) = \exp(-r/r_0). \tag{4}$$

This type of function for the $M(r)$, usually assumed to describe short-range intermolecular interactions resulting from the overlap of the electron shells of interacting molecules, was used in [9, 10] to analyze the experimental data on the temperature dependence of the intensity for several induced absorption bands of oxygen in the spectral region 1260–477 nm.

With a function $M(r)$ of exponential type the best fit for the Herzberg III band data was obtained for $r_0 = 0.61\,\text{Å}$ (Fig. 4). However, one can see that unlike $\mu_{11}^{*H}(T)$, the curve $\mu_{11}^{\nabla}(T)$ leads to a noticeable increase of the BAC values in the temperature range 295–160 K, especially just below room temperature. Besides, it does not describe

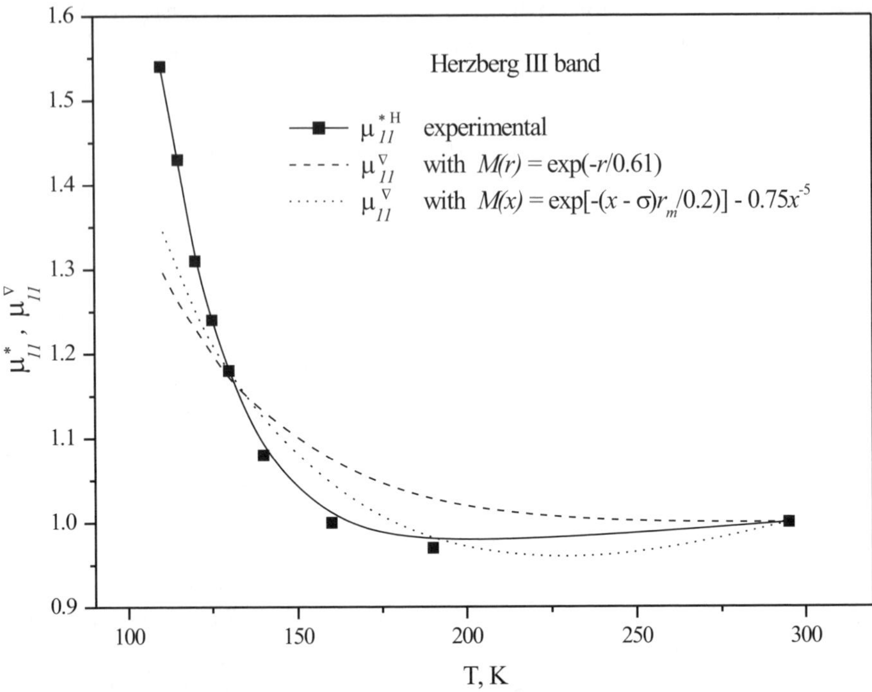

Figure 4.   Experimental and theoretical temperature dependence of the coefficient $\mu^*_{11}(T)$ for the Herzberg III band of oxygen.

the strong temperature dependence of the BAC values for $T \leqslant 120\,\mathrm{K}$, observed in the experiment. As for the Z-band, comparatively good agreement with experimental data was obtained for $r_0 = 0.40\,\text{Å}$ (see Fig. 3).

Since a function $M(r)$ of the exponential type did not allow us to describe the temperature dependence of the BAC values for the Herzberg III band in the whole temperature range under study, we have decided to change the form of the function $M(r)$, based on its behaviour at various intermolecular distances. The new function $M(r)$ potential was represented as:

$$M(r) = M_1(r) + M_2(r), \tag{5}$$

where in addition to a short-range term $M_1(r)$, a long-range term $M_2(r)$, resulting from the interaction of multipole moments of two oxygen molecules in a collision pair was introduced. It can be shown in terms of perturbation theory [1], that the long-range term of the $M(r)$ function for the induced Herzberg III band of oxygen is mainly determined by the interaction of the transient quadrupole moments and therefore depends on the intermolecular distance as r^{-5}. In further calculations we used the following form of the transition dipole moment function:

$$M(x) = M_1(x) + M_2(x) = \exp\left(-(x - \sigma)r_m/r_0\right) - B\,x^{-5}, \qquad (6)$$

where $x \equiv r/r_m$ is the relative intermolecular distance; r_m is the value of the intermolecular distance, corresponding to the minimum of the potential curve ($r_m = 3.94\,\text{Å}$ [8]); σ is the value of the relative coordinate x, corresponding to the zero-value of the potential ($\sigma = 0.89$ [8]); r_0 and B are fitting parameters. Calculations showed that the terms $M_1(x)$ and $M_2(x)$ in equation (6) should be taken with opposite signs, since if not, the curve $\mu_{11}^{\nabla}(T)$ shows considerable increase of the BAC values with the decreasing of temperature, when starting from room temperature.

For a given value of the parameter r_0 in $M_1(x)$, the value of the parameter B in $M_2(x)$ indicates the contribution of the long-range term to the transition dipole moment function. It should be noted that the short-range term $M_1(x)$ is determined by the overlap of the electron shells of the interacting molecules, as well as the repulsive branch of the intermolecular potential. Hence, one can assume that the values of r_0 for these two functions should not differ much. Thus, during the fitting procedure the value of r_0 in $M_1(x)$ was varied around $0.3\,\text{Å}$ (the value of the corresponding parameter r_0 for the repulsive branch of $V(x)$ is equal to $0.31\,\text{Å}$ [8]).

Calculations showed that with the function $M(x)$ described by equation (6) the best fit for the Herzberg III band data is obtained for the interval of r_0 and B values: $r_0 = (0.15\text{–}0.25)\,\text{Å}$, $B = 1.33\text{–}0.58$. The curves $\mu_{11}^{\nabla}(T)$ with r_0 and B values in these intervals have nearly the same shape; one of them is presented in Fig 4. One can see, that the new function $M(x)$ (6) with reasonable values of the parameters allows to improve the theoretical description of the temperature dependence of the BAC values in the whole temperature range studied: in comparison with the pure exponential function, $M(x)$ as given in (6) results in the proper temperature dependence of the BAC values in the temperature range 160–120 K and even predicts a slight decrease of these values around $T = 200\,\text{K}$. However, neither a pure exponential transition dipole moment function, nor the more complicated function $M(x)$ given in (6) describe the steep increase of the BAC values for the Herzberg III band, observed for $T \leq 120\,\text{K}$. This experimental result needs further attention.

References

[1] Robinson, G. W. (1967) Intensity enhancement of forbidden electronic transitions by weak intermolecular interaction, J. Chem. Phys., 46, 572–585.

[2] Yoshino, K., Cheung, A. S.-C., Esmond, J. R., Parkinson, W. H., Freeman, D. E., Guberman, S. L., Jenouvrier, A., Coquart, B., and Merienne, M.-F. (1988) Improved absorption cross sections of oxygen in the wavelength region 205–240 nm of the Herzberg continuum, Planet. Space. Sci., 36, 1469–1475.

[3] Zelikina, G. Ya., Bertsev, V. V., and Kiseleva, M. B. (1994) Absorption of compressed liquid oxygen and its mixtures with Ar, Kr, Xe, N_2, and CF_4 in the 200–280 nm spectral region, Opt. Spectrosc., 77, 513–516.

[4] Zelikina, G. Ya., Bertsev, V. V., Burtsev, A. P., and Kiseleva, M. B. (1996) Spectrum of induced absorption of oxygen in mixtures with various gases in the region of Herzberg photodissociation continuum, Opt. Spectrosc., 81, 685–689.

[5] Zelikina, G. Ya., Kiseleva, M. B., Burtsev, A. P., and Bertsev, V. V. (1998) Spectrum of induced absorption of oxygen in mixtures with various gases in the range of 190–280 nm, Opt. Spectrosc., 85, 520–524.

[6] Saxon, R. P. and Slanger, T. G. (1986) Molecular oxygen absorption continua at 195–300 nm and O_2 radiative lifetimes, J. Geophys. Res., 91, 9877–9879.

[7] Ogawa, M. (1971) Absorption cross section of O_2 and CO_2 continua in the Schumann and far uv region, J. Chem. Phys., 54, 2550–2556.

[8] Brunetti, B., Liuti, G., Luzzati, E., Pirani, F., and Vecchiocattive, V. (1981) Study of the interactions of atomic and molecular oxygen with O_2 and N_2 by scattering data, Can. J. Phys., 74, 6734–6741.

[9] Blickensderfer, R. G. and Ewing, G. E. (1969) Collision-induced absorption spectrum of gaseous oxygen at low temperatures and pressures. I. The $^1\Delta \leftarrow {}^3\Sigma_g^-$ system, J. Chem. Phys., 51, 873–883.

[10] McKellar, A. R. W., Rich, N. H., and Welsh, H. L. (1972) Collision-induced vibrational and electronic spectra of gaseous oxygen at low temperatures, Can. J. Phys., 50, 1–9.

ABSORPTION CROSS-SECTION OF THE COLLISION-INDUCED BANDS OF OXYGEN FROM THE UV TO THE NIR

C. Hermans, A. C. Vandaele

Institut d'Aéronomie Spatiale de Belgique,
3 Av. Circulaire, B-1180 Brussels, Belgium

S. Fally, M. Carleer, R. Colin

Université Libre de Bruxelles,
Laboratoire de Chimie Physique Moléculaire CP160/09,
50 Av. F. D. Roosevelt, B-1050 Brussels, Belgium

B. Coquart, A. Jenouvrier, M.-F. Merienne

Université de Reims Champagne-Ardenne,
Groupe de Spectrométrie Moléculaire et Atmosphérique,
UPRESA 6089, UFR Sciences, Moulin de la Housse,
BP1039, F-51687, Reims, France

Abstract   Collision-induced absorption (CIA) cross-sections of oxygen have been measured in the UV, Visible and near-IR regions from spectra recorded by Fourier Transform Spectroscopy at different pressures and room temperature. An extensive cross-sections dataset from 42000 to 7500 cm^{-1} (238–1330 nm) is presented. The separation procedure of the discrete and diffuse absorption features is described. Pressure and foreign gas effects are discussed, and a comparison with literature data is shown. A preliminary test on the influence of the choice of the dataset on atmospheric retrievals of NO_2, O_3, BrO, and OClO is performed.

Keywords: oxygen / absorption cross-section / collision-induced bands / CIA / O_4 / UV / visible / near IR / Fourier Transform spectroscopy

193

C. Camy-Peyret and A.A. Vigasin (eds.),
Weakly Interacting Molecular Pairs: Unconventional Absorbers of Radiation in the Atmosphere, 193–202.
© 2003 *Kluwer Academic Publishers. Printed in the Netherlands.*

1. Introduction

Although it is the discrete bands of oxygen that are the major contribution of this molecule to the Earth's radiative budget, it is also necessary to take into account of their corresponding diffuse bands whose contribution is estimated to lie between 2.2 and 3.1 Wm^{-2} [1] for an overhead sun at mid-latitude.

The aim of this work is to study the collision-induced absorption (CIA) bands of oxygen in the laboratory under conditions prevailing in the Earth's atmosphere, from the UV to the NIR and the effect of a foreign gas on these bands.

2. Experimental

2.1. Experimental set-up

Absorption spectra of molecular oxygen have been recorded in the spectral region 42000–7500 cm^{-1} (238–1330 nm) under different pressures conditions (100 to 1000 hPa of pure O_2, 400 to 1000 hPa of O_2–N_2 and of O_2–Ar mixtures) with a Bruker IFS120M Fourier transform spectrometer coupled to a multiple reflection path cell. Diffuse absorptions were studied at a resolution of 2 cm^{-1} (OPD$=0.45$ cm) with absorption path lengths of 600 and 1000 m.

2.2. Separation of the discrete and the collisional bands

Figure 1a presents pure O_2 spectra at 600 (left) and 800 hPa (right) at a resolution of 2 cm^{-1}. It shows that the discrete structured bands of O_2 overlap with broader bands attributed to the collisions between O_2 and O_2 and O_2 and X. The discrete bands of molecular oxygen are due to the following electronic transitions: (i) $A^3\Sigma_u^+ \leftarrow X^3\Sigma_g^-$, $c^1\Sigma_u^- \leftarrow X^3\Sigma_g^-$, $A'^3\Delta_u \leftarrow X^3\Sigma_g^-$ in the UV (Herzberg I, II and III band systems), (ii) $b^1\Sigma_g^+(v'=0, 1, 2, 3) \leftarrow X^3\Sigma_g^-$ in the visible (atmospheric A, B, γ and δ bands), and (iii) $a^1\Delta_g(v'=0, 1) \leftarrow X^3\Sigma_g^-$ in the NIR (IR bands). The quality of high resolution spectra (0.12, 0.02 and 0.03 cm^{-1} in the UV, visible and NIR spectral range respectively) has allowed to determine the line parameters [2, 3].[1] A low resolution synthetic spectrum is calculated with the Winprof program written by Hurtmans [4] by applying a Voigt line profile to the values of line positions and intensities and convolving each line with the appropriate instrumental

---

[1] see `http://www.oma.be/BIRA-IASB/Scientific/Topics/Lower/LaboBase/Laboratory.html`

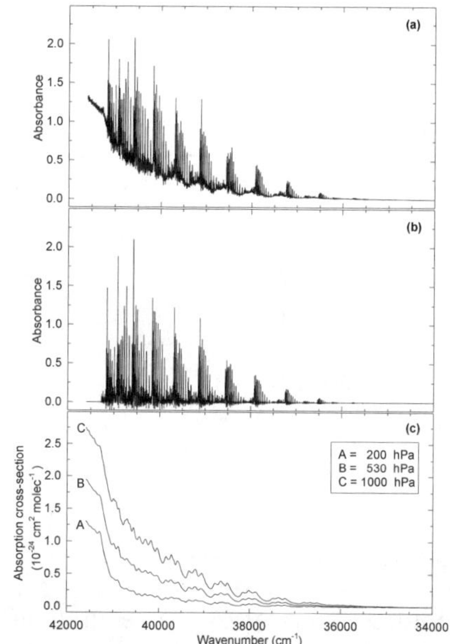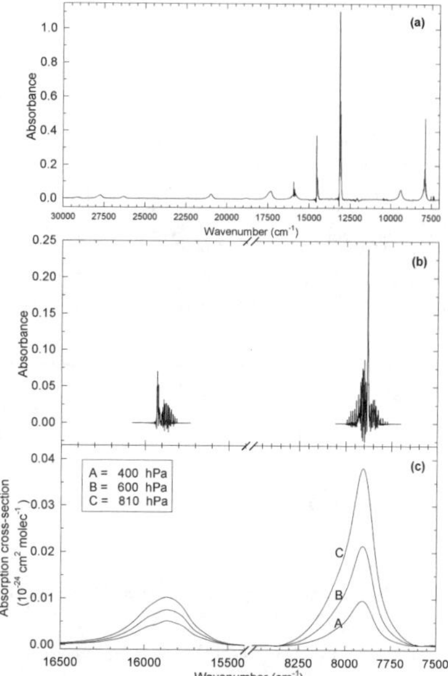

Figure 1. (a) Overview of experimental spectra of pure oxygen at 600 hPa (left) in the UV region and 800 hPa (right) in the visible and NIR regions and a resolution of $2\,\mathrm{cm}^{-1}$. (b) Simulated absorbance spectra at the same pressure and resolution. (c) Examples of absorption cross-sections at three O_2 pressures after the removal of the lines.

function (Fig. 1b). The absorbances of the calculated lines are then subtracted from the experimental ones (Fig. 1c). Collision-induced absorption cross-sections (Fig. 2) are calculated from spectra recorded at several pressures and then averaged. In the UV region, the separation procedure is complicated by the presence of the Herzberg continuum; the entire procedure of removal is described elsewhere [5].

In the following, absorption cross-sections σ and collision-induced absorption cross-sections σ_{O_2-X} are expressed in $\mathrm{cm}^2\ \mathrm{molec}^{-1}$ and in $\mathrm{cm}^5\ \mathrm{molec}^{-2}$ respectively, are related to the absorbance A in the following way:

$$A = \sigma\, P_{O_2}\, k\, l + \sigma_{O_2-X}\, P_{O_2}\, P_X\, k^2\, l,$$

where P_{O_2} and P_X are the oxygen and the foreign gas pressure (hPa) respectively, l is the path length (cm), and k is equal to $N_0 T_0 / P_0 T$ where N_0 is the Loschmidt number ($\mathrm{molec\,cm}^{-3}$), T_0 and P_0 are the normal temperature and pressure (K and hPa), and T is the temperature of the sample (K).

Table 1. Integrated collision-induced absorption cross-section, $10^{-43}\,\mathrm{cm^4\,molec^{-2}}$.

	Band, $\mathrm{cm^{-1}}$										
	29150	27760	26320	22420	20970	18810	17360	15870	9390	7900	All
This work	0.31	1.57	0.77	0.13	1.99	0.29	4.23	2.87	2.96	4.11	19.35
[11]*	0.58	1.87	**0.83**	0.28	**2.09**	0.75	4.79	**3.12**	3.52	5.62	23.61
[10]					2.483	0.431	**4.625**	**2.769**			
[12]					2.4		3.5	1.5			

*calculated

3. Results

3.1. O_2–O_2 collision-induced absorption cross-section

The O_2–O_2 CIA cross section at room temperature is presented in Fig. 2 from 42000 to $7500\,\mathrm{cm^{-1}}$ (238–1330 nm). The 34000–30000 $\mathrm{cm^{-1}}$ (294–333 nm) region is omitted from the figure because the absorption is zero. To our knowledge, the results covering such a large spectral range were never presented before and constitute a new homogeneous and extensive data set.

In the UV, our previous study confirmed that the so-called Wulf bands are related to the spin-orbit substate of the oxygen triplet state $A'^3\Delta_u$ (Herzberg III bands) ([5] and references therein). The assignments of the O_2–O_2 collision-induced bands are shown in Fig. 2 (see e. g. [6]). In the pressure and temperature conditions of our measurements, no bound $(O_2)_2$ discrete absorption bands are observed.

3.2. Comparison with literature data

Figure 3 compares of the O_2–O_2 collision-induced absorption cross-section at room temperature with recent literature data. Despite the different pressure ranges and resolution used in earlier studies, the agreement between the data is good.

Integrated absorption cross-sections for the visible–NIR bands, calculated by integration of the CIA cross-section between the wavenumber limits of each band, are compared with earlier data in Table 1. Our values are generally lower than the literature data, by about 15% for most bands and up to 50% for the three weakest bands. Agreements better than 10% are written in bold characters in Table 1.

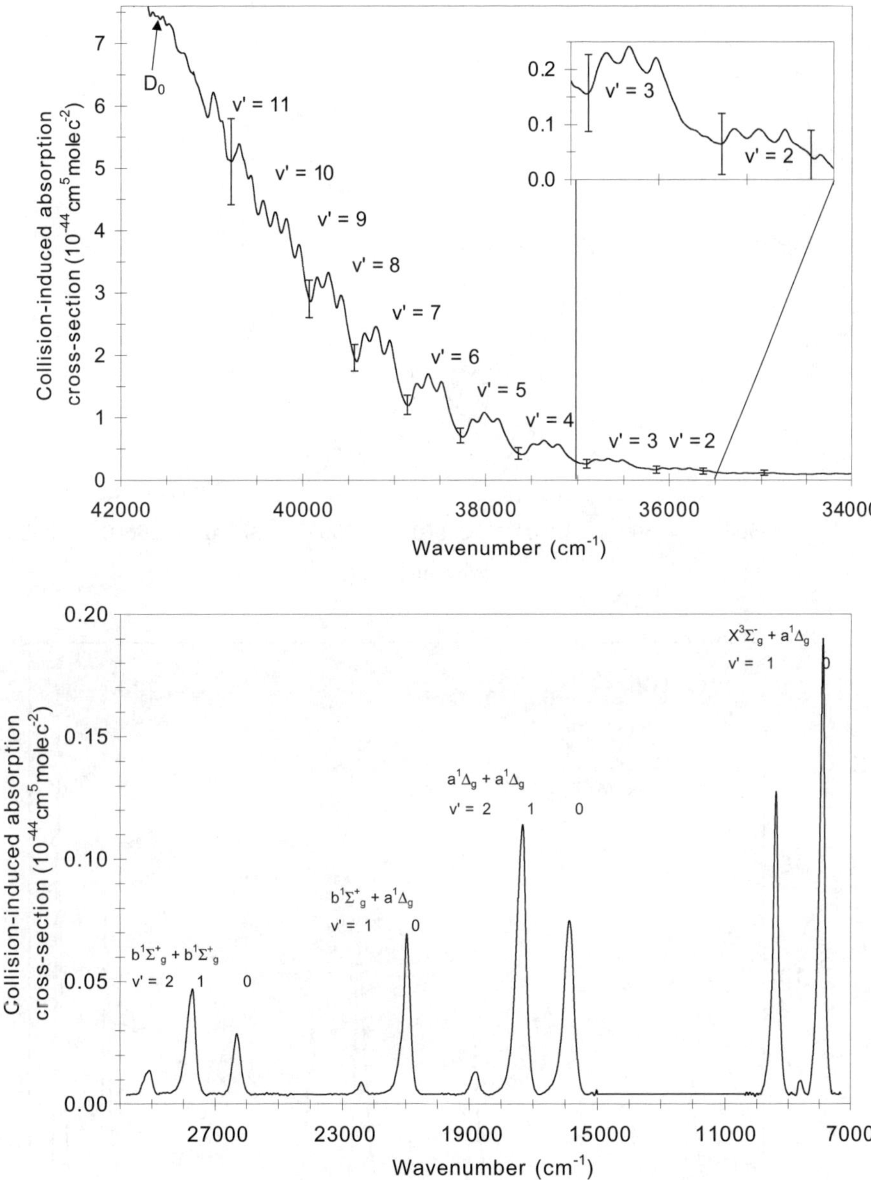

Figure 2. The O_2–O_2 collision-induced absorption cross-section at room temperature from the UV to the NIR. Assignments and vibrational levels of the bands are also shown.

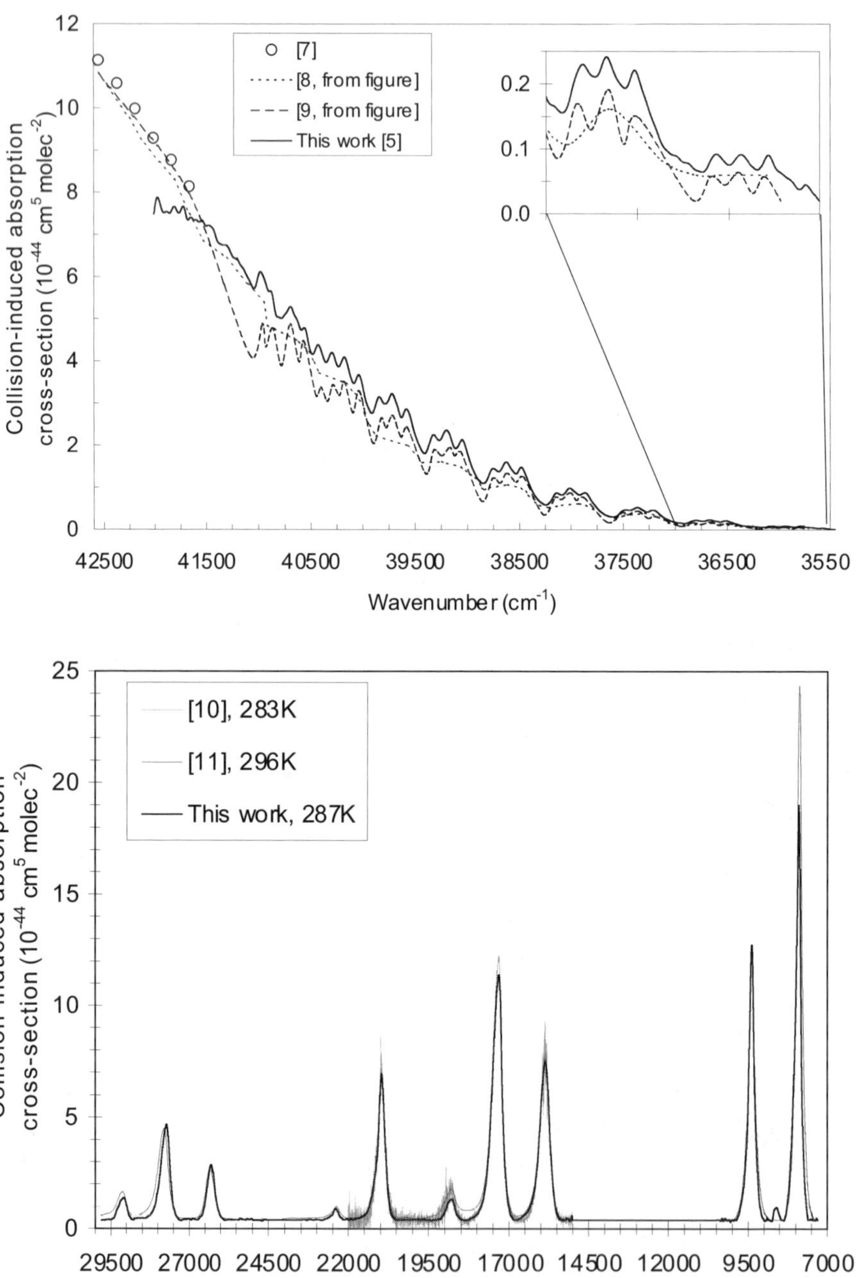

Figure 3. The O_2–O_2 collision-induced absorption cross-section at room temperature from the UV to the NIR and comparison with recent literature data.

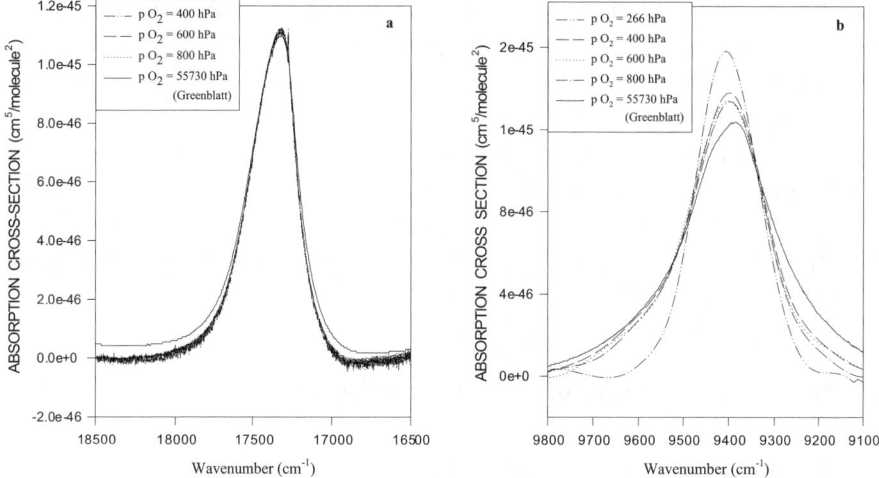

Figure 4. O_2–O_2 CIA cross-sections at different pressures (a) in the visible, (b) in the NIR spectral regions.

3.3. Pressure effect

The pressure effect on the O_2–O_2 CIA cross-sections in the visible region is very weak, as shown in Fig. 4a for a chosen band located at $17335\,\mathrm{cm}^{-1}$. In the pressure range of this study, it is shown that neither a broadening, nor a shift can be observed. Greenblatt's data [11] obtained at $55730\,\mathrm{hPa}$, which are also shown on Fig. 4a, confirm this observation. Similarly, the lack of a pressure effect is observed in the UV region [5]. On the other hand, the two diffuse bands in the NIR range show a more pronounced effect as illustrated in Fig. 4b for the band located at $9390\,\mathrm{cm}^{-1}$.

3.4. Foreign gas effect

The O_2–N_2 and O_2–Ar CIA cross-sections in the UV spectral range were presented and discussed elsewhere [5], and only a brief summary is given here. In agreement with previous studies [9, 13–16], the O_2–N_2 and O_2–Ar CIA cross-sections are almost equal at the pressures used in our experiments, and are approximately twice smaller than the O_2–O_2 CIA cross-section. However, the differences between the O_2–O_2 and the O_2–X cross-sections recorded at the same total pressure are not zero and are directly proportional to the total pressure. In addition, they do not contain any triplet structures (see Fig. 6 in Ref. [5]). These findings led us to suggest that an additional unknown mechanism, involving only O_2–O_2 interactions, might be occurring.

As for the UV bands, the two diffuse bands in the NIR range involve electronic excitation of a single oxygen molecule, and consequently the transition in the complex could be induced by the nature of the foreign gas. Measurements of the O_2 NIR absorption in the presence of N_2 have already been reported in the literature, but the authors greatly disagree concerning the amplitude [1] of this effect. Our measurements show that for the band at $7900\,cm^{-1}$ the O_2–N_2 absorption cross-section is 2.3 times less than the O_2–O_2 value whereas no additional absorption is observed for the $9390\,cm^{-1}$ band.

4. Impact on atmospheric retrievals

Retrievals of NO_2, O_3, BrO, and OClO in the visible region were performed on atmospheric spectra recorded at Harestua, Norway, using the WinDOAS 2.02 software developed at BIRA-IASB [17]. The retrievals were made using the O_2-O_2 cross-section datasets from this work and from the work of Greenblatt *et al.* [11]. The differences in the slant columns vary from 2 to 6% depending on the molecule. The quality of the fit, inferred from the RMS values, varies from one dataset to another, and from molecule to molecule. The best fit is obtained for O_3 and NO_2 when using Greenblatt's data. On the contrary, a better fit is obtained for OClO when using our data. As for BrO, it is not sensitive to the choice of the CIA cross-section.

5. Summary and conclusion

The collision-induced absorption bands of molecular oxygen were studied from 42000 to $7500\,cm^{-1}$ (238–1330 nm) under pressures prevailing in the Earth's atmosphere. A method of lines removal, which was developed for this purpose, is described. The results presented here are in relatively good agreement with earlier results, but have the advantage of being a homogeneous and extensive data set. The O_2–X CIA cross-sections and the O_2 line parameters are available in digital form upon request to the authors or by downloading on the IASB-BIRA website.[2] A sensitivity test of atmospheric NO_2, O_3, BrO and OClO retrievals to O_2–O_2 CIA cross-sections has also been performed.

---

[2]http://www.oma.be/BIRA-IASB/Scientific/Topics/Lower/LaboBase/Laboratory.html

Acknowledgments

This research was supported by the Belgian State — Prime Minister's Service — Federal Office for Scientific, Technical and Cultural Affairs and the Fonds National de la Recherche Scientifique (Belgium). We are grateful for support provided by the Centre National de Recherche Scientifique (France) and Institut National des Sciences de l'Univers (France) through the Programme National de Chimie Atmosphérique.

References

[1] Solomon, S., Portmann, R. W., Sanders, R. W., and Daniel, J. S. (1998) Absorption of solar radiation by water vapor, oxygen, and related collision pairs in the Earth's atmosphere, J. Geophys Res, 103 (D4), 3847–3858.

[2] Jenouvrier, A., Merienne, M.-F., Coquart, B., Carleer, M., Fally, S., Vandaele, A. C., Hermans, C., and Colin, R. (1999) Fourier transform spectroscopy of the O_2 Herzberg bands: I — Rotational analysis, J. Mol. Spec., 198, 136–162.

[3] Jenouvrier, A. (1999) unpublished.

[4] Hurtmans, D. (1999) private communication.

[5] Fally, S., Vandaele, A. C., Carleer, M., Hermans, C., Jenouvrier, A., Merienne, M.-F., Coquart, B., and Colin, R. (2000) Fourier transform spectroscopy of the O_2 Herzberg bands: III - Absorption cross sections of the collision-induced bands and of the Herzberg continuum, J. Mol. Spec., 204, 10–20.

[6] Ellis, J. W. and Kneser, H. O. (1933) Kombinationsbeziehungen im Absorptionsspektrum des flüssigen Sauerstoffs, Z. Phys., 86, 583–591.

[7] Yoshino, K., Cheung, A. S.-C., Esmond, J. R., Parkinson, W. H., Freeman, D. E., Guberman, S. L., Jenouvrier, A., Coquart, B., and Merienne, M.-F. (1988) Improved absorption cross-sections of oxygen in the wavelength region 205–240 nm of the Herzberg continuum, Planet. Space Sci., 36 (12), 1469–1475.

[8] Shardanand and Prasad Rao, A. D. (1977) Collision-induced absorption of O_2 in the Herzberg continuum, J. Quant. Spectrosc. Radiat. Transfer, 17, 433–439.

[9] Zelikina, G. Ya., Bertsev, V. V., and Kiseleva, M. B. (1994) Absorption of compressed liquid oxygen and its mixtures with Ar, Kr, Xe, N_2, and CF_4 in the 200–280 nm spectral region, Optics and Spectroscopy, 77 (4), 579–583.

[10] Newnham, D. and Ballard, J. (1998) Visible absorption cross sections and integrated absorption intensities of molecular oxygen (O_2 and O_4), J. Geophys. Res., 103, D22, 28801–28816.

[11] Greenblatt, G. D., Orlando, J. J., Burkholder, J. B., and Ravishankara, A. R. (1990) Absorption measurements of oxygen beween 330 and 1140 nm, J. Geophys. Res., 95, 18577–18582.

[12] Blickensderfer, R. P. and Ewing, G. E. (1969) Collision-induced absorption spectrum of gaseous oxygen at low temperatures and pressures. II. The simultaneous transition $^1\Delta_g + {}^1\Delta_g \leftarrow {}^3\Sigma_g^- + {}^3\Sigma_g^-$ and $^1\Delta_g + {}^1\Sigma_g^+ \leftarrow {}^3\Sigma_g^- + {}^3\Sigma_g^-$, J. Chem. Phys., 51, 5284–5289.

[13] Dianov-Klokov, V. I. (1966) Absorption by gaseous oxygen and its mixtures with nitrogen in the 2800–2350 Å range, Opt. Spectrosc., 2, 233–236.

[14] Shardanand (1977) Nitrogen-induced absorption of oxygen in the Herzberg continuum, J. Quant. Spectrosc. Radiat. Transfer, 18, 525–530.

[15] Shardanand (1978) Temperature effect on nitrogen-induced absorption of oxygen in the Herzberg continuum, J. Quant. Spectrosc. Radiat. Transfer, 20, 265–270.

[16] Oshima, G. Ya., Yakamoto, Y., and Koda, S. (1995) Pressure effect of foreign gases on the Herzberg photoabsorption of oxygen, J. Phys. Chem., 99, 11830–11833.

[17] Fayt, C. and Van Roozendael, M. (2001) WinDOAS software user manual.

CAVITY RING-DOWN SPECTROSCOPY OF O_2–O_2 COLLISIONAL INDUCED ABSORPTION

M. Sneep, W. Ubachs

Laser Centre, Department of Physics and Astronomy,
Vrije Universiteit, De Boelelaan 1081, 1081 HV Amsterdam,
The Netherlands

Abstract   Absolute cross sections (peak and band-integrated) for collision-induced absorption features in oxygen at 630, 577 and 477 nm are determined using the laser-based technique of cavity ring-down spectroscopy. A wavelength-dependent background, Rayleigh scattering extinction and oxygen-monomer absorption are unraveled from the collision-induced features and its quadratic pressure dependence is demonstrated for sub-atmospheric pressures.

Keywords: Collision induced absorption / oxygen / Cavity Ring-Down Spectroscopy

1. Introduction

The laser based-technique of cavity ring-down spectroscopy (CRDS) has in the past decade grown into an often used method for measuring absorption features of gas-phase molecules. Particularly when cavity mirrors of extreme reflectivity (i. e. exceeding 99.99%) are available, the sensitivity limit for detecting absorption is drastically improved. In this respect it challenges the alternative techniques of long-cell multi-pass Fourier-transform spectroscopy. Here we will describe some general characteristics of CRDS and report on cross section measurements of $(O_2)_2$ features. These collision-induced O_2–O_2 absorptions are associated with electronic resonances of primarily dissociative states, that play a role in the Earth's atmosphere. They contribute to the radiation budget of the atmosphere for $\sim 1\,W/m^2$ and these features with a quadratic pressure dependence can be used for determination of cloud tops in satellite remote sensing.

203

C. Camy-Peyret and A.A. Vigasin (eds.),
Weakly Interacting Molecular Pairs: Unconventional Absorbers of Radiation in the Atmosphere, 203–211.

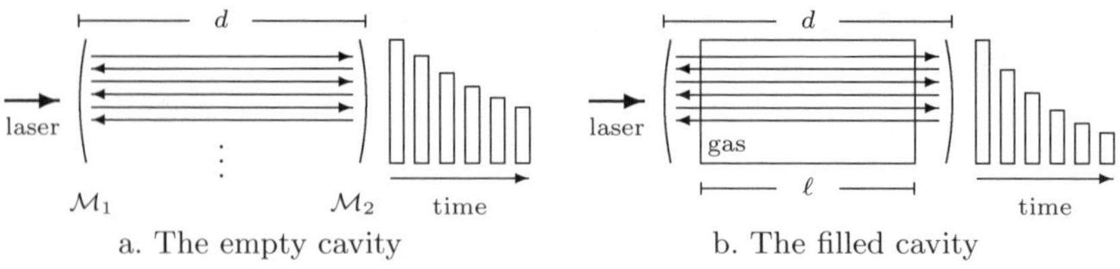

Figure 1. The principle of cavity ring-down.

2. Principles of Cavity Ring-Down Spectroscopy

The principle of CRDS is based upon a measurement of the rate of decay of an optical resonator with a high quality factor. It is schematically illustrated in Fig. 1. A laser pulse of short duration (usually 5 ns) is coupled into a stable but non-confocal cavity, consisting of two highly reflecting (typically $\mathcal{R} > 99.9\%$) curved mirrors. At each mirror a small part $(1 - \mathcal{R})$ of the circulating power will be coupled out, while the remaining part is reflected back in the cavity. The first pulse leaking out of the cavity will have an intensity $I_0 = (1 - \mathcal{R})^2 I_{\text{in}}$. The kth pulse with intensity

$$I_k = I_0 \mathcal{R}^{2(k-1)}$$
$$= I_0 e^{2(k-1)\ln(\mathcal{R})} \tag{1}$$

will leak out $(2d(k-1)n/c)$ seconds after the laser pulse, where c is the speed of light, n the index of refraction and d is the length of the cavity. Since we are dealing with gas-phase environments at low pressure the index of refraction is neglected. We will start with a cavity without absorbers.

A 5 ns pulse has a length of 1.5 m. Practical ring-down cavities have a round-trip length $(2d)$ of less than that, and therefore pulses will overlap inside the cavity. Furthermore the detection system usually has a response time that causes the discrete pulses to blend into a continuous signal as a function of time t,

$$I_{(t)} = I_0 e^{-(c/d)|\ln(\mathcal{R})|t}. \tag{2}$$

Additional losses, usually caused by absorption of light or by Rayleigh scattering on molecules in the cavity, result in a faster decay. When the additional losses satisfy Beer's law, the decay remains exponential, yielding:

$$I_{(t)} = I_0 e^{-(c/d)\left(|\ln(\mathcal{R})|+\kappa_{(\nu)}\ell\right)t}, \tag{3}$$

where ℓ is the length within the cavity that is filled with the absorbing species, see Fig. 1(b), and $\kappa_{(\nu)}$ the frequency dependent extinction coefficient.

The rate at which the ring-down signal decays with $(\beta_{(\nu)})$ or without $(\beta^0_{(\nu)})$ the presence of additional absorbers, is given by:

$$\beta_{(\nu)} = \frac{c\,|\ln(\mathcal{R})| + \kappa_{(\nu)}\ell}{d}; \quad \beta^0_{(\nu)} = \frac{c\,|\ln(\mathcal{R})|}{d}. \tag{4}$$

From $\beta_{(\nu)}$ and $\beta^0_{(\nu)}$, we find an expression for the extinction induced by the medium, where we assume that the absorber fills the complete cavity, so $\ell = d$, yielding a simple equation for the extinction in terms of cavity decay rates

$$\kappa_{(\nu)} = \frac{(\beta_{(\nu)} - \beta^0_{(\nu)})}{c}. \tag{5}$$

The value $\beta^0_{(\nu)}/c$ is the background for the ring-down spectrum. This background is still frequency dependent because the mirror reflectivity depends on the frequency as well.

An important advantage of CRDS is the fact that the intensity of the light does *not* appear in equation 5; all information is contained in the time-evolution of the ring-down signal. The high sensitivity of CRDS is associated with this independence of the fluctuations of the light source. Another advantage is the fact that mirrors can now be made at such high reflectivity, up to $\mathcal{R} = 99.999\%$, that very long effective path lengths can be created, up to $100\,\mathrm{km}$ with a cavity length of only $80\,\mathrm{cm}$. This leads to a very high sensitivity—absorptions as weak as $10^{-9}\,\mathrm{cm}^{-1}$ can be detected.

A crucial issue is that the rate of decay of the cavity should be exponential. Only under that assumption the exponential decay associated with Beer's law for absorption can be added to the decay of the empty cavity (see Eq. 3) thus resulting in mono-exponential decay in a cavity filled with gas. If for some reason the resulting decay function is non-exponential no value of $\beta_{(\nu)}$ can be derived in the fitting procedure. Note that it is mathematically prohibited to design a fitting procedure to an arbitrary sum of exponentials. Multi-exponentiality can be caused by several mechanisms, the most common being laser-bandwidth induced effects. If the bandwidth of the exciting laser is non-negligible with respect to the width (Doppler or collisional) of the spectral line, then the various frequency components within the bandwidth profile are subject to different rates of absorption; hence these frequency components produce a sum of exponentials, which results in a non-exponential decay of the monitored signal. This effect is in fact the well-known "slit-function" problem of all spectroscopy, but its treatment, in the case of CRDS, is more complicated than usual because of its nonlinear nature. Although the problem has been noted in literature it still hampers the use of CRDS for the determination of absolute cross sections of narrow line features.

Another relevant issue for the analysis of ring-down transients is the proper treatment of the decay transients in fitting routines. As was elaborated by Naus *et al.* [4] a weighted non-linear least-squares fit to the data on the transient give the best determination of the decay life time, and in particular produces a reliable estimate of the uncertainty. As for the optical resonator it is important that it is aligned such that its mode structure does not result in wavelength-dependent effects. There is two methods for achieving this. In one method, developed by Van Zee *et al.* [7], the cavity is aligned in confocal position to transmit a single mode adapted to the laser frequency. Although the method yields accurate results it is not well-suited to be used in absorption measurements, where fairly large wavelength ranges are to be scanned. In the alternative method the cavity is aligned far from confocal ($d \approx 0.85\,r_c$, d is the distance between the mirrors and r_c is the radius of curvature), such that a large number of longitudinal and transversal cavity modes is supported, making the cavity a white light transmitter. It is noted that in this condition only a very small fraction ($[1 - \mathcal{R}]^2$) of the initial laser pulse reaches the detector. The optical resonator then acts as a white-light filter and the configuration can be used for wavelength-scanning pulsed CRDS. The relation between mode-structure and transmission, and an experimental method to improve alignment towards single-exponential decay in a non-confocal resonator were discussed in reference [4].

3. Measurements on small samples

As was mentioned in the introduction, both CRDS and Fourier-transform spectroscopy, have developed into sensitive tools for direct absorption measurements. An unambiguous general statement as to the sensitivity of either technique cannot be made, because in CRDS the sensitivity strongly depends on the reflectivity of the mirrors available, while for FT-spectroscopy the sensitivity depends on the geometrical configuration of an absorption path connected to the spectrometer. We mention here that CRDS lacks the so-called multiplex advantage of FT-spectroscopy: data have to be collected at each wavelength position, making the technique time consuming. One noteworthy advantage of

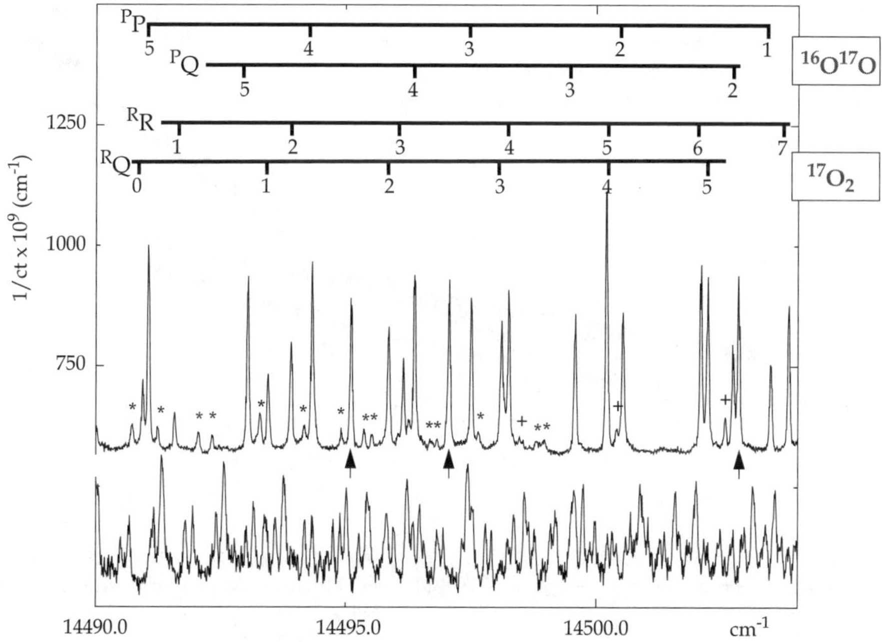

Figure 2. CRDS spectrum of the $b^1\Sigma_u^+ \leftarrow X^3\Sigma_g^- (1,0)$ or B-band of oxygen for a (50% atom) ^{17}O enriched sample, measured at a pressure of 16.5 Torr. Lines pertaining to ^{16}O^{17}O and ^{17}O$_2$ are assigned, while the lines marked with $*$, $+$ and $\uparrow$ belong to ^{17}O^{18}O, ^{16}O^{18}O and ^{16}O$_2$ respectively. The lower panel shows a simultaneously recorded I$_2$ spectrum, used for calibration.

CRDS is that a high sensitivity can be obtained from a compact setup, in which only small amounts of gas are required. This is a particularly useful property for the spectroscopic investigation on isotopically enriched samples, which are only available in small amounts. This is demonstrated in Fig. 2 for measurements of the B-band of molecular oxygen, published before in [3]. The spectral recording is taken from a 50% oxygen-17 enriched sample of oxygen in a cell of 1 cm diameter and 40 cm length.

4. CRD measurements of oxygen collisional complexes

As a result of the binding properties between two ground state oxygen atoms $O(^3P) + O(^3P)$ there are several low-lying electronic states in the O_2 molecule. However, all possible transitions between these molecular states are, in the electric dipole approximation, forbidden by

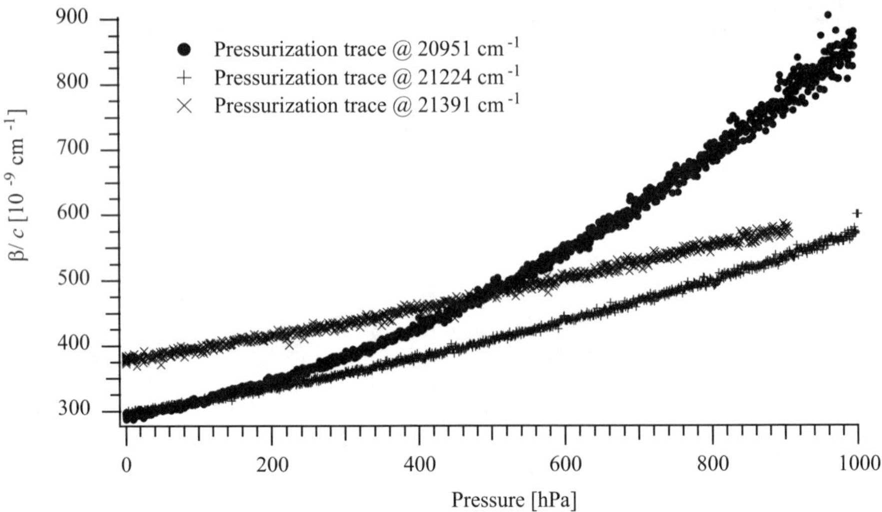

Figure 3.   Three typical pressurization traces. The offset $\beta^0_{(\nu)}/c$ on the vertical axis shows the difference in mirror reflectivity between the three frequency positions.

quantum selection rules. During a collision some of these transitions are allowed and absorption of a photon can occur. The collision pair does not have to be bound and each of the molecules carries away part of the photon energy. Since the number of collision pairs is proportional to the square of the density, the strength of the absorption also increases with the square of the density.

To measure the absorption strength of the oxygen collision complexes, one could scan over the absorption feature at various pressures. A better separation between the background $\beta^0_{(\nu)}/c$—determined by the mirrors and how well they are cleaned, a component with a linear density dependence—Rayleigh scattering, and a component with a quadratic density dependence can be obtained by scanning the density, while maintaining the laser at a fixed frequency. The loss rate as a function of density $\tilde{n}$ is:

$$\beta_{(\nu)}(\tilde{n}) = \beta^0_{(\nu)} + \sigma^R \, \tilde{n} \, c + \alpha^q \, \tilde{n}^2 \, c, \tag{6}$$

where σ^R is the Rayleigh scattering cross section and α^q is the collisional induced absorption cross section.

While oxygen was slowly flushed into the ring-down cavity, the loss-rate $\beta_{(\nu)}(\tilde{n})$ was constantly monitored by a 10 Hz pulsed laser. The pressure ramp spans a range of 0 to 1000 hPa and consists of about 1100 measurements of $\beta_{(\nu)}$. To obtain a spectrum of the entire resonance, the laser is tuned to a different frequency and the procedure is

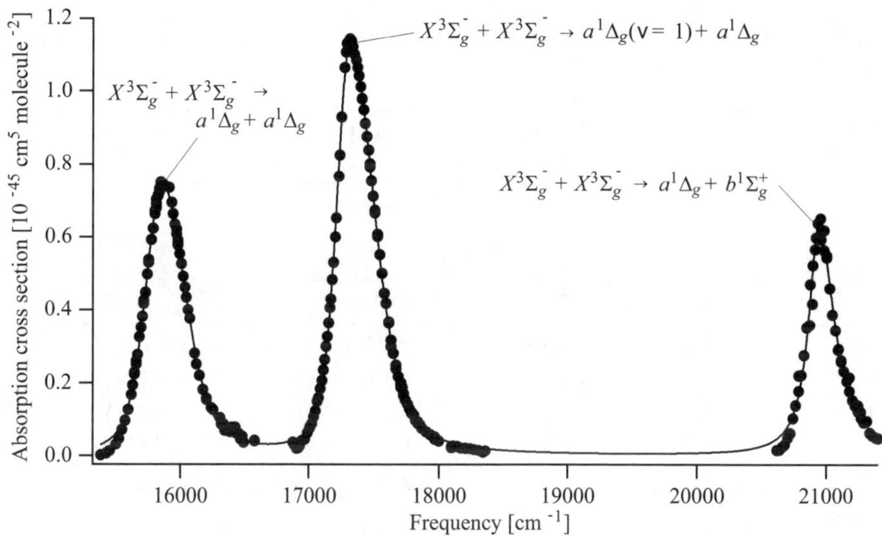

Figure 4. The three visible O_2-O_2 absorption features near 630, 577 and 477 nm. The transitions are indicated in the graph. For all transitions $v = 0$, unless indicated otherwise.

repeated. Three typical pressurization traces can be found in Fig. 3. The parameters of Eq. 6 are fitted to the measured data. The Rayleigh scattering cross section can be calculated from the refractive index and the depolarization ratio. In the final fit, σ^R is fixed to this value.

An explicit advantage of the pressure-ramp method is that at each wavelength position the contributions of baseline, linear extinction and the true collisional-induced quadratic feature can be unraveled in a fitting routine. Extraction of the baseline is crucial if the goal is to derive a cross-section of an absorption feature. Since a major part of the cross-section is near the baseline, the slightest misjudgment of the background causes a strong deviation in the cross-section. In a scanning experiment baseline fluctuations can be interpreted as absorption signal. Note that Newnham *et al.* [5] had to set the baseline to zero at two frequency positions, adding a source of systematic uncertainty, which is avoided in the pressure ramp method.

Another issue is that wavelength positions for pressure ramp scans can be carefully chosen. In the case of the red collisionally induced feature at 630 nm there is oxygen monomer absorption (the γ-band) overlaying the collision-induced feature. The contribution of the monomer and collision feature could be unraveled by choosing wavelengths

Table 1. An overview of the literature value of the $(O_2)_2$ transitions. The values in parentheses are 1σ uncertainties.

Ref.	Position, cm^{-1}	Width, cm^{-1}	Height, $10^{-46}\,cm^{-5}molecule^{-2}$	Integral, $10^{-43}\,cm^{-4}molecule^{-2}$
$[a\,^1\Delta_g(v=0)]_2 \leftarrow [X\,^3\Sigma_g^-(v''=0)]_2$ at 630 nm				
[1]	15869(5)	368(5)	7.2(2)	3.1(1)
[5]	15862(5)	369(5)	7.1(2)	2.8(1)
[2]	15880(1)	367(1)	7.55(5)	3.19(3)
$[a\,^1\Delta_g(v'=0) + a\,^1\Delta_g(v'=1)] \leftarrow [X\,^3\Sigma_g^-(v''=0)]_2$ at 577 nm				
[1]	17320(5)	348(5)	11.0(3)	4.8(1)
[5]	17314(5)	339(5)	11.7(3)	4.7(1)
[2]	17308(1)	340(1)	11.41(5)	4.66(3)
$[b\,^1\Sigma_g^+(v'=0) + a\,^1\Delta_g(v'=0)] \leftarrow [X\,^3\Sigma_g^-(v''=0)]_2$ at 477 nm				
[1]	20951(5)	270	6.3(6)	—
[5]	20951(5)	—	8.34(83)	2.483(48)
[6]	20930(2)	241(7)	6.38(16)	2.18(4)

in between rotational absorption lines in the $b^1\Sigma_g^+(v=2) \leftarrow X^3\Sigma_g^-$ band of $^{16}O_2$ for the construction of the line profile of the broad $[a^1\Delta_g(v=0)]_2 \leftarrow [X^3\Sigma_g^-(v=0)]_2$ collision feature.

Finally we note that effects of non-exponential behaviour were not present in the study of these O_2–O_2 features. The CRDS cavity was carefully aligned towards a condition of purely exponential decay. Moreover laser-bandwidth effects do not play a role: it is below $0.1\,cm^{-1}$, while the width of the oxygen resonances exceeds $100\,cm^{-1}$.

In Fig. 4 an overview is given of three of the visible O_2–O_2 absorption features. Black points represent the quadratic component derived from a pressure ramp scan. The features $[a^1\Delta_g(v=0)]_2 \leftarrow [X^3\Sigma_g^-(v=0)]_2$ at 630 nm and $[a^1\Delta_g(v=0) + a^1\Delta_g(v=1)] \leftarrow [X^3\Sigma_g^-(v=0)]_2$ at 577 nm are described in reference [2], the $[b^1\Sigma_g^+(v=0) + a^1\Delta_g(v=0)] \leftarrow [X^3\Sigma_g^-(v=0)]_2$ feature near 477 nm is described in reference [6]. In this frequency range another feature exists near 532 nm; an overview of the other O_2–O_2 transitions can be found in reference [1].

The data in Table 1 was measured using various methods: Greenblatt *et al.* [1] used a cell at 55 atmosphere. Newnham *et al.* [5] employed a long path FT spectrometer, recording the spectrum at 1000 hPa. Extraction of the quadratic pressure dependence was done from this single pressure alone. In their conclusions it is specifically mentioned that more measurements at a lower pressure are needed for the O_4 absorption bands. An important result of the present CRDS data is, that it is established that the quadratic behaviour and the pressure dependent cross sections also hold for sub-atmospheric pressures.

References

[1] Greenblatt, G. D., Orlando, J. J., Burkholder, J. B., and Ravishankara, A. R. (1990) Absorption measurements of oxygen between 330 and 1140 nm, J. Geophys. Res., 95 (D11), 18577–18582.

[2] Naus, H. and Ubachs, W. (1999) Visible absorption bands of the $(O_2)_2$ collision complex at pressures below 760 Torr, Appl. Opt., 38(15), 3423.

[3] Naus, H., van der Wiel, S. J., and Ubachs, W. (1998) Cavity-Ring-Down Spectroscopy on the $b^1\Sigma_g^+ - X^3\Sigma_g^- (1,0)$ band of oxygen isotopomers, J. Mol. Spectr., 192, 162–168.

[4] Naus, H., van Stokkum, I. H. M., Hogervorst, W., and Ubachs, W. (2001) Quantitative analysis of decay transients applied to a multimode pulsed cavity ringdown experiment, Appl. Opt., 40(24), 4416–4426.

[5] Newnham, D. A. and Ballard, J. (1998) Visible absorption cross sections and integrated absorption intensities of molecular oxygen (O_2 and O_4), J. Geophys. Res., 103(D22), 28801–28816.

[6] Sneep, M. and Ubachs, W. (2003) Cavity Ring-Down Measurements of the O_2–O_2 collision induced absorption resonance at 477 nm at sub-atmospheric pressures, J. Quant. Spectrosc. Radiat. Transfer, accepted for publication.

[7] Van Zee, R. D., Hodges, J. T., and Looney, J. P. (1999) Pulsed, single-mode cavity ringdown spectroscopy, Appl. Opt., 38(18), 3951.

LABORATORY FOURIER TRANSFORM SPECTROSCOPY OF THE WATER ABSORPTION CONTINUUM FROM 2500 TO 22500 cm^{-1}

M. Carleer,[*] M. Kiseleva,[†] S. Fally, P.-F. Coheur,[‡] C. Clerbaux,[§] R. Colin
Laboratoire de Chimie Physique Moléculaire,
Université Libre de Bruxelles, 50 Av. F. D. Roosevelt, B-1050 Brussels, Belgium

L. Daumont,[¶] A. Jenouvrier, M.-F. Merienne
Groupe de Spectrometrie Moléculaire et Atmosphérique,
UFR Sciences, Moulin de la Housse, B.P. 1039, 51687 Reims Cedex 2, France

C. Hermans, A. C. Vandaele
Institut d'Aéronomie Spatiale de Belgique,
Av. Circulaire 3, B-1180 Brussels, Belgium

Abstract The Fourier transform absorption spectrum of unsaturated water is analysed in order to derive the continuous spectral signature of water vapour in the infrared and visible regions. It is shown that the spectrum consists of a broadband continuum, extending over the entire 2500–22500 cm^{-1} spectral range investigated, which underlies the vibrational absorption bands of water vapour. This continuum is tentatively attributed to the extinction of light by small water droplets. A careful analysis of the vibrational bands has failed to prove the existence of the narrow band continuum predicted by theory, due to the far-wings of spectral lines or to water dimers.

Keywords: spectroscopy / water vapour / absorption continuum / extinction

---

[*]Corresponding author (E-mail: `mcarleer@ulb.ac.be`, Fax: +32-2-6504232).
[†]On leave from the Department of Molecular Spectroscopy, StPetersburg University, Russia.
[‡]Scientific Researcher with the Fonds National de la Recherche Scientifique.
[§]Also at Service d'Aéronomie, Université de Paris 6, Paris, France.
[¶]Attaché Temporaire d'Enseignement et de Recherche.

213

C. Camy-Peyret and A.A. Vigasin (eds.),
Weakly Interacting Molecular Pairs: Unconventional Absorbers of Radiation in the Atmosphere, 213–221.
© 2003 *Kluwer Academic Publishers. Printed in the Netherlands.*

1. Introduction

Water vapour plays a key role in the atmosphere, mainly on account of its radiative properties [1]. Being the third most abundant atmospheric species, water has in particular been suspected to be responsible for the excess of atmospheric absorption that models were, until recently [2], unable to reproduce [3, 4]. It has been suggested that weak water lines [5], water dimers [6], clouds [7] or a water continuum, associated with the far-wings of discrete lines [8], might be responsible for this missing absorption. In their recent calculations, Mlawer *et al.* [2] have, however, shown that a new formulation of the scattering by aerosols can account for the discrepancies. Several new investigations of the absorption spectrum have failed to definitively clarify the role of water in the energy balance of the earth. For instance, measurements combining very long absorption paths and Fourier transform spectroscopy have proven to be very valuable in identifying weak lines and measuring their strengths [9, 10], but the results pointed to a small contribution of the latter to the total absorption [10–12]. The more sensitive cavity ringdown laser or Fabry-Perot techniques have provided further information, especially with regard to the water continuum [13–15], but in restricted spectral intervals. This work presents a preliminary analysis of the spectrum of water, recorded at medium resolution along a very long absorption path, from the infrared to the near ultraviolet, with special attention given to the presence of a continuum.

2. Experimental

The spectra have been recorded using the long path multiple-reflection absorption cell facility of the University of Reims. A 600-meter long absorption path was chosen, as in earlier measurements [10, 16]. The spectrometer was a Fourier transform Bruker IFS120M, which was operated at $2\,\mathrm{cm}^{-1}$ spectral resolution ($0.45\,\mathrm{cm}$ OPD) between 2500 and $25000\,\mathrm{cm}^{-1}$. This was achieved using a stabilized Tungsten lamp as the light source, and two different combinations of detectors and optical components: The Indium-Antimony detector with a CaF_2 beamsplitter and the Silicon detector with a quartz beamsplitter were respectively used for the infrared (2500–$12500\,\mathrm{cm}^{-1}$) and the visible (10000–$25000\,\mathrm{cm}^{-1}$) spectral ranges (Fig. 1).

Water was introduced in the cell by heating a recipient containing pure liquid water. The spectra were recorded at room temperature ($\sim 291\,\mathrm{K}$) with water vapour pressures (2 to $16\,\mathrm{hPa}$) well below

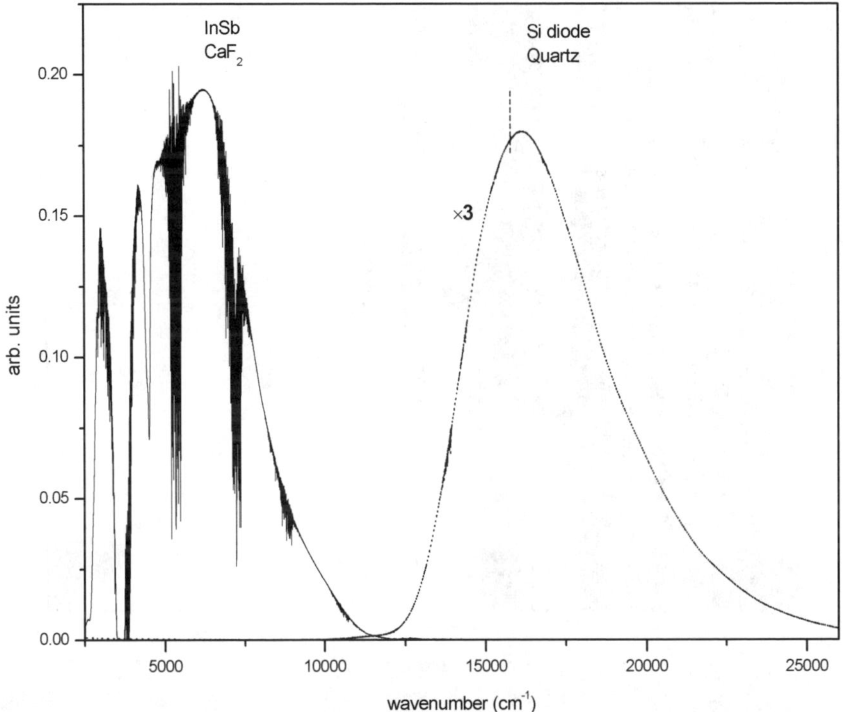

Figure 1. Spectra in the IR (black line) and visible (dotted line) spectral regions.

saturation. The pressures were measured by a MKS Baratron capacitance manometer. Several spectra were also recorded using dry air as a buffer gas.

A typical series of measurements was accomplished as follows: a reference spectrum I_0 was recorded for the empty cell. Water vapour was then introduced in the cell and, when a stable pressure was reached, a spectrum $I(p)$ was recorded. After a series of 8 successive measurements at different pressures, the cell was pumped back and a second reference spectrum I_0' was recorded. The absorbance at a given pressure was calculated by:

$$A(p) = \ln\left(\frac{a\,I_0 + b\,I_0'}{I(p)}\right), \tag{1}$$

where a and b are coefficients weighting the reference spectra. This procedure proved necessary to account for the unavoidable drift of the light source during the experiment, which lasted approximately eight hours. The reference spectra include absorption by water present in the room, which was thus subtracted from the spectra.

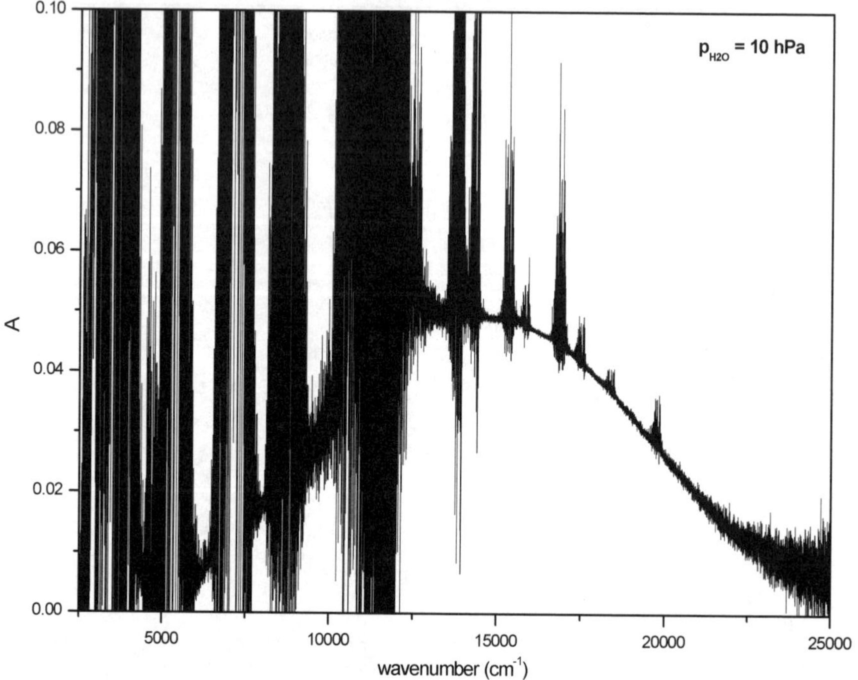

Figure 2.   Absorbance at a water vapour pressure of 10 hPa and 291 K.

3. Results and discussion

A typical spectrum limited to small absorbance is shown in Fig. 2. It is clearly observed that a weak broadband continuum underlies the resolved line spectrum over the entire spectral range investigated. It is maximal at about $12000\,\mathrm{cm}^{-1}$. The continuum is structureless and therefore unlikely to be due to the far-wings of discrete lines or to water dimers, which are both expected to result in a succession of bands, each centered around the water vapour vibrational bands, with a strength about proportional to its integrated intensity [6, 17–19]. In section 3.1 we examine the broadband continuum and its evolution with pressure. In section 3.2 we investigate the vibrational band at $8500\,\mathrm{cm}^{-1}$, which is less saturated than the lower energy bands, though still sufficiently intense to search for a narrow band continuum due to far-wings effects or to dimers.

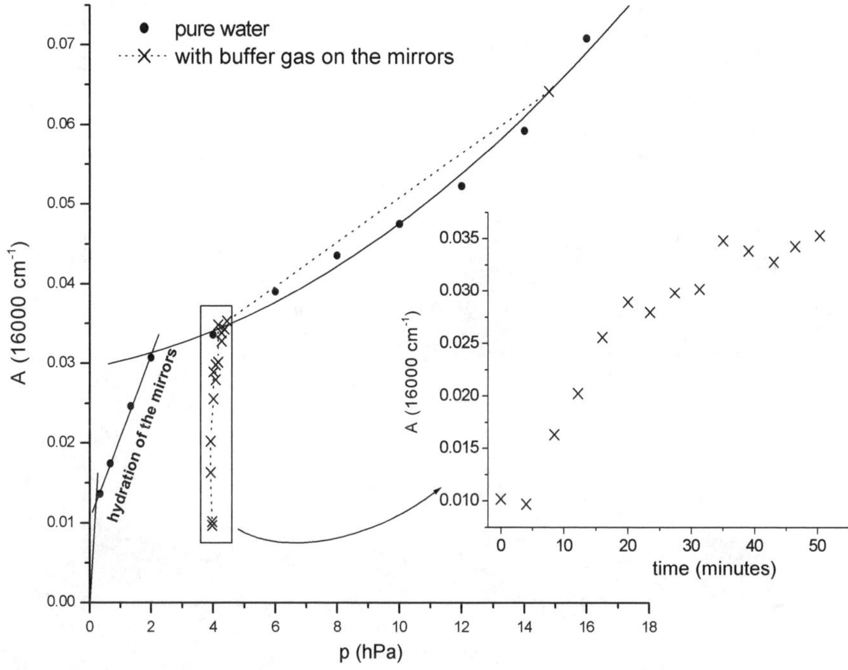

Figure 3.   Evolution of the absorbance of the broadband continuum with water vapour pressure, at $16000\,\mathrm{cm}^{-1}$ (dots). The crosses show the evolution of A with time, at a constant pressure of $4\,\mathrm{hPa}$ and, when a stable value was reached, the result of a single measurement at $15\,\mathrm{hPa}$.

3.1. Broadband continuum

The strength of the broadband continuum increases with increasing water pressure, as illustrated in Fig. 3 at $16000\,\mathrm{cm}^{-1}$. The evolution shows a change in slope at a pressure of $2\,\mathrm{hPa}$. To try to explain this feature, we have carefully followed the evolution of the absorption at $p = 4\,\mathrm{hPa}$, after having first introduced the buffer gas inside the cell, and then water vapour at the centre of the cell. In this situation, the strength of the broadband continuum increases progressively with time, up to a value where it remains constant, and which is similar to that measured without a buffer gas (Fig. 3). These observations suggest that the water introduced in the middle of the cell diffuses towards the mirrors and progressively modifies their reflective properties. A saturation of the mirrors by hydration seems to occur at about $2\,\mathrm{hPa}$. Relying on the above hypothesis, only the measurements made at water vapour pressures above $2\,\mathrm{hPa}$ are relevant for this study and a more realistic

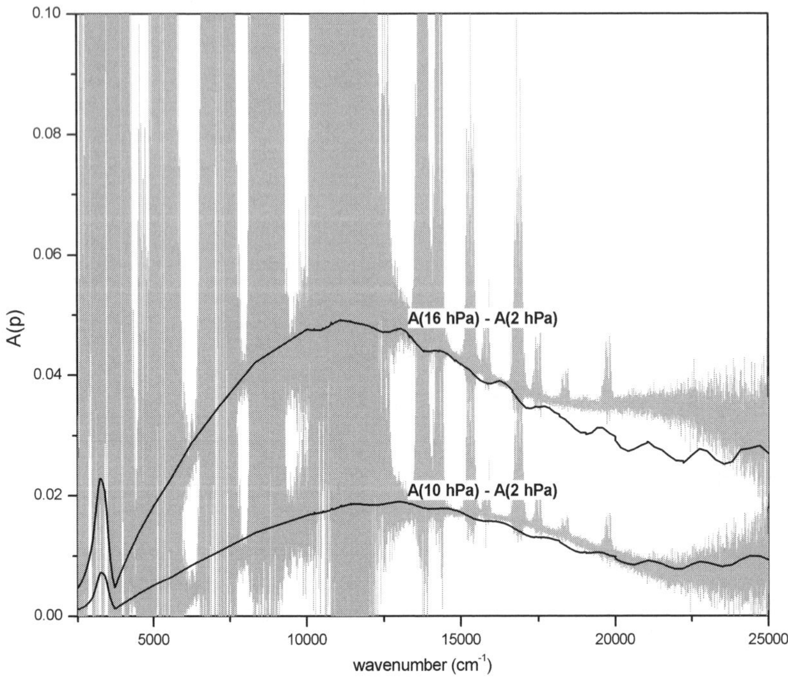

Figure 4.   Broadband continua corrected for the mirror hydration. The black lines are the results of Mie scattering using the optical constants for water of Hale and Querry [20]. For the spectrum at 10 hPa a single water droplet size has been considered ($0.8\,\mu$m radius) whereas for the spectrum at 16 hPa, the contribution of water droplets of 0.8 and $1.0\,\mu$m radius has been used. The curves have been normalized to the maximum of absorbance.

view of the broadband continuum is therefore given by $A(p) - A(2\,\text{hPa})$. Once corrected, the shape of the continua is shown to agree well with the extinction by water droplets of $0.8\,\mu$m radius (Fig. 4). This result is very surprising, as all experiments were performed at pressures well below the saturation vapour pressure (about 20 hPa at 291 K). We have at present very few arguments to support this interpretation and further investigations are required. It should be mentioned, however, that in earlier experiments, performed in similar conditions, the diffusion of light in the cell when using a strong Xenon lamp was observed by the naked eye, suggesting that small droplets were indeed present in suspension in the cell. If these observations are confirmed, several questions remain open: How, where and when are the droplets formed and how do they remain in equilibrium at unsaturated pressures, with a size almost constant for the various pressures investigated.

At this point it is also interesting to note that Mlawer *et al.* have solved the "missing absorber" problem in their atmospheric models [2] by modifying the formulation of the scattering by aerosols. They suggest that the unexplained discrepancies between measurements and calculations in earlier studies were likely due to the use of too large values of the single-scattering albedoes in the atmospheric models. A correlation between their findings and the present observations is therefore tempting, although purely speculative at present.

3.2. Narrow-band continuum

In order to search for a continuum lying beneath the water vapour vibrational bands, as predicted by theory [6, 17–19], we have examined in detail the band at $8500\,\mathrm{cm}^{-1}$, which is intense but does not show a full saturation in the pressure range investigated. The main difficulty of the analysis lies in the complete and accurate removal of all the unbroadened line shapes. To do this, we have used the spectroscopic parameters listed in the HITRAN database [21] and have generated, at each pressure, a synthetic spectrum $I_{\mathrm{synth}}(p)$. The absorbances are then calculated by an expression similar to (1):

$$A'(p) = \ln\left(\frac{a\,I_0 + b\,I'_0}{I(p)/I_{\mathrm{synth}}(p)}\right). \tag{2}$$

Non-vanishing values of $A'(p)$, which increase with increasing pressure, are obtained. However, the removal of the unbroadened line shapes could be in error due to errors in the measurement of the water pressure or to inaccurate line parameters in the HITRAN database. Even by changing the value of the pressure in the simulation, we have not been able to remove the lines sufficiently well to isolate with certainty an underlying continuum. Our results therefore point to the necessity of improving the databases for water vapour. Also, a detailed analysis of the bands at lower energy, where the continuum is expected to be stronger, is obviously needed. This would however require minimizing the saturation and therefore recording spectra at lower pressures or on a shorter absorption path length.

4. Conclusion

We have analysed the absorption spectrum of water vapour between 2500 and $22500\,\mathrm{cm}^{-1}$ and have shown that it consists of a broadband continuum, underlying the water bands and extending over the entire region. The shape of the broadband continuum was shown to match the extinction spectrum of small water droplets (0.8 μm radius) but the presence of the latter at pressures lower than the saturation water vapour pressure remains unexplained. Further studies are needed to assess these observations, especially with regard to the Earth's radiation budget. We have not been able to identify a narrow band continuum, associated to far-wings or to water dimers under the vibrational band at 8500 cm^{-1}, due to the weakness of the continuum and the difficulty of properly removing the water lines.

Acknowledgments

This research was funded by the Belgian State, Federal Office for Scientific, Technical and Cultural Affairs, the Fonds National de la Recherche Scientifique (FNRS, Belgium) and the European Space Agency (contract 14145/00/NL/SFe(IC)). Additional financial support provided by the Centre National de la Recherche Scientifique and the Institut des Sciences de l'Univers (France) through the Programme National de Chimie Atmosphérique is acknowledged.

References

[1] Bernath, P. F. (2002) The spectroscopy of water vapour: Experiment, theory and applications, Phys. Chem. Chem. Phys., 4(9), 1501–1509.

[2] Mlawer, E. J., Brown, P. P., Clough, S. A., Harrison, L. C., Michalsky, J. J., Kiedron, P. W., and Shippert, T. (2000) Comparison of spectral direct and diffuse solar irradiance measurements and calculations for cloud-free conditions, Geophys. Res. Lett., 27(17), 2653–2656.

[3] Stephens, G. L. and Tsay, S.-C. (1990) On the cloud absorption anomaly, Q. J. Roy. Met. Soc., 116, 671–704.

[4] Arking, A. (1996) Absorption of solar energy in the atmosphere: Discrepancy between model and observations, Science, 273, 779–781.

[5] Learner, R. C. M., Zhong, W., Haigh, J. D., Belmiloud, D., and Clarke, J. (1999) The contribution of the unknown weak water vapor lines to the absorption of solar radiation, Geophys. Res. Lett., 26(24), 3609–3612.

[6] Tso, H. C. W., Geldart, G. D. W., and Chylek, P. (1998) Anharmonicity and cross-section for absorption of radiation by water dimer, J. Chem. Phys., 108, 5319–5329.

[7] Pilewskie, P. and Valero, F. P. J. (1995) Direct observations of excess solar absorption by clouds, Science, 267, 1626–1629.

[8] Fu, Q., Lesins, G., Higgins, J., Charlock, T. P. C., and Michalsky, J. (1998) Broadband water vapor absorption of solar radiation tested using ARM data, Geophys. Res. Lett., 25, 1169.

[9] Schermaul, R., Learner, R. C. M., Newnham, D. A., Williams, R. G., Ballard, J., Zobov, N. F., Belmiloud, D., and Tennyson, J. (2001) The water vapor spectrum in the region 8600–15000 cm^{-1}: Experimental and theoretical studies for a new spectral line database: I. Laboratory measurements, J. Molec. Spectrosc., 208, 32–42.

[10] Coheur, P.-F., Fally, S., Carleer, M., Clerbaux, C., Colin, R., Jenouvrier, A., Merienne, M.-F., Hermans, C., and Vandaele, A. C. (2002) New water vapor line parameters in the 26000–13000 cm^{-1} region, J. Quant. Spectrosc. Radiat. Transfer, 74, 493–510.

[11] Chagas, J. C. S., Newnham, D. A., Smith, K. M., and Shine, K. P. (2001) Effects of improvements in near-infrared water vapour line intensities on short-wave atmospheric absorption, Geophys. Res. Lett., 28(12), 2401–2404.

[12] Bennartz, R. and Lohmann, U. (2001) Impact of improved near infrared water vapor line data on absorption of solar radiation in GCMS, Geophys. Res. Lett., 28(24), 4591–4594.

[13] Bauer, A., Godon, M., Carlier, J., and Ma, Q. (1995) Water vapor absorption in the atmospheric window at 239 GHz, J. Quant. Spectrosc. Radiat. Transfer, 53(4), 411–423.

[14] Cornier, J. G., Ciurylo, R., and Drummond, J. R. (2002) Cavity ringdown spectroscopy measurements of the infrared water vapor continuum, J. Chem. Phys., 116(3), 1030–1034.

[15] Kuhn, T., Bauer, A., Godon, M., Buhler, S., and Kunzi, K. (2002) Water vapor continuum: absorption measurements at 350 GHz and model calculations, J. Quant. Spectrosc. Radiat. Transfer, 74(5), 545–562.

[16] Carleer, M., Jenouvrier, A., Vandaele, A. C., Bernath, P. F., Merienne, M.-F., Colin, R., Zobov, N. F., Polyansky, O. L., Tennyson, J., and Savin, S. A. (1999) The near infrared, visible, and near ultraviolet overtone spectrum of water, J. Chem. Phys., 111(6), 2444–2450.

[17] Golovko, V. F. (2000) Dispersion formula and continuous absorption of water vapor, J. Quant. Spectrosc. Radiat. Transfer, 65, 621–644.

[18] Golovko, V. F. (2001) Continuous absorption of water vapor and a problem of the absorption enhancement in the humid atmosphere, J. Quant. Spectrosc. Radiat. Transfer, 69, 431–446.

[19] Tipping, R. H. and Ma, Q. (1995) Theory of the water continuum and validations, Atm. Res., 36, 69–94.

[20] Hale, G. M. and Querry, M. R. (1973) Optical constants of water in the 200 nm–200 μm wavelength region, Appl. Opt., 12, 555–563.

[21] Rothman, L. S., Barbe, A., Benner, C., Brown, L. R., Camy-Peyret, C., Carleer, M., Chance, K., Clerbaux, C., Dana, V., Devi, M., Fayt, A., Flaud, J.-M., Gamache, R. R., Goldman, A., Jaquemart, D., Jucks, K. W., Lafferty, W., Mandin, J.-Y., Massie, S. T., Newnham, D. (in preparation) The HITRAN Molecular Spectroscopic Database: 2000 Edition.

INFRARED SPECTRA
OF WEAKLY-BOUND COMPLEXES
AND COLLISION-INDUCED EFFECTS
INVOLVING ATMOSPHERIC MOLECULES

A. R. W. McKellar

Steacie Institute for Molecular Sciences, National Research Council of Canada, Ottawa, Ontario K1A OR6, Canada

Abstract   Some of the links and gaps between collision-induced spectra and weakly-bound van der Waals complexes are outlined. Even though the spectrum of a complex is in a sense the direct low temperature limit of the corresponding collision-induced spectrum, there is a gap because the spectrum of a dimer at 'real world' (atmospheric) temperatures often cannot easily be determined from its known low temperature (molecular beam) spectrum. The infrared spectra of a 'family' of complexes of atmospheric interest are described: the nitrogen dimer, the carbon monoxide dimer, and the nitrogen-carbon monoxide complex. Although these complexes are isoelectronic with similar intermolecular forces, their spectra are very different due to symmetry effects. The challenge of measuring the spectrum of the water vapor dimer at atmospheric temperatures is described.

Keywords: collision-induced spectra / weakly-bound complexes / dimers / infrared / water vapor / carbon monoxide / nitrogen / van der Waals / supersonic jet / long-path absorption

1. Introduction

Often in the field of molecular spectroscopy, the aim is to study *isolated* molecules which are as free as possible from external influences, so that the true molecular properties can most accurately be determined. This limit can be approached under equilibrium conditions in the gas phase by going to low pressures (to minimize the effects of collisions), and low temperatures (to minimize Doppler line broadening). Alternatively, even more effective isolation of molecules can often be achieved under non-equilibrium conditions by using molecular jets or beams.

223

C. Camy-Peyret and A.A. Vigasin (eds.),
Weakly Interacting Molecular Pairs: Unconventional Absorbers of Radiation in the Atmosphere, 223–232.
© 2003 *Kluwer Academic Publishers. Printed in the Netherlands.*

In nature, the ideal of molecular isolation is closely approached by the environment of interstellar space.

However, in many real world applications, and specifically in the earth's atmosphere, it is necessary to contend with molecular motion and intermolecular collisions, which give rise to such effects as pressure broadening, Doppler broadening, Dicke narrowing, line mixing, and collision-induced absorption (CIA). For the pressures encountered in the earth's atmosphere, it is often sufficient to consider only the effects of binary collisions, which are mediated by an intermolecular potential energy surface for the particular pair of molecules. But in addition to colliding pairs, it is also possible to have bound pairs of molecules, since virtually all intermolecular potentials support one or more bound states labelled by different rotational and vibrational quantum numbers. Such weakly-bound complexes, or *van der Waals molecules*, have been extensively studied in the laboratory during the past two decades, but their manifestations for the earth's atmosphere have traditionally not been widely considered. Such considerations are obviously one of the primary purposes of the present volume.

Some of the factors involved in 'real world' spectroscopy of atmospheric molecules are illustrated in Fig. 1. Almost everything depends ultimately on the intermolecular potential surface (and sometimes a dipole moment surface as well). But while the potential surface is crucial for isolated dimers, in practice it seems to be less relevant for CIA and pressure broadening calculations, where statistical mechanics problems still dominate. In other words, we can and do determine potentials from dimer spectra, but this has never actually been possible from CIA spectra. A distinction is made in Fig. 1 between isolated low temperature dimers (as usually studied in the laboratory) and 'room temperature' dimers (which are more relevant for the atmosphere).

2. Links between weakly-bound dimers and "normal" spectra

This paper will focus on the close link between weakly-bound complexes and real world collisional effects such as CIA and pressure broadening. But at the same time, it should be realized that binary (and larger) clusters are themselves ideally studied under isolated molecule (collision-free) conditions in molecular jets. An example of a truly isolated dimer is shown in Fig. 2, part of the infrared spectrum of H_2O–CO [1] with an effective rotational temperature of about $5\,\mathrm{K}$. This species is closely related to H_2O–N_2, which might be called one of the most significant atmospheric van der Waals molecules.

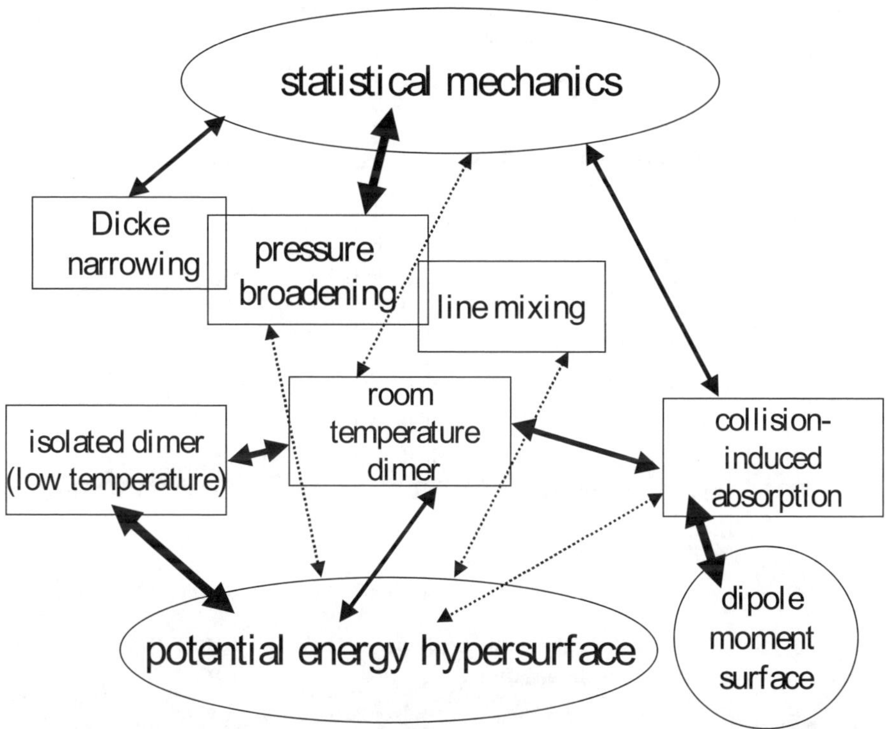

Figure 1. Schematic representation of some relations among factors affecting atmospheric spectra.

But although quite a bit is known about the spectra, dynamics, and structures of H_2O–CO and H_2O–N_2 [2], no link has been made between this knowledge and its possible atmospheric consequences. This relates to the gap in Fig. 1 between the 'isolated low temperature dimer' and the 'real world room temperature dimer'. No current experiment or calculation can reliably provide the spectrum of H_2O–N_2 under atmospheric conditions.

An example of the intimate link between dimer and CIA spectra is provided by the case of H_2–Ar. Part of the infrared spectrum [3] of a mixture of para-hydrogen and argon under conditions of 77 K temperature and 1.2 amagat density is shown in Fig. 3. We observe a characteristic broad ($\sim 80\,\mathrm{cm}^{-1}$) collision-induced line centered at the $S_1(0)$ transition frequency of H_2, together with a 'vibrational band' consisting of sharp lines due to bound H_2–Ar sitting on top of the CIA continuum. However, when we try to separate the CIA and dimer components, we run into an interesting problem! If we fit all of the smooth portion of the CIA line, then we predict (dashed curve) much more CIA intensity

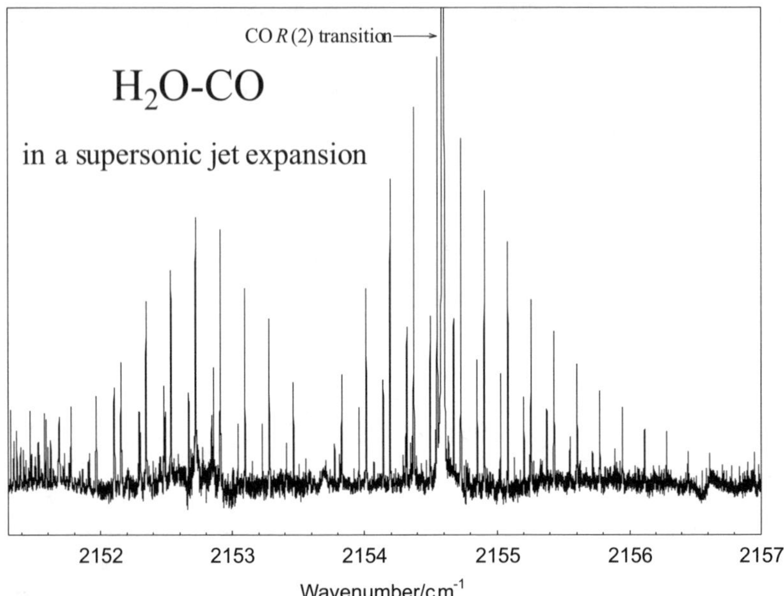

Figure 2.   Infrared spectrum of the weakly-bound complex H_2O–CO in the C-O stretching region [1]. An 'isolated' dimer.

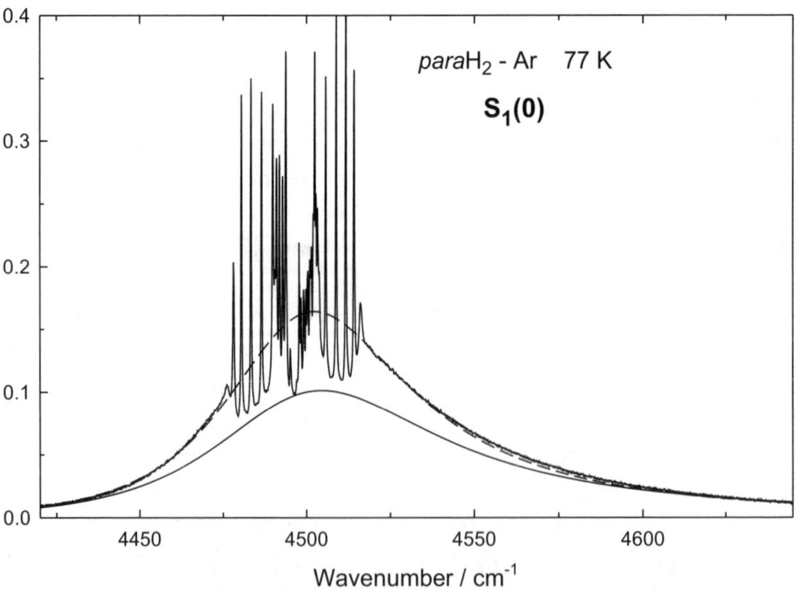

Figure 3.   The $S_1(0)$ line in the infrared spectrum of the weakly-bound complex *para*H_2–Ar [3]. The dashed and solid curves represent two attempts to fit the collision-induced background, but neither is satisfactory (see text).

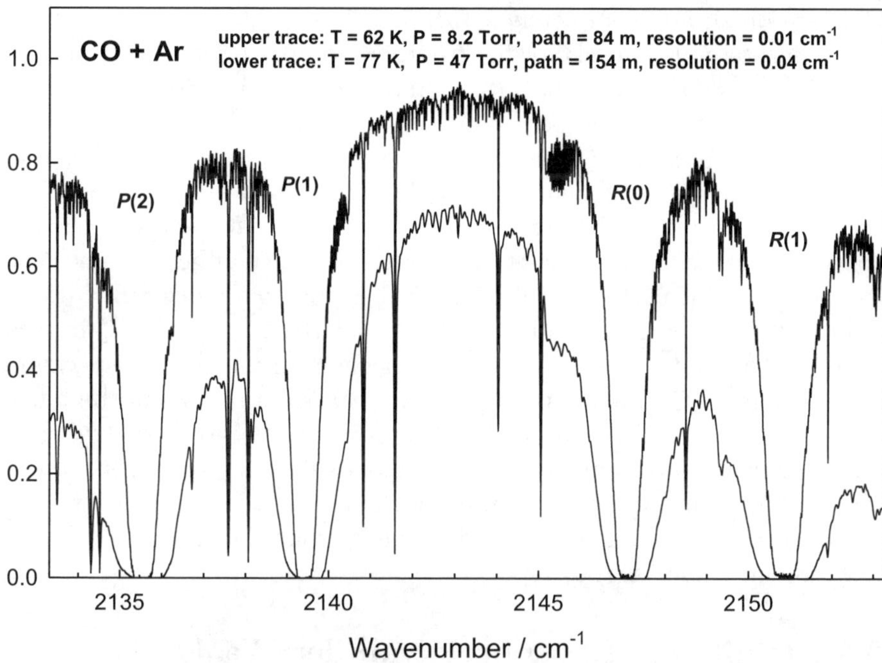

Figure 4.   Infrared spectra of long-path low temperature mixtures of CO and Ar, showing transitions of CO monomers together with those of CO–Ar van der Waals complexes.

near the line center than is observed. But if we constrain the CIA curve to equal the observed intensity near line center, then the prediction (solid curve) seriously underestimates the smooth CIA absorption at $\pm 25\,\mathrm{cm}^{-1}$ from line center. The problem is that the decomposition of the profile into bound-bound, bound-free, and free-free components is not at all obvious. Reasonably rigorous calculations of such effects have perhaps been done only by J. Schaefer [4], for the case of H_2–H_2.

Turning now to normal allowed spectra, as opposed to collision-induced spectra, the link with weakly-bound dimers is not quite as direct and dramatic. Nevertheless, there are significant connections, as illustrated in Fig. 4 which shows a small part of the infrared spectrum of carbon monoxide near the center of its fundamental band at $2143\,\mathrm{cm}^{-1}$. The experimental conditions are (moderately) close to 'real world' ones: small amounts of carbon monoxide (0.4 or 0.6 Torr partial pressure) mixed with argon to give total pressures of 8.2 or 47 Torr, and temperatures of 62 or 77 K. With the long absorption paths used here, the normal $^{12}C^{16}O$ rotation-vibration lines $P(2)$, $P(1)$, $R(0)$, and

$R(1)$ are broad and completely saturated. A few moderately strong but sharp lines are also observed due to other isotopes of carbon monoxide ($^{13}C^{16}O$, $^{12}C^{18}O$, etc.) in natural abundance. But, in addition, there is a veritable 'forest' of weaker sharp lines which are especially evident in the lower pressure (top) trace: these are all due to the weakly-bound van der Waals complex CO–Ar [5,6]. In the higher pressure (bottom) trace, the CO–Ar lines are themselves subject to pressure broadening, and so they start to merge with each other and become less evident. But they are still present: note for example the $K = 1 \leftarrow 0$ Q-branch between 2145 and 2146 cm^{-1} and the $K = 2 \leftarrow 1$ Q-branch near 2149.3 cm^{-1}. The separation or decomposition of this spectrum into monomer (CO) and dimer (CO–Ar) contributions would be much easier than in the CIA case of H$_2$–Ar above. But if we were trying to study pressure broadening of carbon monoxide lines by argon at low temperatures, and we were unaware of the presence of the CO–Ar complexes, then we would have certainly have trouble in trying to fit the CO monomer lines shapes.

3. A family of atmospheric van der Waals molecules, (N$_2$)$_2$, N$_2$–CO, and (CO)$_2$

Nitrogen is, of course, the dominant component of the earth's atmosphere, and carbon monoxide is an important trace constituent. These molecules are isoelectronic, and so some of their properties are rather similar (for example the normal boiling points are 77 K for N$_2$ and 82 K for CO). We can consider the 'family' of closely related van der Waals molecules, (N$_2$)$_2$,N$_2$–CO,, and (CO)$_2$, which are isoelectronic, have the same total and reduced masses (for the normal isotopes), and have roughly similar intermolecular potentials. There are important differences as well, namely symmetry and dipole moment properties, which have crucial effects on their spectra. Most importantly, N$_2$–CO and (CO)$_2$ have permanent dipole moments and significant vibrational transition moments, which give rise to allowed spectra in the millimeter wave and infrared regions, whereas (N$_2$)$_2$ has only a small induced dipole moment and consequently only very weak spectra.

During the past six years, both N$_2$–CO and (CO)$_2$ have been successfully studied by millimeter wave and infrared spectroscopy, and detailed knowledge has been gained regarding their lower energy levels. For example, with the CO dimer, we now have model-independent relative positions for 47 energy levels below 10 cm^{-1} with a precision of about 100 kHz (0.000003 cm^{-1}). Due to the difficulty of the necessary calculations, this spectroscopic information has not yet been translated

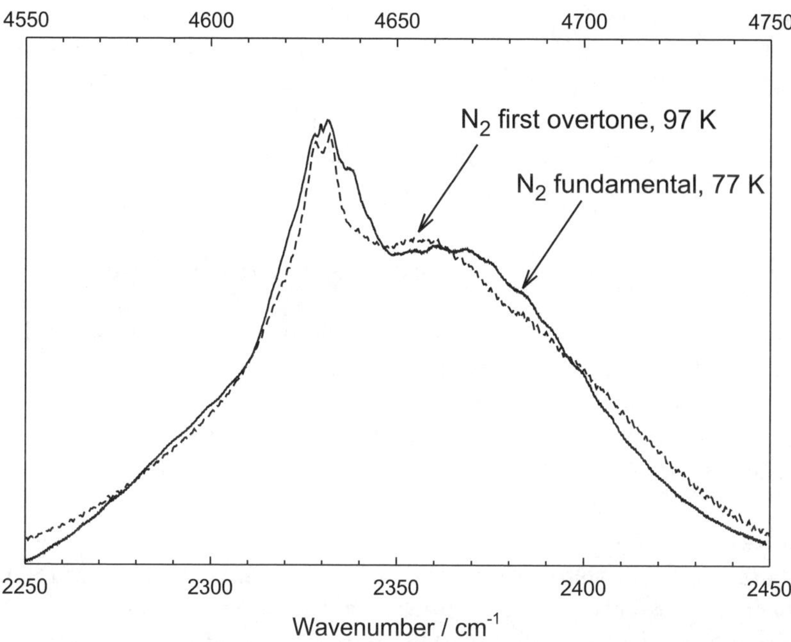

Figure 5. Low temperature collision-induced spectra of N$_2$ in the fundamental (solid line, lower scale) and first overtone (dashed line, upper scale) regions.

into improved intermolecular potentials or even definite structures, but such developments should only be a matter of time. The reader is referred to recent publications by Tang *et al.* [7] for (CO)$_2$ and Xia *et al.* [8] for N$_2$–CO, as well as to the references in these papers.

Due to the absence of allowed dipole transitions, the N$_2$ dimer can so far only be studied at moderately high pressures in a long-path absorption cell, conditions where we would hope to observe sharp dimer features sitting on top of a broad collision-induced spectrum, similar (we hope!) to the H$_2$–Ar result in Fig. 3. Unfortunately, the best existing result [9] shows only slight hints of sharp dimer structure. The problem is that (N$_2$)$_2$ likely has thousands of populated rotation-vibration levels at the temperature of the experiment, giving rise to tens of thousands (at least!) of transitions with significant intensity in the infrared spectrum. This myriad of lines is then hopelessly congested, with no hope of being individually resolved.

These observed low-temperature N$_2$ collision-induced spectra are illustrated in Fig. 5, where the shape of the fundamental band [9] ($v = 1 \leftarrow 0$, 2330 cm^{-1}) is compared with that of the (much weaker) overtone band [10] ($v = 2 \leftarrow 0$, $4630\,\mathrm{cm}^{-1}$) under similar conditions. Only the somewhat sharp features at the band center can be attributed to bound $(N_2)_2$ dimers, that is, for the fundamental, the features at 2327.3, 2329.5, and $2331.5\,\mathrm{cm}^{-1}$ (see also Fig. 3 of [9]). Accurate measurements of the N_2 fundamental CIA band at atmospheric temperatures have been reported by Lafferty *et al.* [11].

4. The water dimer, a spectroscopic challenge

Although $(N_2)_2$ might be considered the ultimate atmospheric dimer in terms of abundance, it is unlikely to have very much practical spectroscopic importance in the earth's atmosphere. Of greater potential significance is the water dimer, $(H_2O)_2$. The nature and importance of water collisional and dimer effects in the atmosphere have been the subject of a great deal of recent interest and controversy. This field is the subject of other contributions in this volume, and will not be covered here. As well, the infrared and microwave spectroscopy of $(H_2O)_2$ has been extensively studied under low temperature (supersonic jet) conditions (see for example, a recent paper by Goldman *et al.* [12] and references therein).

Here I would like to consider an important aspect of this problem, the challenge contained in the following question: *What is the infrared spectrum of water dimers at room temperature?* This is a meaningful and well-defined question, which could be answered experimentally, theoretically, or by a combination of approaches. Our already extensive knowledge of the dimer spectrum at low temperatures is helpful, but does not even begin to answer the challenge. Some calculations of room temperature water dimer spectra have already been reported, but (for the present author) they are not sufficiently reliable and detailed to answer the challenge.

One key component of our question is whether or not water dimers absorb significantly in spectral regions other than where water monomers absorb. If they do, then the dimers are obviously more likely to play a role in the earth's solar energy budget.

A direct experimental approach to our challenge is to fill a long path absorption cell with water vapor at various temperatures and pressures, and record the resulting infrared spectra. Since we already know the water monomer spectrum reasonably well (?!), it might then be possible to observe effects due to dimers, keeping in mind the difficulties discussed above in separating monomer and dimer contributions. One experimental problem is the limited vapor pressure of H_2O at room temperature (~ 0.02 atmospheres), which means that high pressure

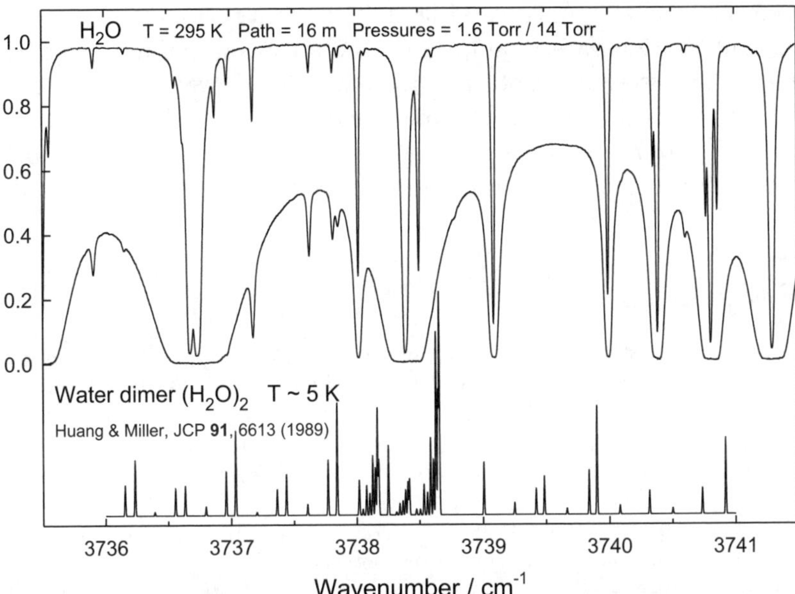

Figure 6.   Long path infrared spectra of water vapor at room temperature (top two traces) in the region where water dimer transitions (bottom trace) have been observed in low temperature supersonic jet experiments [13].

spectra have to be obtained at elevated temperatures. This complicates the sample handling part of the experiment, which is already difficult since water tends to adsorb on cell walls, windows, mirrors, etc.

As a small contribution to this challenge, I have tried to observe evidence of discrete room temperature water dimer spectral features in the far infrared (200–$800\,\mathrm{cm}^{-1}$) and mid-infrared (3600–$3900\,\mathrm{cm}^{-1}$) regions. But so far I have had no success! An example is shown in Fig. 6, which presents experimental long-path spectra of room temperature water vapor near $3740\,\mathrm{cm}^{-1}$ for pressures of 1.6 (top trace) and 14 (middle trace) Torr. In this spectral region, we know for sure that water dimers absorb, because well-resolved transitions have been observed under low temperature ($5\,\mathrm{K}$) supersonic jet conditions. These H_2O,dimer transitions are shown in the bottom trace of Fig. 6, which reproduces the spectrum observed by Huang and Miller [13].

Unfortunately, there are no features in these room temperature spectra that can be attributed to water dimers! Of course, we do not expect the exact $5\,\mathrm{K}$ spectrum to appear unchanged at $295\,\mathrm{K}$. But we might have hoped that here, near the origin of the dimer band, some features like a Q-branch might survive and make their presence known, much as in the CO–Ar case, Fig. 4. In reality, however, any water dimer contributions are swamped by the strong water monomer transitions. The awesome problem in linking the 'low temperature' and 'room temperature' spectra of dimers can be summed up in the term *partition function*. A dimer like $(H_2O)_2$ has an *enormous* number of occupied rotation-vibration quantum states at real world (atmospheric) temperatures!

References

[1] Brookes, M. D. and McKellar, A. R. W. (1998) Infrared spectrum of the water–carbon monoxide complex in the CO stretching region, J. Chem. Phys., 109, 5823–5829.

[2] Leung, H. O., Marshall, M. D., Suenram, R. D., and Lovas, F. J. (1989) Microwave spectrum and molecular structure of the N_2–H_2O complex, J. Chem. Phys., 90, 700–712.

[3] McKellar, A. R. W. (1998) High resolution infrared spectra of H_2–Ar, HD–Ar, and D_2–Ar van der Waals complexes between 160 and $8620\,cm^{-1}$, J. Chem. Phys., 105, 2628–2638.

[4] McKellar, A. R. W. and Schaefer, J. (1991) Far-infrared spectra of hydrogen dimers: Comparisons of experiment and theory for $(H_2)_2$ and $(D_2)_2$ at $20\,K$, J. Chem. Phys., 95, 3081–3091.

[5] McKellar, A. R. W., Zeng, Y. P., Sharpe, S. W., Wittig, C., and Beaudet, R. A. (1992) Infrared absorption spectroscopy of the CO–Ar complex, J. Mol. Spectrosc., 153, 475–485.

[6] Xu, Y. and McKellar, A. R. W. (1996) The infrared spectrum of the Ar–CO complex: comprehensive analysis including van der Waals stretching and bending states, Mol. Phys., 88, 859–874.

[7] Tang, J., McKellar, A. R. W., Surin, L. A., Fourzikov, D. N., Dumesh, B. S., and Winnewisser, G. (2002) Millimeter wave spectra of the CO dimer: Three new states and further evidence of distinct isomers, J. Mol. Spectrosc., 214, 87–93.

[8] Xia, C., McKellar A. R. W., and Xu, Y. (2000) Infrared spectrum of the CO–N_2 van der Waals complex: Assignments for CO–*para*N_2 and observation of a bending state for CO–*ortho*N_2, J. Chem. Phys., 113, 525–533.

[9] McKellar, A. R. W. (1990) Infrared spectra of the $(N_2)_2$ and N_2–Ar van der Waals molecules, J. Chem. Phys., 88, 4190–4196.

[10] McKellar, A. R. W. (1990) Low temperature infrared absorption of gaseous N_2 and $N_2 + H_2$ in the 2.0–2.5 μm region: application to the atmospheres of Titan and Triton, Icarus, 80, 361–369.

[11] Lafferty, W. J., Solodov, A. M., Weber, A., Olson, W. B., and Hartmann, J.-M. (1996) Infrared collision- induced absorption by N_2 near 4.3 μm for atmospheric applications: measurements and empirical modeling, Appl. Opt., 35, 5911–5917.

[12] Goldman, N., Fellers, R. S., Brown, M. G., Braly, L. B., Keoshian, C. J., Leforestier, C., and Saykally, R. J. (2002) Spectroscopic determination of the water dimer intermolecular potential-energy surface, J. Chem. Phys., 116, 10148–10163.

[13] Huang, Z. S. and Miller, R. E. (1989) High-resolution near infrared spectroscopy of water dimer, J. Chem. Phys., 91, 6613–6631.

THE FAR-INFRARED CONTINUUM
IN THE SPECTRUM OF WATER VAPOR

X. Wang, A. Senchuk, G. C. Tabisz

Department of Physics and Astronomy, University of Manitoba,
Winnipeg MB, R3T 2N2, Canada

Abstract   The origin of the continuum underlying the infrared spectrum of water vapor has been controversial. Measurements in the far infrared region for H_2O-N_2 mixtures are compared with the theory of Ma and Tipping which attributes the continuum to the cumulative effect of extremely far wings of individual spectral lines. Theory and experiment agree for the behavior of the shape of the continuum with increasing perturber density.

Keywords: water vapor continuum / atmospheric windows / spectral line wings / HITRAN database

1. Introduction

The far-infrared pure rotational spectrum of water vapor has been a subject of great interest for many years, both as a topic of fundamental study and in its importance to the atmospheric sciences. The recent intense interest in the Earth's radiation budget and global warming has placed great emphasis on the absorption characteristics of water vapor, since it is still the most important greenhouse gas despite the recent publicity attached to other gases such as carbon dioxide and methane. In addition, the operation of many communications and remote sensing systems in the infrared and millimeter- wave windows of the atmosphere requires the capability to predict the attenuation in these regions, at-tenuation that is mainly due to absorption from water vapor.

The water molecule is an asymmetric top belonging to the point group C_{2v}. The rotational energy level scheme in the ground vibrational state is obtained through application of the formula:

$$E_{J\sigma} = \frac{\hbar^2}{2}\left(S\,J\,(J+1) + W_\sigma\right),$$

233

C. Camy-Peyret and A.A. Vigasin (eds.),
Weakly Interacting Molecular Pairs: Unconventional Absorbers of Radiation in the Atmosphere, 233–237.
© 2003 *Kluwer Academic Publishers. Printed in the Netherlands.*

where J is the angular momentum quantum number, σ takes on the values $-J$, $-J+1,\ldots$, $J-1$, J and S depends of the three principle moments of inertia. The selection rules for J namely, $\Delta J = 0$, ± 1 and those of σ which are more complicated, yield a host of spectral lines in the far-infrared region.

Usually, the parameters for the spectral lines of atmospheric gases are known well enough to permit calculation of the absorption using line-by-line methods. However, the absorption by water vapor measured in the laboratory or in the atmosphere has usually been greater than that predicted on the basis of theoretical line shapes and theoretical or empirical intensities and widths, frequently by a factor of two or three.

This excess absorption has several characteristics that seem to hold in the infrared and millimeter regions of the spectrum. It takes the form of a continuum that stretches from $0.2\,\mathrm{cm}^{-1}$ to $10,000\,\mathrm{cm}^{-1}$, varying slowly with frequency. It has a quadratic pressure or density dependence and negative temperature dependence. It is much greater for self-broadening (pure water vapor) than for nitrogen broadening of the water vapor lines (mixtures of water vapor and nitrogen). The discrepancy between theory and experiment is greater in regions of weak absorption than in regions of strong or medium absorption.

There has been considerable disagreement as to the magnitude of this continuum and physical mechanisms responsible for it. A few of the proposed sources for the absorption are the extreme wings of absorption lines of water molecules or monomers [1, 2], H_2O–H_2O dimers [3, 4], ionic or uncharged clusters of water molecules and collision-induced effects [5, 6]. In 1990, Ma and Tipping [7–9] developed a unified theory of the far wings of spectral transitions based on the quasi-static formalism of Rosenkranz [10, 11] and offered strong evidence that the dominant source of continuum absorption is due to monomers. The lines are super-Lorentzian out to approximately $200\,\mathrm{cm}^{-1}$ from line center, after which they fall off rapidly with frequency. The quasi-static approximation is, however, not valid below $20\,\mathrm{cm}^{-1}$ from line center. By excluding contributions from line centers, one can calculate the total absorption coefficient due to the far wings.

2. Experimental considerations

The experiments were performed using a Michelson interferometer operating at a theoretical resolution of $0.06\,\mathrm{cm}^{-1}$. The sample was contained in a single pass stainless steel cell, one meter in length. The cell had a volume to surface ratio of 7.5 to reduce the rate of adsorption on the walls. Stainless steel was chosen to avoid chemical reactions between the sample and the walls of the cell. Provisions were made to allow the cell to be filled with either pure water vapor or mixtures of water vapor and nitrogen. The range of the experimental region was $80\,\mathrm{cm}^{-1}$ to $400\,\mathrm{cm}^{-1}$.

3. Continuum analysis

The HITRAN database parameters were used to calculate the line spectrum out to $\pm 25\,\mathrm{cm}^{-1}$ from the line centers. This calculated spectrum was then subtracted from the experimental one, and the result was numerically smoothed to obtain the measured continuum. The results of this procedure are displayed in Figs. 1 and 2 together with the theoretical results from the Ma-Tipping formalism for increasing nitrogen density. The spectra are placed on an arbitrary intensity scale because of uncertainties in the normalization of the experimental results. It is clear nevertheless that the behavior of the continuum found in experiment is predicted by the theory. In particular, the increased absorption at low frequencies (100–$250\,\mathrm{cm}^{-1}$) and the change in shape at high frequency as perturber density increases.

4. Concluding remarks

Initial results indicate a good agreement between the Ma-Tipping theory of water vapor continuum absorption and experiment with regard to the shape of the continuum with increasing perturber density. Further experiments are underway to refine the value of the absorption coefficient.

References

[1] Davies, R. W., Tipping, R. H., and Clough, S. A. (1982) Dipole autocorrelation function for molecular pressure broadening: A quantum theory which satisfies the fluctuation-dissipation theorem, Phys. Rev., 26, 3378–3394.

[2] Clough, S. A Kneizys, F. X. Davies, R. Gamache, R., and Tipping, R. H. (1980) Theoretical line shape for H_2O vapor: Application to the continuum. In: *Atmospheric Water Vapor*, edited by Deepak, A. Wilkerson, T. D., and Runke, L. H., Academic, New York.

[3] Penner, S. S. and Varanasi, P. (1967) Spectral absorption coefficients in the pure rotation spectrum of water vapor, J. Quant. Spectrosc. Radiat. Transfer, 7, 687–690.

[4] Varanasi, P. (1988) Infrared absorption by water vapor in the atmospheric window, Proc. Soc. Photo Opt. Instrum. Eng., 928, 213.

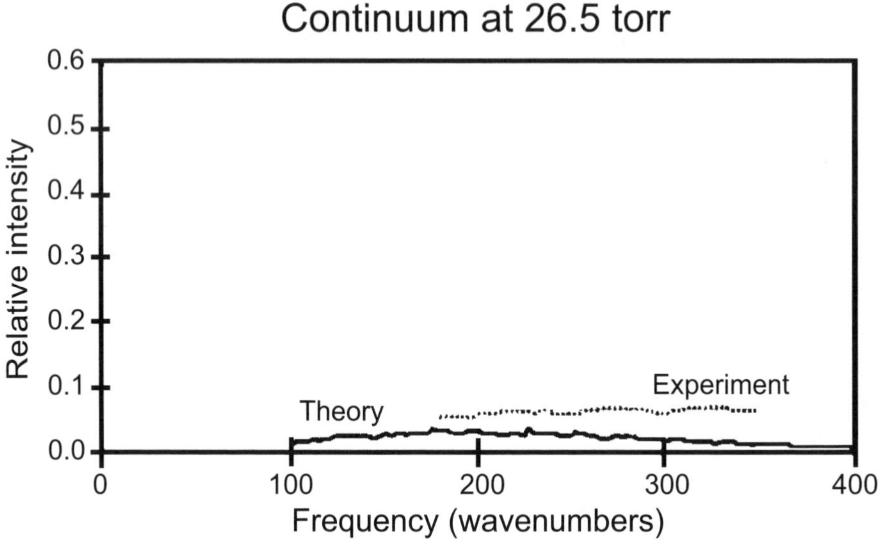

Figure 1. The H_2O–N_2 continuum in the far-infrared region. The intensity is in arbitrary units. The initial water vapor pressure is 0.5 torr and the total pressure is 26.5 torr.

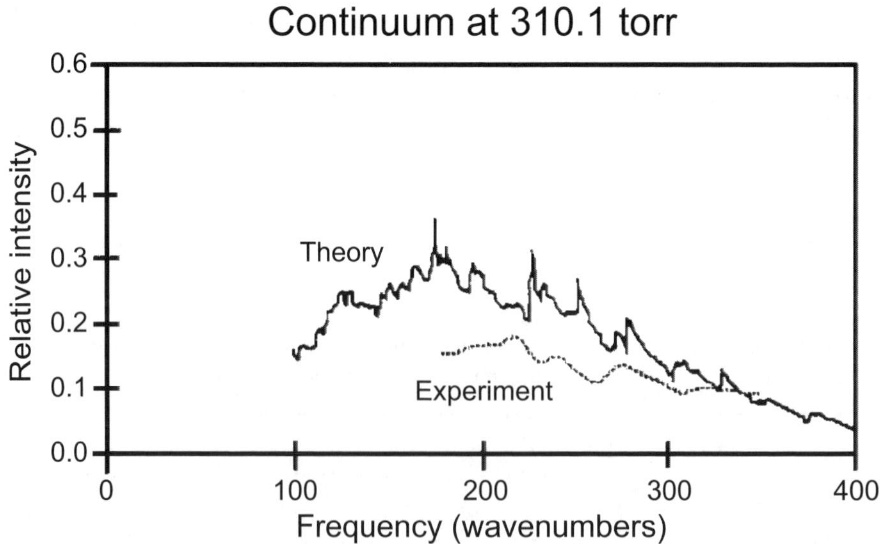

Figure 2. The H_2O–N_2 continuum in the far-infrared region. The intensity is in arbitrary units. The initial water vapor pressure is 0.5 torr and the total pressure is 310.1 torr.

[5] Birnbaum, G., editor. (1985) *Phenomena Induced by Intermolecular Interactions*, Plenum Press, New York.

[6] Tabisz, G. C. and Neuman, M. N., editors. (1995) *Collision- and Interaction-Induced Spectroscopy*, Kluwer Academic Publishers, Dordrecht, The Netherlands.

[7] Ma, Q. and Tipping, R. H. (1990) Water vapor continuum in the millimeter spectral region, J. Chem. Phys., 93, 6127–6139.

[8] Ma, Q. and Tipping, R. H. (1990) The atmospheric water continuum in the infrared: Extension of the statistical theory of Rosenkranz, J. Chem. Phys., 93, 7066–7075.

[9] Ma, Q. and Tipping, R. H. (2002) The frequency detuning correction and the asymmetry of line shapes: The far wings of H_2O–H_2O, J. Chem. Phys., 116, 4102–4115.

[10] Rosenkranz, P. W. (1985) Pressure broadening of rotational bands I: A statistical theory, J. Chem. Phys., 83, 6139–6144.

[11] Rosenkranz, P. W. (1987) Pressure broadening of rotational bands II: Water vapor from 300–1100 cm^{-1}, J. Chem. Phys., 87, 163–170.

RESONATOR SPECTROSCOPY AS A NEW METHOD OF INVESTIGATION OF UNCONVENTIONAL MILLIMETER-WAVE ATMOSPHERIC ABSORBERS

A. A. Shvetsov, M. Y. Tretyakov,
M. A. Koshelev, A. F. Krupnov, V. V. Parshin
Institute of Applied Physics, Russian Academy of Sciences,
46 Uljanova Street, Nizhnii Novgorod, 603950 GSP-120, Russia

Abstract   Principles of millimeter wave resonator spectroscopy of broad lines are presented. Block-diagram and parameters of high sensitive resonator spectrometer are described. Results of investigations of millimeter water and oxygen rotation lines in laboratory atmosphere demonstrate great potential of this technique for the measurements both broad atmospheric absorption lines and continuum absorption.

Keywords: millimeter wave absorption of water vapor and oxygen / resonator spectroscopy / millimeter waves / water dimer / rotational lineshape / collision-induced absorption spectra

1. Introduction

Water vapor and molecular oxygen are among main species in the Earth's atmosphere absorbing radiation in the millimeter and submillimeter wave ranges. They have strong rotational transitions over broad spectral region. In the gaps between these lines, called atmospheric windows, the absorption is due to the sum of the line wings and continuum. Conventional lineshapes, describing an isolated resonant transmission are the Van Vleck-Weisskopf (VVW) lineshape and the kinetic lineshape (Gross) rewritten by Zhevakin-Naumov (ZN). It is recognized, however, that the absorption far from the centers of lines, where the impact approximation fails, cannot be accounted for by the above lineshapes. Physical mechanisms leading to the formation of a continuum are still a matter of controversies. Several mechanisms have been proposed besides that of far line wings contribution. Excess absorption

239

C. Camy-Peyret and A.A. Vigasin (eds.),
Weakly Interacting Molecular Pairs: Unconventional Absorbers of Radiation in the Atmosphere, 239–246.
© 2003 *Kluwer Academic Publishers. Printed in the Netherlands.*

may arise from the water dimer H_2O–H_2O and/or from such weakly-bound molecular complexes as H_2O–N_2 and H_2O–O_2. It could also be caused by collision-induced absorption. The measurements of absorption were made in different atmospheric windows both in the field and in the laboratory environments, see for example [1]. The amount of measurements in the mmw and sub-mmw atmospheric windows is still very limited and no agreement is achieved as to the origin of anomalous temperature and water vapor pressure dependence. Our belief is that significant new perspectives in the solution of this problem could be offered through the use of broadband resonator spectrometry. This method is especially suitable for investigations of broad line shapes, in particular continuum absorption, under high and normal atmospheric pressure.

There are two main classes of resonator spectrometry, depending on the ratio of absorption linewidth to the cavity resonance width: (i) the line is narrower than the cavity resonance and (ii) the line is much broader. The first case includes high-resolution spectroscopy while the second one relates to highly pressure-broadened and atmospheric spectral lines as well as to absolute measurements of absorption at one fixed frequency. Our paper deals with the spectrometers belonging to the second class only.

2. A principle of the wide range microwave resonator spectroscopy

Resonator spectrometer consists of an open high quality Fabry-Perot resonator, precisely controlled and powerful source of radiation, and sensitive detector. Fabry-Perot resonator uses the fundamental $TEM(00q)$ mode, where q is the longitudinal mode number, i.e. the number of half-wavelengths between mirrors. The sample to be studied is placed inside the cavity. In case of atmospheric measurements the cavity is filled with a gas. The measurements of absorption in this version of resonator spectroscopy reduces to comparison of the quality factors Q characteristic to the empty and loaded resonator. The sensitivity of resonator spectroscopy increases with an increase in both the resonator quality factor and in the accuracy of the measuring the resonance width. The resonator quality factor Q is defined as the ratio of the frequency of the resonance f_0 to the width of the resonance (FWHM) Δf

$$\Delta f = \frac{c\,P_t}{2\,\pi\,L}, \tag{1}$$

where c is the velocity of light in a substance, L is the resonator length, and P_t stands for the total relative losses of radiation energy during one traversal in the resonator. We can evaluate the total resonator losses by measuring the width of the resonance curve. Total relative losses in the Fabry-Perot cavity P_t consist of

$$P_t = P_r + P_c + P_d + P_a. \tag{2}$$

The reflection P_r, coupling P_c, and diffraction P_d losses can be calculated or evaluated experimentally. The absorption coefficient γ and the atmospheric losses P_a are related by

$$P_a = 1 - e^{-\gamma L}. \tag{3}$$

Such measurement can be performed at any resonator mode under constant resonator length. If the resonator length is about some tens of a cm, the distance between two consecutive longitudinal modes correspond approximately to some hundreds MHz. The measurement of γ reduces to finding the resonance width. Techniques and instruments for such measurements are described below.

3. Experimental set-up

The block-diagram and parameters of the real broad band resonator spectrometer developed at the Institute of Applied Physics RAS [2] are shown in Fig. 1. The Fabry-Perot resonator used the fundamental TEM$(00q)$ mode. The quality factor of a Fabry-Perot resonator having 25–42 cm length, spherical silver-plated mirrors 12 cm in diameter and 24 cm in curvature radius, coupled with a source and detector by 6-mm Teflon film placed at $45°$ to the resonator axis, is determined by unavoidable reflection losses and equals approximately 600 000. Expressions for all the losses are known more or less accurately. Diffraction losses can be calculated as a fraction of total energy accumulated inside the resonator which is cut by the resonator mirrors' aperture and can be done much smaller than all other losses; coupling losses and reflection losses can be calculated from the coupling film parameters and from the dc conductivity of bulk silver, respectively.

The synthesized frequency radiation source employs a backward wave oscillator (BWO) [3] which is stabilized by a phase lock-in loop (PLL) with the use of two reference synthesizers in this case: one microwave (MW) synthesizer (8–12 GHz) defining the central frequency of the BWO, and one fast radio frequency (RF) synthesizer (20–40 MHz)

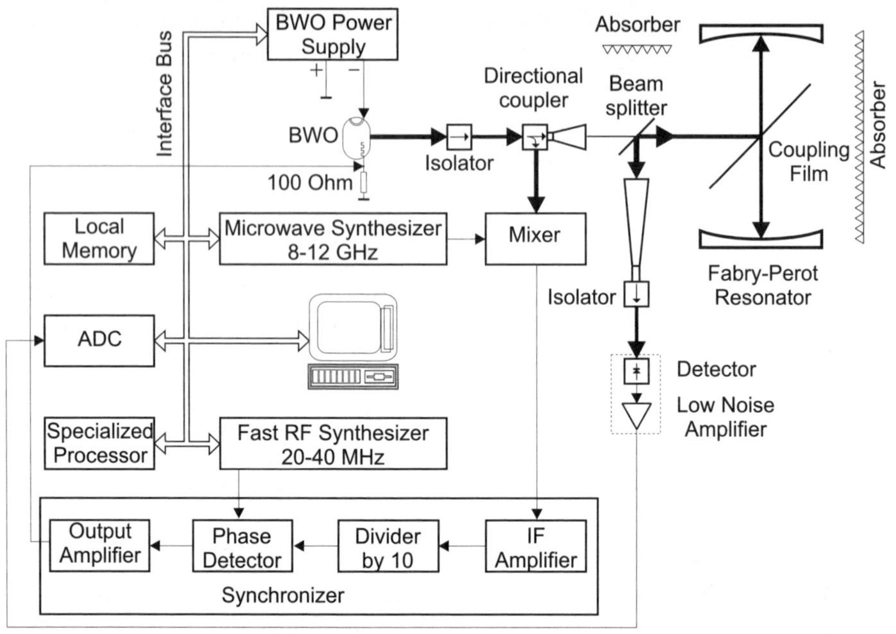

Figure 1.   Block-diagram of the millimeter-wave broad band resonator spectrometer.

for precision fast scanning of the BWO frequency around the chosen central frequency. The radio frequency synthesizer provides frequency scanning without loss of the phase of oscillations (without phase jumps). Both synthesizers are computer-controlled. As a result, the BWO frequency is defined as

$$f_{\mathrm{BWO}} = n\,f_{\mathrm{MW}} - 10\,f_{\mathrm{RF}}, \qquad (4)$$

where n varied from 4 to 20, and a factor of 10 before f_{RF} appears because phase detection was done at 10 times the digitally divided intermediate frequency (IF), which was 350 MHz. The main source of uncertainty in the measurement of the resonance width is the drift of the central frequency of a resonance during the time of measurement. To minimize this error one has to measure the resonance curve as fast as physically possible. Response time of the resonator itself equals $\tau \sim 1/\pi\Delta f$. In our case it amounts $\sim 2\,\mathrm{ms}$. For precision measurement the observation time should be increased, say, 10 times, i.e. up to $\geq 20\,\mathrm{ms}$. The microwave and the millimeter-wave synthesizers commonly used in spectroscopy employ indirect frequency synthesis and have therefore $\sim (10\text{--}50)\,\mathrm{ms}$ switching time, thus preventing fast scanning of the resonance curve. A fast direct radio frequency synthesizer with switching time 200 ns and time between switching 58 ms was used in this work as a source of a reference signal for the phase detector in the lock-in loop. Thus both precision and fast scanning of the BWO radiation frequency within 200 MHz around the central frequency defined by the microwave frequency synthesizer were reached. Scanning without loss of a phase permits the physical limits of the resonance observation time to be approached and reduces the source phase noise. The passed-through resonator radiation was received then by a low-barrier Schottky diode detector. The precision frequency control, signal acquisition, and processing were done by a computer. The results of each scan were recorded and processed separately. The automation system consisted of an IBM PC and a module containing an RF synthesizer and a data acquisition system. The main part of the module is a microprocessor with external memory. The RF synthesizer is based on direct digital synthesizer (DDS) and is able to generate a harmonic signal in the 20–40 MHz range with 0.03 Hz discreteness and without phase jumps at the switching.

A microprocessor controls the frequency of the synthesizer and synchronizes the data-acquisition process with the frequency steps. The frequency was changed by a triangle law, meaning forward and backward scan. The maximum number of points per one scan was 512. At each frequency point, the microprocessor collected the data several times from the ADC, averaged the obtained results, and stored the average into data memory. In such a way 32 of the 512 point triangle scans may be put into a local memory. The data were transferred then to a PC for further processing.

4. Procedure and results of the measurements

The basic procedure in resonator spectroscopy is to measure the width of the resonance. The experimentally observed resonant curve of the Fabry-Perot resonator at 85 GHz is presented in Fig. 2, as an example. The curve is a combination of 500 scans with a duration of 30 ms each, i.e. corresponding to the total averaging time 15 s. Each fast scan was processed separately, then resonant curves were combined so that their centers coincided to allow for obtaining the averaged curve in Fig. 2. The residual of the fit, shown in the lower part of the figure, indicates the adequacy of the fitting model. The increased noise on the line slopes corresponds to transformation of the phase noise of radiation into the amplitude noise. The width of the resonance (FWHM) was defined then as $\Delta f = 164728(20)$ Hz. The actually achieved accuracy of the resonance width measurements is 20 Hz. It corresponds to

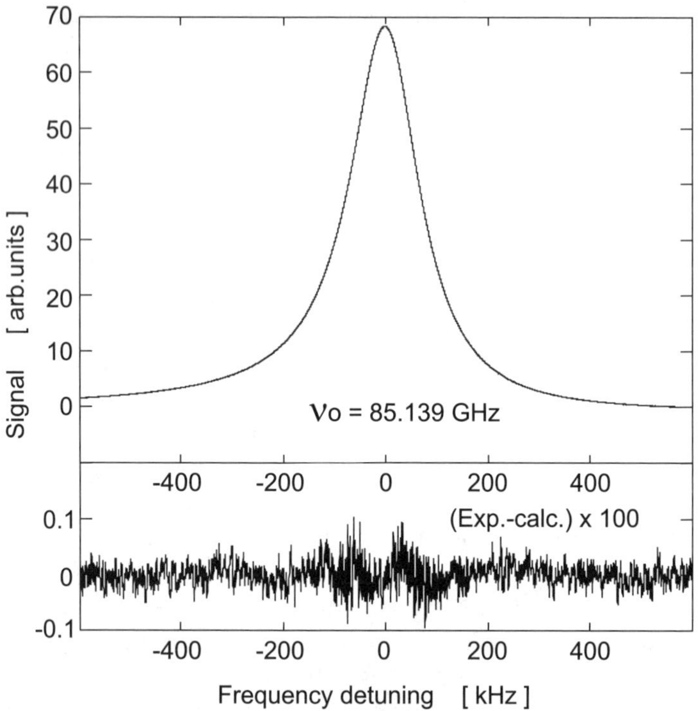

Figure 2.   Observed Fabry-Perot resonator curve and its fitting to Lorentzian profile.

$4 \times 10^{-9}\,\mathrm{cm}^{-1}$ or 0.002 dB/km sensitivity limit (in absorption coefficient) of the spectrometer that exceeds by an order of magnitude the
values characteristic to previously known analogous schemes [4].

The capabilities of the spectrometer were demonstrated with in *in
situ* analysis of the laboratory air. Scans covering 44–98 GHz and 113–
200 GHz were recorded in experiment for each range. The results are
shown in Fig. 3. The spectra shown are the absorption band in atmospheric oxygen produced by magnetic dipole transitions between fine
structure rotational energy levels [6] and [5], and electric dipole transition between rotational energy levels of water [2]. Broad scanning
in these experiments was produced by jumping from one longitudinal
mode to another without mechanical tuning of the resonator. The circles in Fig. 3 represent experimental points (after subtraction of the
calculated resonator losses); solid line in the low frequency range corresponds to calculation according to [4]; residuals of the fit multiplied
by 10 is presented at the bottom. The solid line in the higher frequency
range is the Van Vleck-Weisskopf profile with an addition of the linear and quadratic frequency terms fitted to the experimental points to

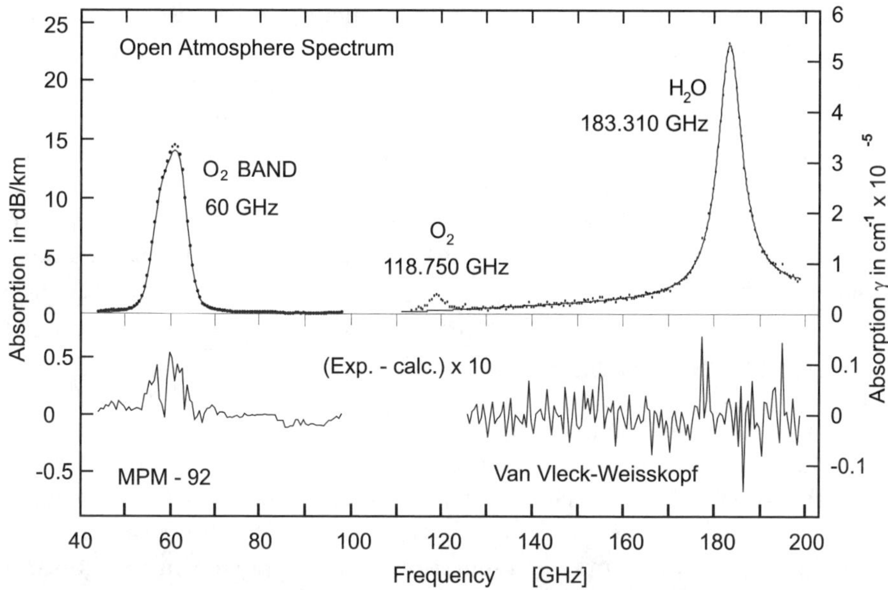

Figure 3.   Experimental and calculated spectra of laboratory atmosphere.

account for the possible inaccuracy of the calculated resonator losses, nonresonance absorption in the air filling the resonator, and the wings of strong higher frequency water lines.

The departure of the calculated profile from the observed one is clearly seen in the residuals in Fig. 3 in the range of atmospheric oxygen band absorption. The nature of these regular deviations is not obvious as for now. It was found that the 183 GHz water line profile at atmospheric pressure fits the Van Vleck-Weisskopf shape with experimental accuracy over twenty halfwidths down and six halfwidths up from the line center. The measured integrated absorption coefficient for the water line coincides with the calculated one within 2%. Dry air broadening parameter for this line was found equal 3.985(40) MHz/Torr. This result is more accurate than those previously reported in the literature. First measurement of the 119 GHz oxygen line allowed for obtaining the air broadening parameter of 2.14(9) MHz/Torr. Within 10 halfwidths the line profile can be fitted nicely by the VVW lineshape [5].

5. Conclusion

Millimeter-wave resonator spectrometer have the highest presently available sensitivity among similar devices operating in that region.

The spectrometer allows for very wide frequency tuning range provided the BWO is used a radiation source. Similar techniques apply to the frequency range in excess of 1 THz and the submillimeter-wave resonator spectrometer can be constructed soon. Such instruments could be very useful for investigations of millimeter and submillimeter lines of conventional molecules under elevated pressure. These may play an important role in the studies of broad spectral features appropriate to unconventional atmospheric absorbers like dimers and other weakly-bound complexes.

Acknowledgments

The authors are thankful to Svetlana Tretyakova for her assistance in producing the camera-ready typescript and to the organizer of the NATO Advanced Research workshop "Weakly Interacting Molecular Pairs: Unconventional Absorbers of Radiation in the Atmosphere" for their interest to our work. Present work was supported in part by the Russian Foundation for Basic Research (RFBR) Grant and the Contract from the Ministry of Industry, Science and Technology of Russian Federation. The authors express their deep gratitude to these sources of support.

References

[1] Bauer, A., Godon, M., Carlier, J., and Ma, Q. (1995) Water vapor absorption in the atmospheric window at 239 GHz, J. Quant. Spectrosc. Radiat.Transfer, 53, 4, 307–318.

[2] Krupnov, A. F., Tretyakov, M. Yu., Parshin, V. V., Shanin, V. N., and Myasnikova, S. E. (2000) Modern millimeter-wave resonator spectroscopy of broad lines, J. Mol. Spectrosc, 202, 107–115.

[3] Golant, M. B., Vilenskaya, R. L., Zyulina, E. A., Kaplun, Z. F., Negirev, A. A., Parilov, V. A., Rebrova, E. B., and Saveliev, V. S. (1965) A series of low power broad-band oscillators operating in millimeter and submillimeter wave ranges, Pribory i Technika Eksperimenta, 4, 136–139 (in Russian, available in English as Sovjet Physics-Pribory).

[4] Liebe, H. J., Rosenkranz, P. W., and Hufford, G. A. (1992) Atmospheric 60-GHz oxygen spectrum: new laboratory measurements and line parameters, J. Quant. Spectrosc.Radiat. Transfer, 48, 629–643.

[5] Tretyakov, M. Yu., Parshin V. V., Shanin, V. N., Myasnikova, S. E., Koshelev, M. A., and Krupnov, A. F. (2001) Real atmosphere laboratory measurements of the 118.75 GHz oxygen line: Shape, shift and broadening of the line, J. Mol. Spectrosc, 208, 110–112.

[6] Krupnov, A. F., Tretakov, M. Yu., Parshin, V. V., Shanin, V. N., and Kirilov, M. I. (1999) Precision resonator microwave spectroscopy in millimeter and submillimeter range, International Journal of Infrared and Millimeter Waves, 20(10), 1731–1737.

III

ATMOSPHERIC APPLICATIONS

LOOK-UP TABLE
AND INTERPOLATION METHODS
FOR RADIATIVE TRANSFER CALCULATIONS
IN THE INFRARED APPLICATION
TO BALLOON AND SATELLITE SPECTRA

M. Eremenko*, S. Payan,
Y. Té, G. Dufour, V. Ferreira, P. Jeseck, C. Camy-Peyret
Laboratoire de Physique Moléculaire et Applications,
Université Pierre et Marie Curie & CNRS, Paris, France

A. Mitsel
Tomsk State University of Automated Control Systems and Radioelectronics,
Tomsk, Russia

Abstract   Remote sensing of atmospheric state parameters (such as temperature
and mixing ratio profiles, trace gas column amounts) requires fast and
accurate forward model radiative transfer calculations. Application
of commonly used line-by-line (LBL) models to calculate atmospheric
spectra and simulate actual measurements provides high accuracy but
is often too much time consuming. One way to decrease the calculation time is to use pre-calculated absorption cross sections performing
a satisfactory accuracy on an optimized grid of wavenumber, temperature and pressure. This pre-calculated database method (known as
Look-up table method) is becoming more and more widely used for
processing spectra of satellite-borne instruments like IMG and ILAS,
of the balloon-borne instruments like LPMA (Limb Profile Monitor of
the Atmosphere), and of future satellite instruments (IASI, ACE, etc.).
However, the use of Look-up table on a wide wavenumber interval and
for a large number of trace gases is efficient when a good interpolation
scheme and an appropriate storage of the data are chosen. Comparison
of data processing results obtained using Look-up table technique with
results obtained by direct line-by-line LPMA algorithm is presented
and discussed. Retrievals of CFC-12 and CO_2 from solar occultation
limb FTIR balloon spectra are exemplified.

*Permanent address: Tomsk State University of Automated Control Systems and Radioelectronics,
Tomsk, Russia

249

C. Camy-Peyret and A.A. Vigasin (eds.),
Weakly Interacting Molecular Pairs: Unconventional Absorbers of Radiation in the Atmosphere, 249–258.
© 2003 *Kluwer Academic Publishers. Printed in the Netherlands.*

1. Introduction

The retrieval of atmospheric state parameters requires sufficiently high level of both accuracy and efficiency of the radiative transfer algorithms. While various line-by-line (LBL) models exist to calculate accurately IR radiances, they are slow when a large number of spectra have to be analyzed (satellite or balloon measurements). It is possible to pre-calculate absorption cross-sections (ACS) and store them in a set of files known as Look-up tables for a given grid of wavenumber, temperature and pressure. ACS can then be quickly derived for any wavenumber, temperature and pressure by interpolation. However, this interpolation must be performed with no significant loss of accuracy. The choice of an optimal interpolation scheme and data storage form is thus of first importance and must be carefully study.

2. Look-up table description

The ACS database is a three dimensional table: wavenumber, temperature and pressure — ACS (σ, T, P). The spectral range as well as the spectral step is depending on the species and pressure range. Indeed, a given species, absorbing in a finite spectral domain around electronic and/or ro-vibrational transitions energies, can have continuous or discrete absorption depending on the corresponding set of lines with variable intensities and width. The form of our data storage is binary file with records. Each record corresponds to ACS (σ, T_i, P_i) for a given wavenumber and contains all the information needed to interpolate ACS for any value of P and T. The use of a fast index search allows to access quickly the data needed for the calculation. To define ranges by temperature and by pressure, a documented set of real atmospheric soundings (from satellite and balloon profiles) was analyzed. The selected ranges to cover 0–120 km atmospheric altitude range are presented below.

Before calculating ACS for each temperature and associated pressure range, it is necessary to define a pressure and temperature step. These steps are depending on the interpolation scheme. The steps in pressure and temperature are a critical point determining the size of the final database and interpolation accuracy.

Temperature grid is linear ($T = 150$ to $380\,$K). Pressure grid (P is in hPa) is Napierian logarithmic ($\ln P = -11$ to 7).

Table 1. Nominal temperature and pressure domains for Look-up table generation.

	Minimum value	Maximum value
Temperature	158 K	380 K
Pressure	2×10^{-5} hPa	1026 hPa

2.1. Logarithmic scheme

To interpolate the $f(x)$ value by two tabulated values $f(x_i)$ and $f(x_{i+1})$ the logarithmic scheme uses the following equation:

$$\ln f(x) = \ln f(x_i) - \frac{x - x_{i+1}}{H}, \quad H = \frac{x_i - x_{i+1}}{\ln f(x_{i+1})/f(x_i)}. \tag{1}$$

For a two-dimensional function, the corresponding interpolation used is:

$$\ln f(x, y) = (1 - A)(1 - B)\ln f(x_i, y_j) + (1 - A)B\ln f(x_{i+1}, y_j)$$
$$+ A(1 - B)\ln f(x_i, y_{j+1}) + AB\ln f(x_{i+1}, y_{j+1}), \tag{2}$$

where

$$A = \frac{x - x_i}{x_{i+1} - x_i}, \quad B = \frac{x - x_i}{x_{i+1} - x_i}. \tag{3}$$

To study the logarithmic scheme, several temperature and pressure steps have been used and tested in term of impact on the relative interpolation error. Using the logarithmic interpolation scheme, a good (error on the integrated ACS less than 1%, see Section 3) compromise between results accuracy and database size is obtained with the following steps:

- 10 K for temperature: 25 grid points;

- 0.25 for logarithm of pressure: 73 grid points.

Thus, the total number of ACS needed in a database is $25 \times 73 = 1825$ by wavenumber and by species. Each value of ACS in simple precision format uses four memory bytes. It means that for one wavenumber and a given species, 7300 bytes are necessary to interpolate ACS for any temperature and any pressure.

It can be seen that logarithmic scheme generates a large database size. However, the size could be reduced by using the Singular Value Decomposition (SVD) method as describe below.

Table 2. Look-up table properties for the logarithmic scheme interpolation.

	Step by T	Grid points, T	Step by $\ln P$	Grid points, $\ln P$	Number of PT sets	Bytes for one wavenumber
Logarithmic scheme	10	25	0.25	73	1825	7300

2.2. Singular Value Decomposition method

A Singular Value Decomposition (SVD) can be used to transform calculated Look-up tables into a compressed representation. We tested this approach in order to decrease the size storage of our Look-up tables. SVD is a well-known technique [1] for orthogonal decomposition of data sets. The initial Look-up table matrix A could be represented as:

$$A = U\Sigma V^T, \tag{4}$$

where $A[m \times n]$ is initial matrix, $U[m \times n]$ and $V[n \times n]$ are orthonormal matrices, $\Sigma[n \times n]$ is a diagonal matrix, m and n are the number of temperature and pressure, respectively.

The diagonal elements of Σ are named the singular values of A and they are ordered from the largest to the smallest. If many of the singular values are small, we can suppose that only the L first singular values are significant and we can drop the rest. The size of matrices is thus decreased. U becomes $[m \times L]$, Σ becomes $[L \times L]$, and V becomes $[n \times L]$. For a small value of L the decomposition requires smaller storage size than A itself. For optimization, the line number m has been chosen to be the succession of temperatures for 3 wavenumbers and A is the stack of the Look-up tables for 3 wavenumbers

$$A = \begin{vmatrix} (T_1,P_1) \ \dots \ (T_1,P_n) \\ \dots\dots\dots\dots\dots \\ (T_m,P_1)\dots(T_m,P_n) \\ (T_1,P_1) \ \dots \ (T_1,P_n) \\ \dots\dots\dots\dots\dots \\ (T_m,P_1)\dots(T_m,P_n) \\ (T_1,P_1) \ \dots \ (T_1,P_n) \\ \dots\dots\dots\dots\dots \\ (T_m,P_1)\dots(T_m,P_n) \end{vmatrix} \begin{matrix} \left.\vphantom{\begin{matrix}a\\b\\c\end{matrix}}\right\}\sigma_1 \\ \left.\vphantom{\begin{matrix}a\\b\\c\end{matrix}}\right\}\sigma_2 \\ \left.\vphantom{\begin{matrix}a\\b\\c\end{matrix}}\right\}\sigma_3 \end{matrix} \ . \tag{5}$$

Table 3. Truncated Look-up table properties (after a singular value decomposition in logarithmic scheme).

Species	Selected number of singular values	Bytes for one wavenumber
O_3, CFC-12	5	654
HNO_3, CH_4, N_2O, CO_2	6	784
H_2O	7	915

As a result, the table to be truncated has the dimension $[75 \times 75]$, and the total number of singular values is then 75. The minimum number of singular values varies between 5 and 7. This number depends on the species and on the spectral range being considered. For CFC-12 in $920\,\mathrm{cm}^{-1}$ region it equals 5, for CO_2 in $1900\,\mathrm{cm}^{-1}$ region this number is 6. SVD truncated Look-up table properties are summarized in Table 3. The application of SVD to Look-up table allows decreasing database at least by a factor 8.

3. Comparison of Look-up tables interpolation and Line-by-line calculations

In order to test our Look-up tables, we first compare (for a representative set of temperature and pressure) ACS calculation with LBL and Look-up tables interpolation. Calculations have been compared in term of integrated ACS:

$$I_{\mathrm{ACS}} = \int_\sigma \mathrm{ACS}(\sigma)\, d\sigma \tag{6}$$

and the relative error is defined as:

$$\mathrm{ERR}_{\mathrm{ACS}} = 100 \times (I_{\mathrm{LBL}} - I_{\mathrm{LUT}})/I_{\mathrm{LBL}},$$

where: $I_{\mathrm{LBL}} = I_{\mathrm{ACS}}$ for LBL calculation $I_{\mathrm{LUT}} = I_{\mathrm{ACS}}$ for Look-up tables.

The second test has been made on the retrieved vertical column amounts from atmospheric spectra recorded by the LPMA balloon-borne FTIR experiment [2]. The current LBL LPMA retrieval algorithm and the new Look-up tables retrieval algorithm have been used for these calculations. The relative difference of column amounts retrieved using LBL algorithm and using Look-up tables were calculated:

$$\mathrm{ERR}_{\mathrm{col}} = \frac{\sqrt{\Sigma\,(\mathrm{col}_{\mathrm{LBL}} - \mathrm{col}_{\mathrm{LUT}})^2}}{N}, \tag{7}$$

where: col_{LBL} = column amount, obtained using LBL calculation, col_{LUT} = column amount, obtained using Look-up tables calculation, N = number of retrieved column amounts. Interpolation schemes were tested in several spectral regions (and for two LPMA flights, see Fig. 1 and Fig. 2).

- FLIGHT LPMA09: spectral region: 1289.618–1295.378 cm^{-1}, species: HNO_3, CH_4, and N_2O.

  As an example, Fig. 1 presents spectra in the selected microwindows. Observed and calculated spectra are plotted. It can be seen that simulation of the observed spectra are very close using LBL or Look-up tables.

  As it can be seen in Tables 4 and 5, the difference introduced by Look-up tables in the column retrieval is small compared to column retrieval accuracy from LPMA measurements (between 5 and 20% depending on the species). The maximum relative difference of 1% is found for N_2O. These differences are smaller than the accuracy of the line intensities parameters. Thus, the use of Look-up tables produces acceptable inversion compare to LBL.

- FLIGHT LPMA13: spectral region: 3037.751–3043.130 cm^{-1}, species: H_2O, O_3 and CH_4.

  The set of comparisons between retrieval using LBL and using Look-up tables (see Fig. 2 and Tables 6 and 7) indicates that Look-up tables introduce errors less than 1% on the retrieved column with a significant gain in term of calculation time. The new inversion algorithm can then be successfully used for operational analysis of a large number of atmospheric spectra.

4. Calculation time

The radiative transfer calculation to simulate observed spectra needs as input the calculation of ACS for each atmospheric layer crossed by the observational line of sight (i. e. temperature and pressure in the layer), for all species having significant absorption contribution. This calculation is time-consuming using direct LBL calculations (strongly dependent on the number of absorbing lines in the given spectral interval). Using the LBL algorithm, this calculation had to be done for each spectrum analysis because the associated line of sight is different from a spectrum to an other one. Using Look-up tables, calculation time does not depend on the spectral region. If the generation of Look-up tables is time consuming, the calculation is done only one time, and resulting table can be used to process any measurements. To make a

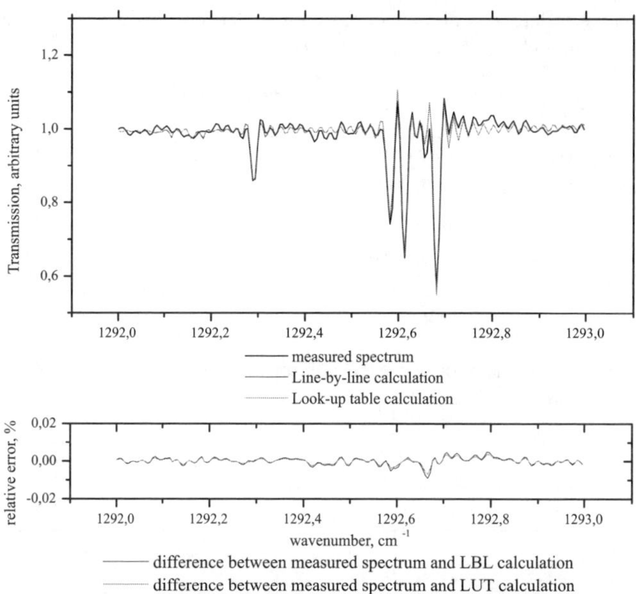

Figure 1.   Comparison of measured spectrum and simulations using LBL and Look-up tables and relative differences (for flight LPMA09, spectrum recorded at float, balloon altitude = 31.9 km).

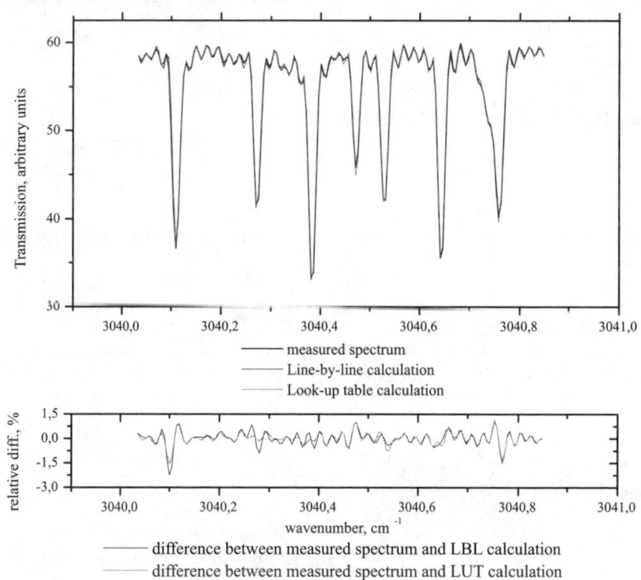

Figure 2.   Comparison of spectrum calculated using LBL and Look-up tables and relative differences (for flight LPMA13, spectrum recorded during ascent, balloon altitude = 14.7 km).

Table 4. Relative differences of wavenumber integrated cross sections between calculation using LBL algorithm and calculation using Look-up tables.

	HNO$_3$	CH$_4$	N$_2$O
Relative difference (%) in ACS, first layer ($T = 195.77$ K, $P = 28.21$ hPa)	0.06	0.24	0.25
Relative difference (%) in ACS, last layer ($T = 314.93$ K, $P = 0.3836 \times 10^{-4}$ hPa)	0.01	0.02	0.02

Table 5. Relative differences of retrieved column amounts obtained using LBL algorithm and using Look-up tables.

	HNO$_3$	CH$_4$	N$_2$O
Relative difference (%) in column amounts	0.10	0.08	1.02

Table 6. Relative differences of wavenumber integrated absorption cross-sections between calculation using LBL algorithm and calculation using Look-up tables.

	H$_2$O	O$_3$	CH$_4$
Relative difference (%) in ACS, first layer ($T = 228.13$ K, $P = 125.4$ hPa)	0.35	0.20	0.34
Relative difference (%) in ACS, last layer ($T = 348.42$ K, $P = 3.018 \ 10^{-5}$ hPa)	0.08	0.09	0.002

Table 7. Relative differences of retrieved column amounts obtained using LBL algorithm and using Look-up tables.

	O$_3$
Relative difference (%) in column amounts	0.08

Table 8. Calculation time summary.

LPMA flight	Time of measurement processing, min	
	Line-by-line	Look-up table
LPMA09	50:00	00:42
LPMA13	04:13	00:15

comparison of calculation time, two LPMA flights were processed using the two techniques: LPMA09 — measurements in spectral interval 1289.618–1295.378 cm^{-1}, absorbing species are HNO_3, CH_4, and N_2O, number of atmospheric layers: 44; LPMA13 — measurements were in spectral interval 3037.751–3043.130 cm^{-1}, absorbing species are H_2O, O_3, and CH_4, number of atmospheric layers: 50. The use of Look-up tables allows decreasing of the calculation time by a factor of 70 in the case of LPMA09 measurements processing, and by a factor of 16 for LPMA13 measurements processing (see Table 8).

5. Application of wide spectral region analysis

CO_2 absorption lines can be used to verify and study possible solar pointing problems. Indeed, CO_2 volume mixing ratio is reasonably known in the stratosphere and can be fixed in the radiative transfer model. The Look-up table algorithm has been used to simulate a wide spectral region. It allows to cover several CO_2 lines and to improve the analysis. Figures 3 and 4 present example of simulations compared to atmospheric measured spectra.

6. Conclusions

Results indicates that the use of Look-up tables allow to decrease greatly calculation time without significant loss of accuracy. A logarithmic interpolation scheme has been chosen and successfully tested. The size of Look-up tables is truncated using SVD technique. 130 Mbytes of memory are needed to operate a spectral region of 40 cm^{-1}. The new algorithm has been tested successfully on balloon-borne FTIR measurements. It has been used to process wide spectral window (that made possible to retrieve vertical profile of species like CFC-12. Wide spectral region around CO_2 absorption lines has been analyzed using a new algorithm in order to study possible solar pointing problems.

References

[1] Goulb, G. H. and Van Load, C. F. *Matrix Computations*, 2nd edn., Johns Hopkins University Press, Baltimore, 1989.

[2] Camy-Peyret, C. (1995) Balloon-borne infrared Fourier transform spectroscopy for measurements of atmospheric trace species, Spectrochim. Acta, 51A, 1143–1152.

[3] Payan, S., Camy-Peyret, C., Jeseck, P., *et al.* (1998) First direct simultaneous HCl and ClONO$_2$ profile measurements in the Arctic vortex, Geophys. Research Letters, 25, 2663–2666.

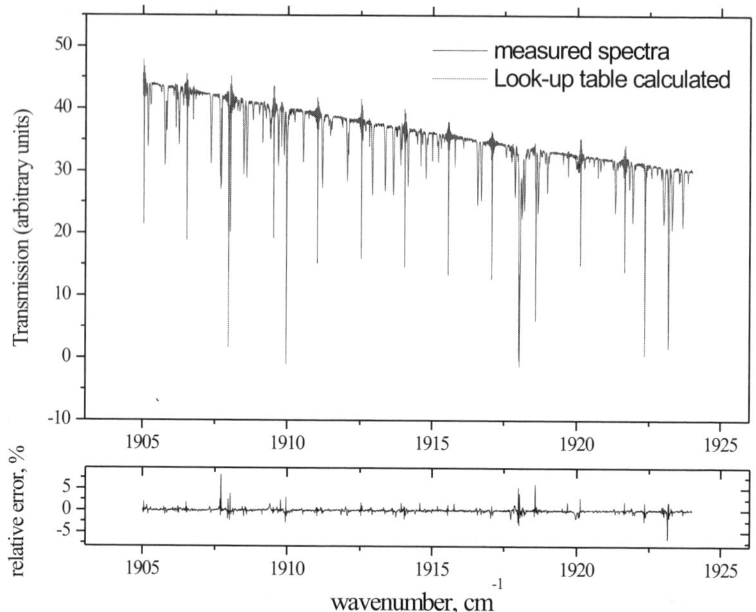

Figure 3.   Measured and calculated spectrum (spectrum number 185, balloon altitude 38.82 km).

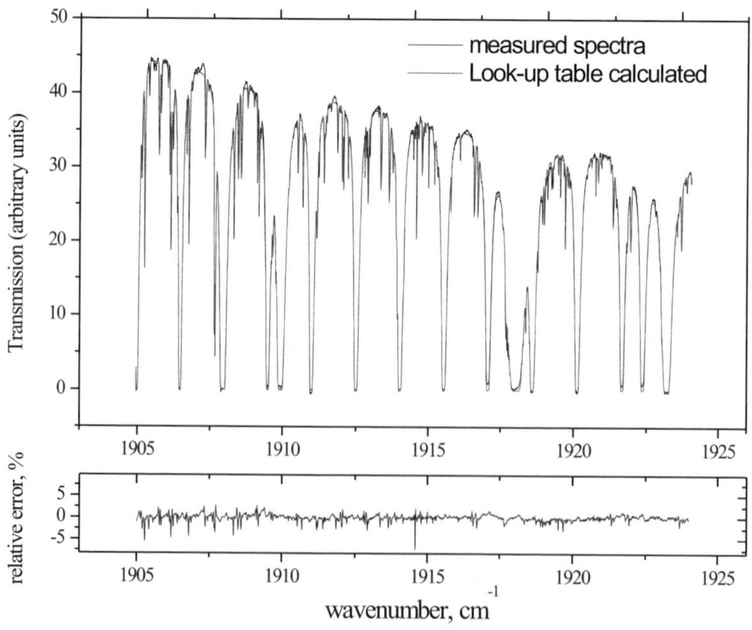

Figure 4.   Measured and calculated spectrum (spectrum number 271, tangent altitude 10.13 km).

THE IMPACT OF NEW WATER VAPOR SPECTROSCOPY ON SATELLITE RETRIEVALS

A. N. Maurellis
SRON National Institute for Space Research,
Sorbonnelaan 2, 3584 CA Utrecht, The Netherlands

R. Lang, J. E. Williams, W. J. van der Zande
FOM-Institute for Atomic and Molecular Physics,
Kruislaan 407, 1098 SJ Amsterdam, Amsterdam, The Netherlands

K. Smith, D. A. Newnham
CCLRC Rutherford Appleton Laboratory, Oxon, OX11 0QX, The United Kingdom

J. Tennyson, R. N. Tolchenov
Department of Physics and Astronomy, University College London,
Gower Street, London WC1E 6BT, The United Kingdom

Abstract   Water vapor, arguably the most important trace gas constituent of Earth atmospheric physics, is also both a retrieval goal and a hindrance in the retrievals of other trace gases from nadir-measuring satellite spectrometers. This is because the atmospherically-attenuated solar spectrum in the visible and shortwave infrared is littered with water vapor bands. The recent plethora of water vapor spectroscopy databases in this spectral region has prompted us to study their utility in satellite retrievals. We consider water vapor spectroscopy compiled from four sources including new spectroscopy due to University College London and Imperial College London. Radiative transfer models of satellite measurements, in combination with accurate retrieval techniques, are quite sensitive to the accuracy and completeness of the water vapor spectroscopy. Notwithstanding the high degree of variability of a number of different factors in satellite measurements we show that retrievals are sensitive to database differences which suggests that our knowledge of water vapor spectroscopy is not as yet complete. In addition, new laboratory measurements indicate that the role of both the far-line wings of water vapor and the cumulative effect of many weak lines each have an important role to play in forming the so-called continuum.

259

C. Camy-Peyret and A.A. Vigasin (eds.),
Weakly Interacting Molecular Pairs: Unconventional Absorbers of Radiation in the Atmosphere, 259–272.
© 2003 *Kluwer Academic Publishers. Printed in the Netherlands.*

1. Introduction

Oddly enough, a quantification of broadband absorption effects in atmospheric radiative transfer modeling is closely linked to an accurate knowledge of line spectra, especially those of water vapor which absorbs significantly in the atmosphere at most wavelengths longward of about 400 nm.

Many instances of water vapor line signatures may be found in visible and shortwave infrared spectra taken by nadir-sounding instruments such as the European Space Agency's Global Ozone Monitoring Experiment (GOME) on board the ERS-2 platform [12] and the Scanning Imaging Absorption Spectrometer for Atmospheric Chartography (SCIAMACHY) on board ESA's recently-launched ENVISAT platform [1]. In this work we present retrieval simulations for these instruments using rovibrational overtone bands of water vapor in the visible and shortwave infrared and spectroscopy from four different sources:

- new data (hereafter UCL-IC) due to a collaboration between University College London and Imperial College, and based on new laboratory measurements at the Molecular Spectroscopy Facility, Rutherford Appleton Laboratory (RAL) [17];

- new data (hereafter ULB-UFR) due to a collaboration between Laboratoire de Chimie Physique Moléculaire at the Université Libre de Bruxelles, the Institut d'Aéronomie Spatiale de Belgique, and the Groupe de Spectrométrie Moléculaire et Atmosphérique, UFR Sciences, Reims [2, 4];

- the HITRAN-1996 [15] and HITRAN-2000 [7] databases;

- the ESA-WV (B) and (R) databases [9, 16].

We show here that synthetic spectra generated from these line parameter databases and a knowledge of the background atmosphere may be used in combination with satellite-measured spectra to fix requirements for line intensity accuracies in spectral databases. Thus it becomes possible to use the atmosphere as a long pathlength laboratory in order to place limits on the levels of spectroscopic accuracy required for water vapor retrieval. This has additional, important consequences for our understanding of the water vapor continuum which are supported by other modeling efforts as well as by recent laboratory measurements of the water vapor continuum at RAL.

2. The atmosphere as laboratory

In order to compare modeled and satellite-measured earth-backscattered solar radiances we define a correction to the ratio between modeled and measured reflectivity spectra in terms of uncertainties in cross-section line intensities, averaged over the spectral region covered by each detector pixel in the focal plane of a grating spectrometer array. First we assume that the relative error in the cross-section corresponding to the spectral region covered by the satellite detector pixel, $\hat{\sigma}$, is dominated by the relative error in the pixel-averaged line intensity so that

$$\frac{\Delta\hat{\sigma}}{\hat{\sigma}} = \frac{\Delta S}{S} \equiv -\Delta\phi. \tag{1}$$

The quantity $\Delta\phi$ represents the correction required to bring the cross-section using the line intensity values of the spectral database into agreement with the satellite-measured average cross section. Thus [8]

$$R_{\text{sat}} = R_{\text{mod}} \exp\left(-\frac{\Delta S}{S}\int_s \hat{\sigma}_{\text{mod}}(s)\,n(s)\,ds\right), \tag{2}$$

where R_{sat} is the measured reflectivity. The modeled reflectivity and cross-section are given by R_{mod} and σ_{mod}, respectively.

Figures 1 and 2 show how this quantity may be used to distinguish directly between spectral databases. There is a caveat; this approach does not exclude the possibility that detector pixel electronics offsets could also contribute a residual signature which is comparable to that of the spectral database uncertainties. There is no way to exclude this possibility in the case of GOME. However SCIAMACHY has the capability to make so-called "dark" measurements which can be used to "flatfield" each detector pixel to high precision.

3. Retrieval theory

Standard retrieval theory [6], begins with a forward model $\mathcal{T}$ which dictates the passage of radiation through the atmosphere and instrument, viz.

$$y = \mathcal{T}(x; b, S) + \varepsilon, \tag{3}$$

where y is the measured spectrum, x reflects the true state of the atmosphere, ε is the shotnoise on the measurement, and the two other arguments of the forward model constitute a set of instrumental parameters b and spectroscopic parameters S. Retrieval is the process

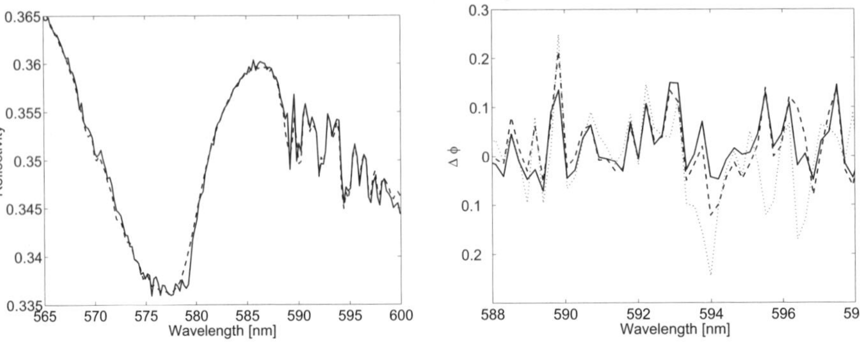

Figure 1.   **Left Panel:** comparison of a GOME-measured reflectivity (solid curve) with
the results of a line-by-line calculation (dashed curve). The background consists of the
$(O_2)_2$ collision-induced absorption complex (broad feature centered at 577 nm), part of the
O_3 Chappuis band throughout the spectral window and a sharp feature at about 590 nm due
to the sodium lines. **Right Panel:** pixel-to-pixel variation of the residual mismatch $\Delta\phi$
calculated for three different GOME measurements using a line-by-line forward model [8]
corresponding to a high (solid line), medium (dashed line) and low (dotted line) water vapor
column. The measurements were taken at significantly different geolocations and very different solar zenith angles ($37°$, $23°$, $73°$, respectively). Notwithstanding differences in surface
albedo, aerosol content and the contribution of multiple scattering to the measurements the
three curves show similar patterns which could imply either detector biases or spectroscopic
biases are present in the modeling.

of formally inverting Eq. (3) in order to determine a set of trace gas
profiles, notated by $\hat{x}$, which satisfy a modeled spectrum

$$\hat{y} = \mathcal{T}(\hat{x}; b, S). \tag{4}$$

The problem is generally ill-posed, since the number of elements
in the measurement vector is usually much larger than the number of
retrievable profile points (see DFS below). Thus even though $|\hat{y} - y| \leq \varepsilon$
may result from a good inversion, $\hat{x}$ will almost never be identical to
x simply because there may be insufficient information in the measurement or because information has been lost due to the spatial and
spectral discretization of the model as well as any spectral smoothing
caused by the spectrometer (so-called nullspace errors). In addition,
there may be uncertainties in the radiation transport model and errors
in instrumental parameters b which, at least in the latter case, are pre-
dictable through careful calibration of the instrument. Both of the last
two errors are avoided in what follows by carrying out retrievals from
simulated spectra which are generated using the same atmospheric and
instrument physics as the retrieval. This allows us to focus primarily
in this work on the error in the forward model due to spectroscopic un-

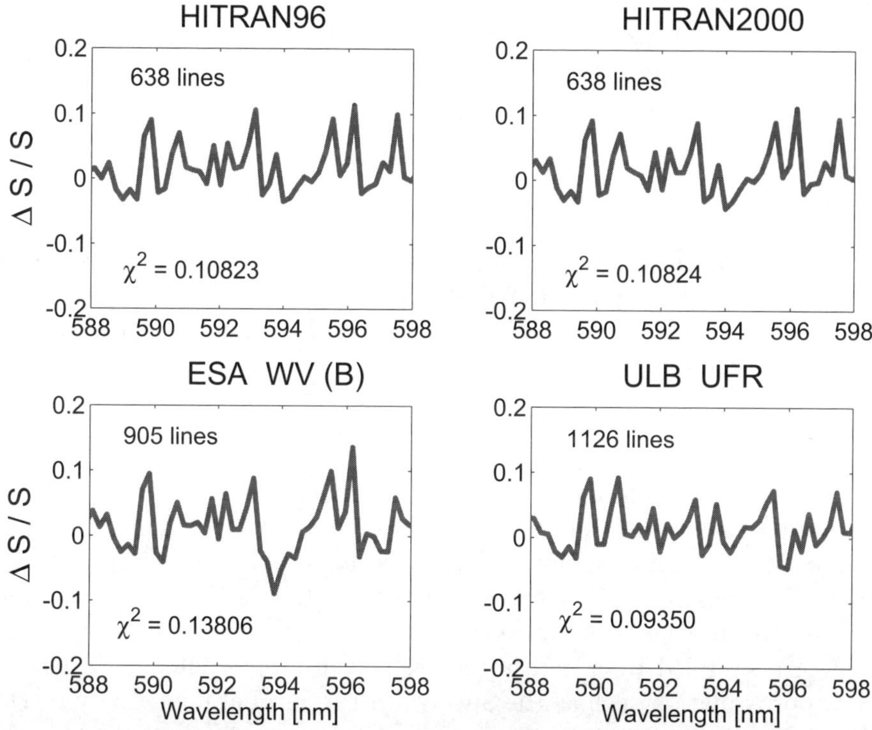

Figure 2. Determination of cross-section correction for different spectral databases in the region of the 5ν polyad using real GOME data, as per Figure 1 and Eq. (2). This confirms the sensitivity of the retrieval to small spectroscopic differences on the order of 10%. Each plot shows the correction derived from a model fit to the data which uses a different source of line parameters: HITRAN-1996; HITRAN-2000; ESA-WV (B only); ULB-UFR. The number of water vapor lines contained in each database in this spectral window is also shown. *At least for the 5ν polyad* HITRAN data appears to fare reasonably well, though the ULB data comes out better. The ESA-WV(B) data fares poorly, as expected, since this portion of the ESA-WV database mixes theoretical values and data derived from other studies independent of the ESA-WV study. In addition, it should be noted that the databases are not entirely independent of one another. ULB-UFR lines have intensities from [2, 4] but are assigned from [13, 14, 20] and ESA-WV (R). ESA-WV (B) line assignments are made using [20] but intensities are taken from HITRAN-2000 with some assignments from [2].

certainties, in particular, the effects of uncertainty (δS) in S (mean line intensity per wavelength grid cell as in Eq. (1)). The net contribution to the total retrieval errors may then be written as

$$\delta x = \left[A - I \right] (\hat{x} - x) + \frac{\partial \hat{x}}{\partial y} \varepsilon + \frac{\partial \hat{x}}{\partial y} \frac{\partial T}{\partial S} \delta S. \tag{5}$$

Here the first term encompasses the contribution to retrieval uncertainty of the nullspace errors, the second the shotnoise error and the third the mean line intensity parameter error. The first term includes the so-called averaging kernel of the inversion

$$A = \frac{\partial \hat{x}}{\partial y} \frac{\partial \mathcal{T}}{\partial x}. \qquad (6)$$

A wavelength grid cell is assumed, for the purposes of this discussion, to be the wavelength region covered by a detector pixel. Matrix A can also be used to calculate the degrees of freedom for signal in the retrieval (abbreviated to DFS), via

$$\mathrm{DFS} = \mathrm{trace}(A). \qquad (7)$$

The profile weighting function (or dependence of the forward model on the profile) $\partial \mathcal{T}/\partial x$ may be calculated trivially as a function of altitude if one assumes a simple Beer's law for atmospheric extinction of light. The spectroscopic parameter weighting function $\partial \mathcal{T}/\partial S$ is harder to calculate in general. However, if one assumes that the total spectral structure sampled by a detector pixel may be represented by a giant-line approximation such as the Spectral Structure Parameterization [11] then one may express the mean transmittance corresponding to each detector pixel as

$$\mathcal{T} = 1 + w \exp(-SN) - w, \qquad (8)$$

where S is the line intensity of the giant line for a layer of absorbing gas with column N and w is the line width. The spectroscopic parameter weighting functions follow immediately on partial differentiation with respect to S or w. Finally, the mean transmittance must be further smoothed in order to simulate the spectrometer slit and so to obtain the modeled reflectivity, $\hat{y}$.

4. Results of simulations

Figure 3 summarizes some nadir profile retrieval simulation results for the different databases. The implicit assumption is that each database is complete for the simulated atmosphere. Thick lines show the effect on the retrieved profile, in percentage of the true profile and as a function of altitude, of the nullspace and shotnoise uncertainties induced by the errors included in Eq. (5). Lighter lines correspond to the effect of a 10% average line intensity S uncertainty for each detector pixel wavelength grid cell in the $3\nu + \delta$ polyad (11,600–12,750 cm^{-1})

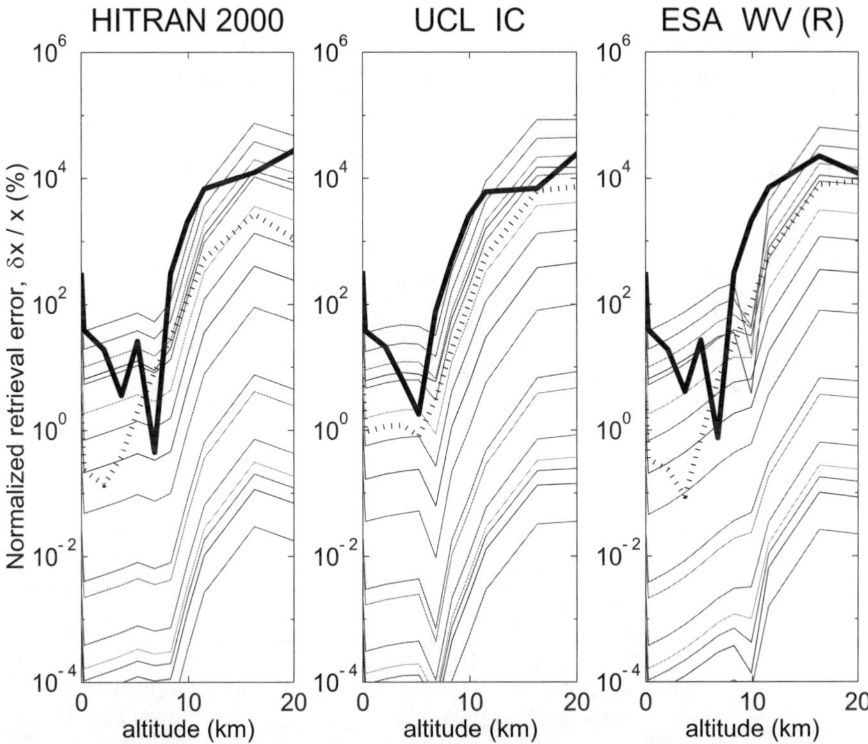

Figure 3.   Profile retrieval errors (cf. Eq. (5)) derived from simulations assuming a single nadir-gazing measurement and using the spectroscopy in three of the databases for the $3\nu + \delta$ polyad. Although a single nadir measurement in a single absorption band would seldom be deemed sufficient for a good profile retrieval it is nonetheless useful to simulate the resulting errors as a function of altitude. The solid line indicates the effect of nullspace errors, the dotted line the effect of the detector shotnoise error and the many colored lines the effect of line intensity uncertainties on the order of 10% for each of an array of simulated detector pixels. The effects of spectroscopic uncertainties are still more significant than other sources of error, especially for altitudes below 15 km, where the bulk of the water vapor column is situated. The much higher relative uncertainties at altitudes above 15 km derive from the much lower water vapor densities at these altitudes.

(i. e. $\delta S = 0.1S$, cf. Eq. (1) and the maximum uncertainties in Figure 2). The forward model assumes simple Beer's law extinction, a surface albedo of 30%, a solar zenith angle of 30° and a mid-latitude summer water vapor profile, with total column of 1.24×10^{23} molecules cm^{-2}. The simulated spectra are smoothed with a Gaussian function with FWHM corresponding to a slit of 0.2 nm, typical for satellite grating spectrometers in this wavenumber range. Clearly the effects of spectroscopic uncertainty (light lines) are frequently comparable to and

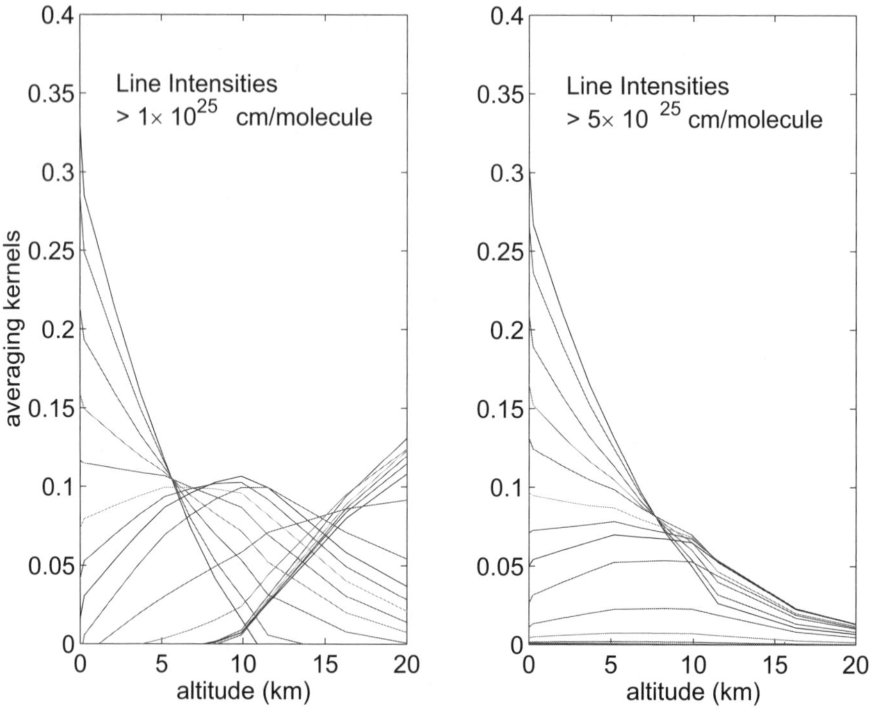

Figure 4. Two sets of averaging kernels (row vectors of the matrix A) for a set of simulations using the ESA-WV database. (The averaging kernels give an indication of the sensitivity of the retrieved profile to the true profile.) Successive levels of weak line intensity were removed from the ESA-WV database starting with all the lines in the database until a significant change in the averaging kernels resulted (a drop in the retrieval sensitivity above 7 km). This change dictates a level in line intensity above which line intensities have to be known to high accuracy and below which the completeness of the spectral database will contribute to a continuum signal but for which a high accuracy of line intensity parameters is probably not crucial.

occasionally exceed the other sources of error, especially at the important lower altitudes which correspond to the bulk of the water vapor column.

An increase in the requirement on spectral uncertainty from 10% to 1% would place all the colored curves a factor of 10 lower, with the result that the retrieval would be optimal, i.e. nullspace- and shotnoise-limited. There is also a change in the shape of the averaging kernels (see Figure 4) as lines are removed from the retrieval (assuming a full set of lines in each simulated measurement). The sensitivity of the retrieved water vapor profile to the true profile at higher altitudes drops dramatically when lines with intensity levels less than 5×10^{-25} cm/molecule are excluded from the retrieval (but not from the forward model used to calculate the simulated measurement). Together these indicate that line intensities greater than about 5×10^{-25} cm/molecule should be known to an accuracy on the order of 1% or better and the completeness of the databases below this level should be well established since this will contribute to the overall level of the assumed broadband or continuum signal. We note in passing that this level corresponds roughly to the range, 1–6×10^{-24} cm/molecule, below which line intensity differences between UCL-IC and HITRAN-2000 become significant [17]). In addition the increase in total number of weak lines in the new databases yields an increase in DFS value from approximately 3 (when using the HITRAN-2000 database) to 5–7 (for the ESA-WV and UCL-IC databases) which implies that the new databases have the potential to substantially enhance retrievals. Figure 5 shows the results of doing column retrievals from real GOME data using two of the databases, HITRAN-1996 and UCL-IC. Notwithstanding difficult remote sensing issues to be dealt with in this wavelength region (not least of which is a considerable change in the surface albedo of vegetation at approximately 720 nm) the effects on column retrieval of using different spectral databases are quite significant. However it is still too early to decide whether or not using the new database constitutes an improvement.

5. New laboratory measurements of the water vapor continuum

The distinction between broadband effects due to large numbers of low intensity lines on the one hand and true continuum effects on the other has been alluded to, above, in terms of completeness of the spectral line list used in a given radiative transfer model. Recently this distinction was made in the laboratory using a combination of a high-resolution Fourier transform spectrometer and two variable path-length absorption cells at RAL. Comparisons between modeled and measured laboratory transmittance spectra at two temperatures (296 and 342 K) in the spectral region 4400–6000 cm^{-1} (1.6–2.3 μm) indicate that absorption by pure water vapor is underestimated between 4900 cm^{-1} and 5700 cm^{-1}. Figure 6 shows results from the short path-length absorption cell (SPAC) operated at an optical path length of 9.74 m and a temperature of 342.1 K. The measurements were modeled using the Reference Forward Model line-by-line code [5] in conjunction with one or the other of the CKD version 2.4 continuum (cf. Clough, this volume

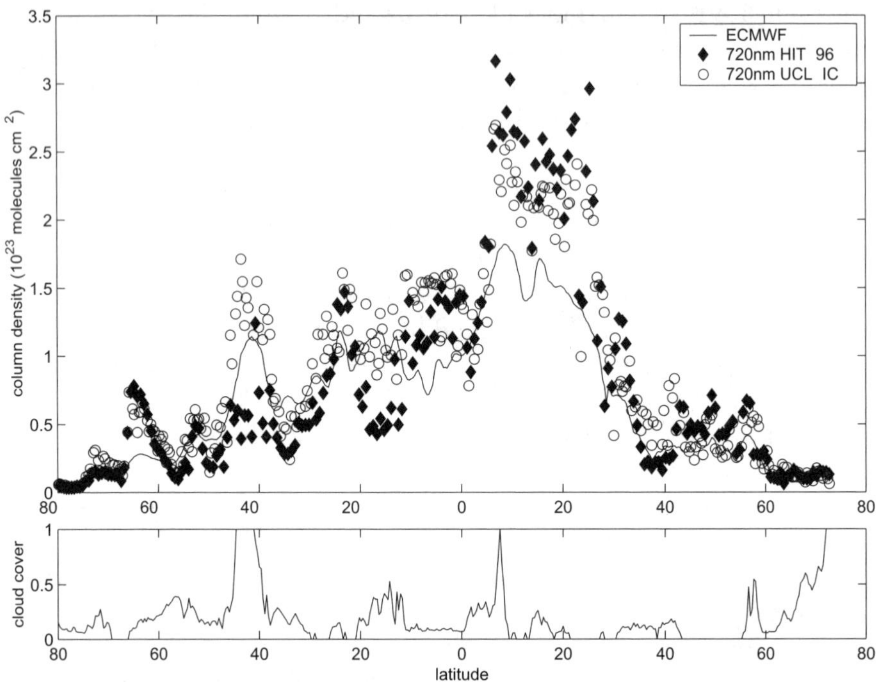

Figure 5. Water vapor column retrievals (**upper panel**) using the HITRAN-1996 (filled diamonds) and UCL-IC (circles) databases for the 4ν polyad (centered at approximately 720 nm). The columns have been retrieved from measurements across a single north-south track of GOME data (see horizontal axis). The solid line shows water vapor columns taken from the European Center for Medium-Range Weather Forecasting (ECMWF) and is provided purely for comparison purposes, as is the measure of cloud cover (**lower panel**) (from 0 for no cloud cover to 1 for fully-clouded) inferred from other GOME data.

and [3]) and the continuum cross sections from new far-line wing calculations for water vapor [10]. Observed and calculated absorbances were compared by subdividing each absorption band into $5\,\mathrm{cm}^{-1}$ wide intervals. In each interval a baseline offset was determined by a non-linear least squares fit of the modeled spectrum to the measurement. Only those data where the calculated and observed Napierian absorbances were below 1.0 were used in the comparisons to avoid introducing spectral saturation errors. As shown in Figure 6 neither of the two water vapor continuum treatments fully accounts for the observed additional absorption. The residual absorption is ostensibly reduced the most by the use of the CKD continuum in the forward model. Future investigations at RAL will center on the role of meta-stable water vapor dimers as a possible cause for the discrepancy between observations and calculations. However, since the CKD continuum is an empirically-adjusted broadband offset rather than a continuum calculation derived from first principles, *one likely conclusion is that CKD accounts for the net contribution of many missing water vapor lines, the cumulative effect of which swamps any true continuum effect.*

6. Conclusions

Thus it turns out that the addition of weak lines to the current databases is probably crucial for explaining inconsistences between laboratory measurements and models as well as improving the accuracy of retrievals in general. The role of weak lines in contributing, effectively, to a continuum was further corroborated recently by a study of the radiative forcing effect due to different water vapor spectral databases [21]. This study concluded that large numbers of weak water vapor lines such as those included in the ESA-WV database (but not in the HITRAN databases) could account for as much as 90% of the CKD continuum.

Retrievals, both real and simulated, have shown that sufficient levels of cross-section accuracy have not yet been attained. The indication is that strong lines should be known to accuracies of 1% or better. This will hopefully be met by future improvements to the UCL-IC database due to the availability of new, more accurate potential surfaces and dipole moment calculations. The temperature dependence of air-broadening coefficients, although not discussed here, is also important and will be addressed in a future study.

Thus it is possible to verify spectroscopic databases using high-resolution laboratory as well as medium-resolution satellite measurements. Another way is to use high-resolution ground-based measurements of atmospheric absorption [18, 19]. In any case, we have at our disposal a unique long pathlength laboratory setup — the atmosphere — providing we can use it correctly!

Acknowledgments

We thank Ilse Aben (SRON) for being among the first to recognize the importance of spectral residual analysis based on satellite measurements. ESA is acknowledged for providing GOME data processed by DFD/DLR. NERC and EPSRC are also acknowledged for additional support.

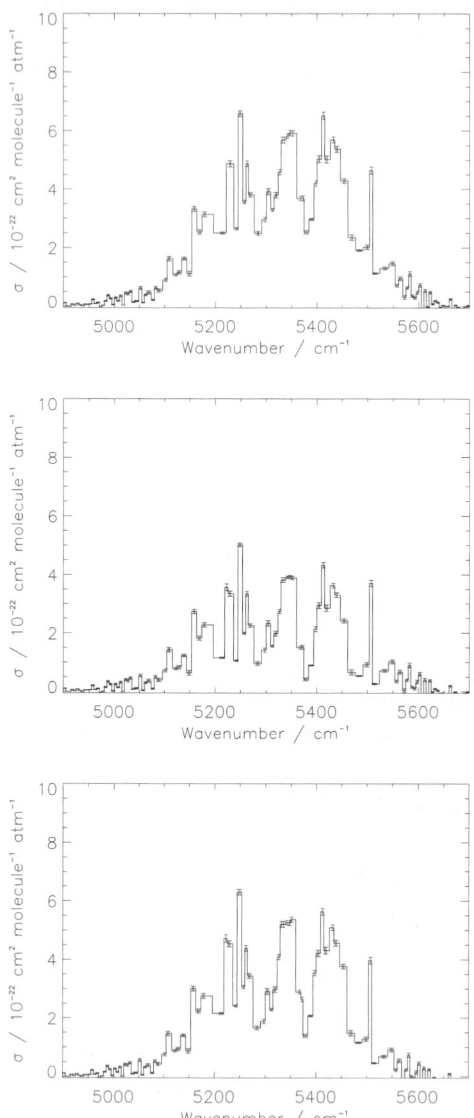

Figure 6.   Residuals in absorption cross-sections between RAL SPAC measurements and spectra generated by the RFM/HITRAN line-by-line model in $5\,cm^{-1}$ intervals between $4900\,cm^{-1}$ and $5700\,cm^{-1}$ for $100\,hPa$ of pure water vapor at $342\,K$. **Top panel:** results from a pure line-by-line calculation. Error bars indicate all known experimental uncertainties. **Center panel:** results from a RFM/HITRAN calculation assuming a Voigt line shape up to $25\,cm^{-1}$ on either side of the line centre and the CKD 2.4 water vapor continuum. **Lower panel:** as in the other panels except that the RFM/HITRAN calculation assumed a Voigt line shape up to $10\,cm^{-1}$ on either side of the line centre and the far-line wing continuum due to Tipping and Ma beyond $10\ cm^{-1}$.

References

[1] Bovensmann, H., Burrows, J. P., Buchwitz, M., Frerick, J., Noel, S., Rozanov, V. V., Chance, K. V., and Goede, A. P. H. (1999) SCIAMACHY: Mission objectives and measurement modes, J. Atmospheric Sciences, 56(2), 127–150.

[2] Carleer, M., Jenouvrier, A., Vandaele, A. C., Bernath, P. F., Merienne, M.-F., Colin, R., Zobov, N. F., Polyansky, O. L., Tennyson, J., and Savin, V. A. (1999) The near infrared, visible, and near ultraviolet overtone spectrum of water, J. Chem. Phys., 111(6), 2444–2450.

[3] Clough, S. A., Kneizys, F. X., and Davies, R. W. (1989) Line shape and the water vapor continuum, Atmospheric Research, 23, 229–241.

[4] Coheur, P. F., Fally, S., Carleer, M., Clerbaux, C., Colin, R., Jenouvrier, A., Merienne, M.-F., Hermans, C., and Vandaele, A. C. (2002) New water vapor line parameters in the 26 000–13 000 cm^{-1} region, J. Quantitative Spectrosc. Radiative Transfer, 74(4), 493–510.

[5] Dudhia, A. (1997) Reference forward model version 3: Software user's manual. *European Space Technology Centre (ESTEC)* Document PO-MA-OXF-GS-0003, European Space Agency, Paris, France.

[6] Eriksson, P. (2000) Analysis and comparison of two linear regularization methods for passive atmospheric observations, J. Geophys. Research, 105(D14), 18,157–18,167.

[7] Giver, L. P., Chackerian, C., and Varanasi, P. (2000) Visible and near-infrared H$_2^{16}$O line intensity corrections for HITRAN-96, J. Quaintitative Spectrosc. Radiative Transfer, 66(1), 101–105.

[8] Lang, R., Maurellis, A. N., van der Zande, W. J., Aben, I., Landgraf, J., and Ubachs, W. (2002) Forward modelling and retrieval of water vapor from the global ozone monitoring experiment: Treatment of narrowband absorption spectra, J. Geophys. Research, 107(D16), 4300, doi:10.1029/2001JD001453.

[9] Learner, R. C. M., Schermaul, R., Tennyson, J., Zobov, N. F., Ballard, J., Newnham, D., and Wickett, M. G. Measurement of Water Absorption Cross-Sections for the Exploitation of GOME data. ESTEC Contract No. 13312/9/NL/SF.

[10] Ma, Q. and Tipping, R. (2002)The frequency detuning correction and the asymmetry of line shapes: The far wings of H$_2$O–H$_2$O, J. Chem. Phys., 116(1), 4102–4115.

[11] Maurellis, A. N., Lang, R., and van der Zande, W. J. (2000) A new DOAS parameterization for retrieval of trace gases with highly-structured absorption spectra, Geophys. Research Letters, 27(24), 4069–4072.

[12] Noël, S., Buchwitz, M., Bovensmann, H., and Burrows, J. P. (2002) Retrieval of total water vapor column amounts from GOME/ERS-2 data, Advances in Space Research, 29(11), 1697–1702.

[13] Partridge, H. and Schwenke, D. W. (1997) The determination of an accurate isotope dependent potential energy surface for water from extensive *ab initio* calculations and experimental data, J. Chem. Phys., 106(11), 4618–4639.

[14] Polyansky, O. L., Zobov, N. F., Viti, S., and Tennyson, J. (1998) Water vapor line assignments in the near infrared, J. Molec. Spectrosc., 189(2), 291–300.

[15] Rothman, L. S., Rinsland, C. P., Goldman, A., Massie, S. T., Edwards, D. P., Flaud, J.-M., Perrin, A., Camy-Peyret, C., Dana, V., Mandin, J.-Y., Schroeder, J., Mccann, A., Gamache, R. R., Wattson, R. B., Yoshino, K., Chance, K. V., Jucks, K. W., Brown, L. R., Nemtchinov, V., and Varanasi, P. (1998) The HITRAN molecular spectroscopic database and HAWKS (HITRAN Atmospheric Workstation): 1996 Edition, JQSRT, 60, 665–710.

[16] Schermaul, R., Learner, R. C. M., Newnham, D. A., Ballard, J., Zobov, N. F., Belmiloud, D., and Tennyson, J. (2001) The water vapor spectrum in the region 8600–15 000 cm^{-1}: Experimental and theoretical studies for a new spectral line database. I. Linelist Construction, J. Mol. Spectrosc., 208(1), 43–50.

[17] Schermaul, R., Learner, R. C. M., Canas, A. A. D., Brault, J. W., Polyansky, O. L., Belmiloud, D., Zobov, N. F., and Tennyson, J. (2002) Weak line water vapor spectra in the region 13 200–15 000 cm^{-1}, J. Molec. Spectrosc., 211(2), 169–178.

[18] Smith, K. M. and Newnham, D. A. (2001) High-resolution atmospheric absorption by water vapor in the 830–985 nm region: Evaluation of spectroscopic databases, Geophys. Research Letters, 28(16), 3115–3118.

[19] Veihelmann, B., Lang, R., Smith, K. M., Newnham, D. A., and van der Zande, W. J. (2002) Evaluation of spectroscopic databases of water vapor between 585 and 600 nm, Geophysical Research Letters, 29(15), 1752, doi: 10.1029/2002GL015330.

[20] Zobov, N. F., Belmiloud, D., Polyansky, O. L., Tennyson, J., Shirin, S. V., Carleer, M., Jenouvrier, A., Vandaele, A.-C., Bernath, P. F., Mérienne, M.-F., and Colin, R. (2000) The near ultraviolet rotation-vibration spectrum of water, J. Chem. Phys., 113(4), 1546–1552.

[21] Zhong, W., Haigh, J. D., Belmiloud, D., Schermaul, R., and Tennyson, J. (2001) The impact of new water vapor spectral line parameters on the calculation of atmospheric absorption, Quarterly J. Royal Meteorological Society, 127, 1615–1626.

SPECTROSCOPIC AND THERMOCHEMICAL INFORMATION ON THE O_2-O_2 COLLISIONAL COMPLEX INFERRED FROM ATMOSPHERIC UV/VISIBLE O_4 ABSORPTION BAND PROFILE MEASUREMENTS

K. Pfeilsticker,* H. Bösch, R. Fitzenberger
Institut für Umweltphysik,
INF 229, University of Heidelberg, D-69120 Heidelberg, Germany

C. Camy-Peyret
Laboratoire de Physique Moléculaire et Applications,
Université Pierre et Marie Curie & CNRS, Paris, France

Abstract   Atmospheric profiles of the ultraviolet/visible (UV/vis) absorption bands of the collision complex O_2–O_2, or O_4 in brief, are reported. The O_4 absorption profiles are inferred from direct Sun spectra observed from the LPMA/DOAS (Laboratoire de Physique Moléculaire et Applications/Differential Optical Absorption Spectroscopy) balloon gondola. Seven O_4 absorption bands — centered at $\sim$ 360.7, 380.2, 446.7, 477.1, 532.2, 577.2, and 630.0 nm — are investigated for atmospheric pressures (p) ranging from $\sim$ 500 hPa to $\sim$ 40 hPa and temperatures (T) ranging from 203 K to 250 K. For the encountered atmospheric conditions, it is found that, (a) the band shapes do not change with T and p and (b) the peak collision pair absorption intensities (α_i) concurrently increase with decreasing T (by about 11% over a $\Delta T = 50$ K). That result is in agreement with previous laboratory O_4 studies mostly conducted at high O_2 partial pressures (up to several hundred bars). Furthermore, by reasonably assuming that the O_4 absorption cross sections are T-independent, the inferred T-dependence of $\alpha_i(T)$ suggests a thermally averaged enthalpy change $<\Delta H> = -(1207 \pm 83)$ J/Mol involved in the formation of O_4. Our inferred ΔH is in reasonable agreement with the orientation and spin averaged O_4 well depth $D_e(O_4)$ $(= -(1130 \pm 80)$ J/Mol) measured in a recent O_2–O_2 collision experiment, when accounting for the rovibrational energy change during O_4 formation (189 J/Mol).

---

e-mail: `klaus.pfeilsticker@iup.uni-heidelberg.de`

273

C. Camy-Peyret and A.A. Vigasin (eds.),
Weakly Interacting Molecular Pairs: Unconventional Absorbers of Radiation in the Atmosphere, 273–284.
© 2003 *Kluwer Academic Publishers. Printed in the Netherlands.*

1. Introduction

The physics of the collision complex O_2–O_2 is of vital interest in many fields, such as combustion chemistry, molecular, solid state, and atmospheric physics, and atmospheric chemistry. In atmospheric physics, O_2–O_2 has recently attracted much interest as (a) an absorber of solar radiation [e. g. Pfeilsticker *et al.*, 1997; Solomon *et al.*, 1998], (b) a tool to infer atmospheric photon path lengths and cloud heights [e. g. Erle *et al.*, 1995], and (c) a marker to locate the absorption of atmospheric absorbers in remote sensing studies [Wagner and Platt, 1998].

The UV/vis absorption of the O_4 collision complex was first reported by [Janssen, 1885]. Since then the p and T-dependence, spectroscopy, and nature of the UV/vis, and near-IR transitions of O_4 have been subject to many studies [e. g. Ellis and Kneser, 1933; Salow and Steiner, 1936; Herman, 1939; Dianov-Klokov, 1964; Blickensderfer and Ewing, 1969a, b; Tabisz *et al.*, 1969; McKellar *et al.*, 1972; Long and Ewing, 1973; Perner and Platt, 1980; Rinsland *et al.*, 1982; Johnston *et al.*, 1984; Horowitz *et al.*, 1998; Greenblatt *et al.*, 1990; Orlando *et al.*, 1991; Newnham *et al.*, 1998; Mlawer *et al.*, 1998; Campargue *et al.*, 1998].

While previous studies have shown that the O_4 absorption results from a relaxation of the selection rules for the dipole forbidden O_2 transitions by perturbing neighbor molecules [Blake and McCoy, 1987], contradictory conclusions on the nature of O_4 have been drawn. It has been discussed whether O_4 has to be regarded either as (1) a covalently bonded dimer [e. g. Shardanand, 1969], or (2) an unbonded collision complex where the absorption may take place because of the distorted O_2 'Hamiltonian' (or electron orbitals) during the collision [e. g. Tabisz *et al.*, 1969; Greenblatt *et al.*, 1990], or (3) a metastable dimer with molecular and spectroscopic properties intermediate between the characteristics (1) and (2), where mutual bondings (without pairing of electrons necessarily) of the O_2 molecular orbitals—depending on the orientation of individual O_2 molecules during collision—may exist [e. g. Aquilanti *et al.*, 1999a, b]. By solely applying spectroscopy, the nature of O_4 could not be totally unraveled until recently when a series of new studies appeared in the literature. These include (a) O_2–O_2 orientation controlled collision experiments [Aquilanti *et al.*, 1999a, b], where it has been shown that the O_2–O_2 bonding, depending on the orientation of each O_2 molecule, is due to electrostatic forces (van der Waals), but chemical (spin-spin) contributions are not negligible, and (b) spectroscopic studies of O_2 at very high p (10 GPa), where the formation of an O_4 lattice and a concurrent 'metalization' was found [Gorelli *et al.*, 1999]. The study of Aquilanti *et al.* [1999a] indicates a O_4 well depth

$D_e(O_4) = -(1130 \pm 80)$ J/mol when averaging over the mutual molecular orientations and spin contributions [Aquilanti *et al.*, 1999a,b].

Another problem related to the O_4 collision complex is due to its relatively weak UV/vis absorption at atmospheric T's and p's (at most 10^{-6}/m for one standard atmosphere of O_2). Therefore, previous laboratory work was either conducted at high O_2 pressures (with a possibility of a p-modification of the absorption bands), or at extremely long optical path lengths to measure the weak O_4 absorption features at the expense of a restricted T-range.

Atmospheric measurements of the O_4 absorption band intensities have a good potential to overcome these difficulties. Accordingly, we present here first measurements of absorption profiles of 7 UV/vis O_4 absorption bands measured by direct Sun spectroscopy employed during two balloon flights of the azimuth angle controlled LPMA/DOAS gondola [e.g. Ferlemann *et al.*, 2000]. Our observations are interpreted with respect to the spectroscopy, and the formation enthalpy of the O_4 collisional complex.

2. Observations

Details of the spectroscopic measurements conducted from aboard the LPMA/DOAS gondola have already been reported on several occasions [e.g. Ferlemann *et al.*, 2000] and, therefore, they need only to be briefly revisited here. The azimuth angle controlled LPMA/DOAS gondola carries a tracker that continuously directs a parallel beam of Sun light into a FT-IR, and into a grating UV/vis spectrometer as well. The optical set-up allows us to analyze the line of sight absorption of many atmospheric constituents for wavelengths ranging from the near UV into the mid-IR. For the present study the UV/vis O_4 absorption features were monitored in direct Sun light with two grating spectrometers (320 to 420 nm, and 416 to 680 nm) and analyzed using the DOAS technique [Platt, 1994]. For the spectral retrieval the same set of reference spectra than in previous studies was used [e.g. Ferlemann *et al.*, 2000]. The O_4 absorption signature were inferred with reference spectra taken from Greenblatt *et al.* [1990], or alternatively by [Newnham and Ballard, 1998]. The amount of absorbing O_4 in the background, or reference spectra was determined by extrapolating the measured O_4 absorption features as a function of the air mass to zero air mass. Relative errors of the spectral retrieval range from 1% for the strong absorption bands at high ambient pressures to 30% for the weak absorption bands at lower pressures, respectively. Spectral retrievals from observations at

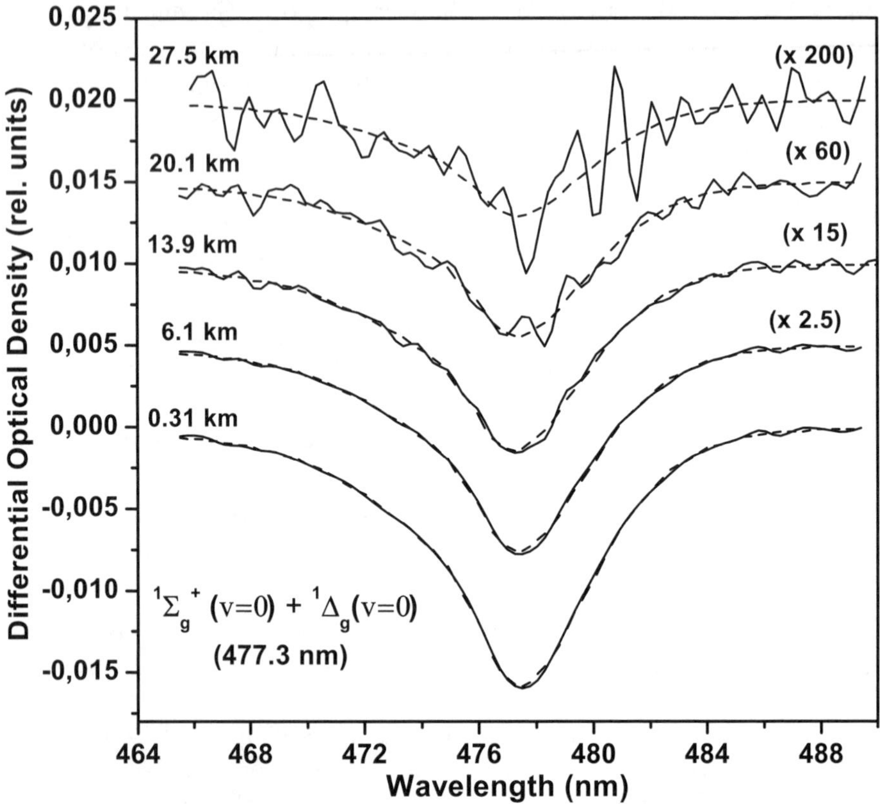

$$^1\Sigma_g^+ (v=0) + {}^1\Delta_g(v=0)$$
$$(477.3 \text{ nm})$$

Figure 1.   Absorption spectra (full line) for the $^1\Sigma_g^+ + {}^1\Delta_g$ transition (centered at 477.3 nm) as inferred from direct Sun observation taken at different altitudes (0.31, 6.1, 13.9, 20.1, and 27.5 km) during balloon ascent from Kiruna on 14 Feb. 1997. For comparison, the O_4 band shape recorded at high O_2 pressure (55 atm) by [Greenblatt *et al.*, 1990] is also shown (dashed line). For better illustration, the spectra are vertically shifted by bins of 0.005 optical densities units, and multiplied by factors given at the right side of the panel.

low ambient p's exceeding relative errors of 30% were discarded in the analysis.

Figure 1 shows the 477.3 nm O_4 absorption band in the direct Sun monitored from different atmospheric altitudes during the Kiruna balloon flight on Feb. 14, 1997. It is apparent that, the absorption is monotonically decreasing with height but the band shape does not change with ambient p, and in particular not when compared with the higher p measurements (e. g. conducted at 55 atm [Greenblatt *et al.*, 1990]). For our instrumental resolution and observational conditions (500 hPa $> p > 40$ hPa and 250 K $> T > 203$ K), no change in absorption band shapes is observed for the 477.3 nm band as of any for the other

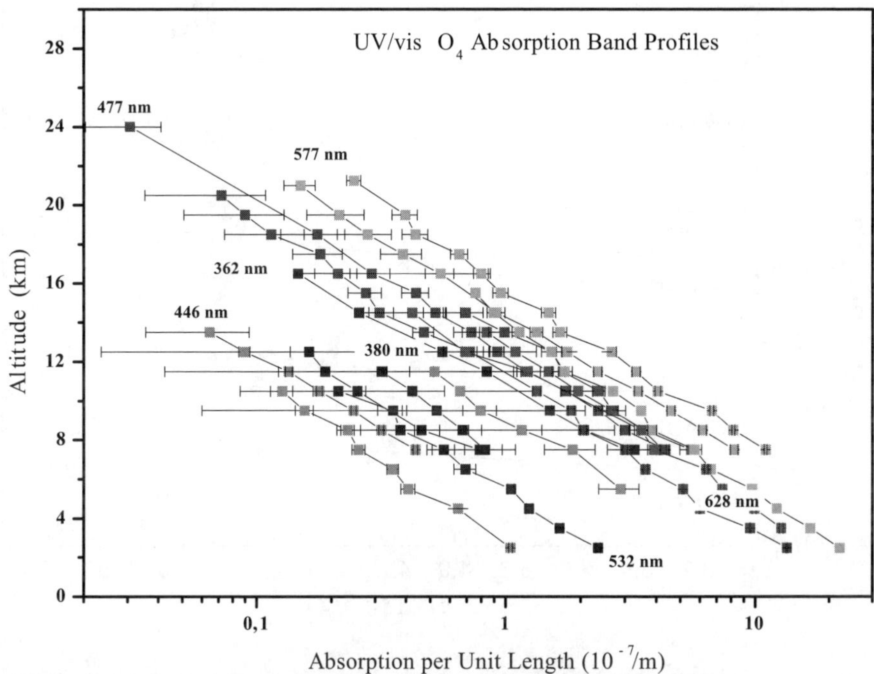

Figure 2.   Atmospheric absorption profiles for the UV/vis O$_4$ absorption bands inferred for the balloon flights from León/Spain on Nov. 23, 1996, and Kiruna/Sweden on Feb. 14, 1997.

UV/vis O$_4$ absorption bands, and in particular the O$_4$ absorptions bands have no fine structure at our spectral resolution.

From the measured O$_4$ band absorption as a function of line of sight, O$_4$ absorption profiles are inferred using the onion peeling technique [e. g., Rodgers, 1976] (Figure 2).

It is found that, for a particular atmospheric height the individual peak O$_4$ absorption intensities change by as much as 50% between the individual balloon flights, mainly because of a change in atmospheric p (and less because of T). The measured atmospheric O$_4$ absorption per unit length (ϵ_i) for band i, T and p profiles allows us to infer the peak collision pair absorption cross section $\alpha_i(T)$ (sometimes also called peak binary absorption cross section or binary absorption coefficient) as a function of T

$$\alpha_i(T) = \frac{\epsilon_i(T)}{[O_2]^2} = \sigma_i\, K_{\mathrm{eq}}(T)$$

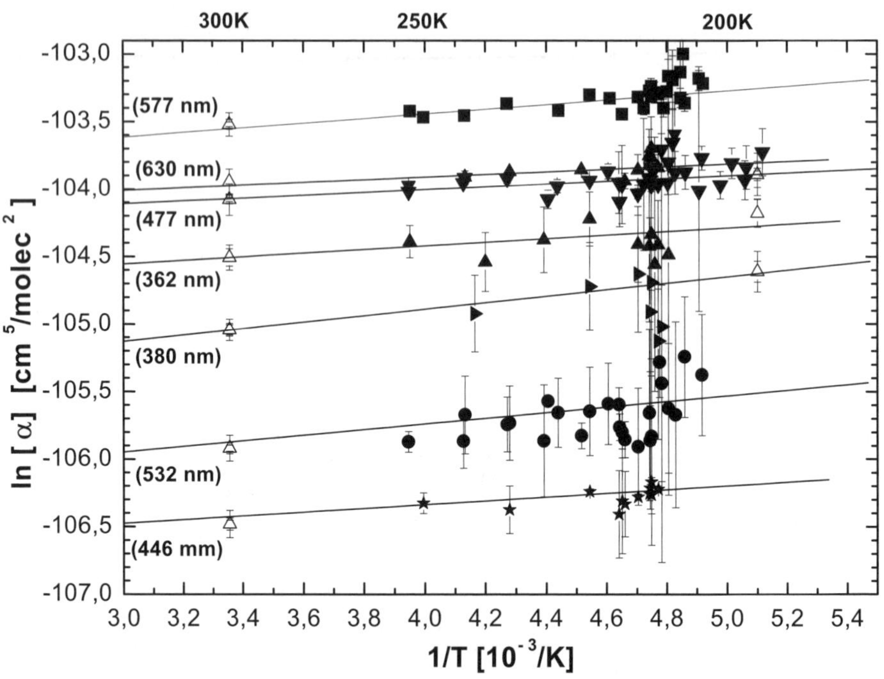

$\ln [\alpha]$ [cm ^{5}/molec ^{2}]

1/T [10^{-3}/K]

Figure 3. T-dependence of the O_4 collision pair absorption cross section ($\alpha_i(T)$). Regression lines to the measured data are also shown. The open triangles are the peak collisional absorption cross section by Greenblatt *et al.*, 1990.

with $[O_2]$ being the oxygen number density $(= 0.2094\, p/k_B\, T)$, σ_i the absorption cross section, $K_{eq}(T)$ is the O_4 equilibrium constant and all other quantities are denoted as in the standard physical literature. Assuming as [Johnston *et al.*, 1984] that σ_i is T-independent, then the T-dependence of α_i is solely due to the T-dependence of $K_{eq}(T)$. Even not totally justified for collisional complexes at sufficiently high T's [e. g. Vigasin, 2001] $K_{eq}(T)$ may be approximated by

$$\ln K_{eq}(T) = -\frac{<\Delta H>}{R}\left(\frac{1}{T}\right) + \frac{\Delta S}{R}.$$

A thermally averaged O_4 formation enthalpy, $<\Delta H>$ (see below), can therefore be derived from the slope of the plotted (α_i) against $1/T$ (Fig. 3).

We find that though the errors are large for some of the weaker absorption bands (at 446, 532, and 380 nm), mainly because of their weakness and interfering absorption by other gases like O_3, H_2O, NO_2, the observed slopes cluster around a $<\Delta H>$ value of $-(1207 \pm 83)$ J/Mol.

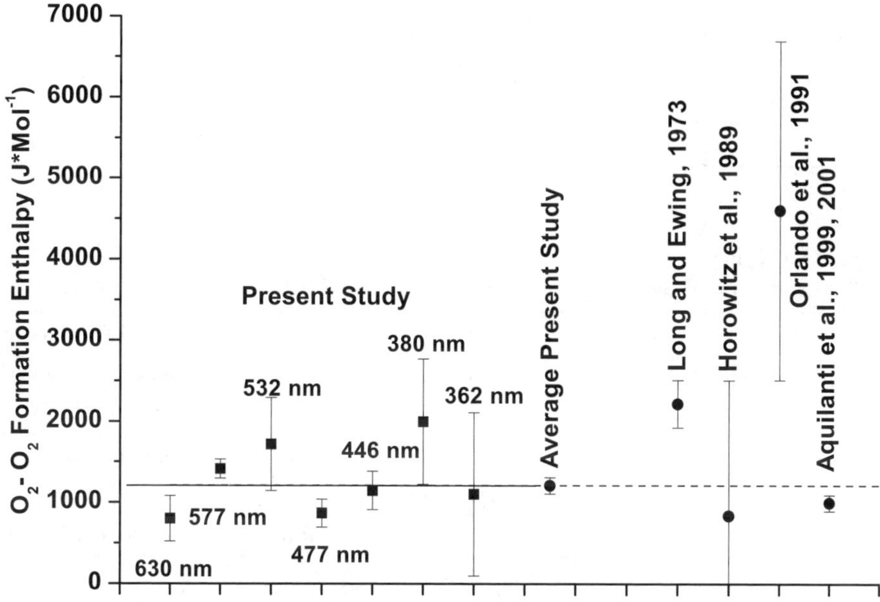

Figure 4.   Compendium of inferred O$_4$ formation enthalpies.

This small value indicates a weak bonding of O$_4$, with a $<\Delta H>$ corresponding to $T = 145 \pm 11$ K, being however, significantly smaller than typical atmospheric T's.

3. Discussion and conclusion

It is clear that the UV/vis O$_4$ absorption bands are forbidden O$_2$ electronic transitions $(^3\Sigma_g^+, \, ^1\Delta_g \leftarrow \, ^3\Sigma_g^-, \, ^3\Sigma_g^-)$ induced by weak intermolecular interactions [e. g. Blake and McCoy, 1987; for the spectroscopic nomenclature see Greenblatt *et al.*, 1990]. Therefore, in first order of approximation the intensity for the pair absorption is due to the exchange interaction of the $(^3\Sigma_g^-, \, ^1\Delta_g)$ and the $(^3\Sigma_g^-, \, ^3\Sigma_u^+)$ states. Accordingly, it may be expected that the O$_4$ band intensities should depend somewhat on the collision duration and energy involved in the interacting pair (for velocity/energy dependent O$_2$–O$_2$ collision cross section see Figure 1 in Aquilanti *et al.*, [1999a]), but the band shapes should not strongly depend on p, a behaviour that is actually observed (see Fig. 1).

In order to make our result comparable to the previously measured $D_e(O_4)$ [Aquilanti *et al.*, 1999a, b], we now consider the various terms

Table 1.   Compendium of energies involved in the $O_2 + O_2 \rightarrow O_4$ collision.

Contribution	J/Mol	cm^{-1}	Ref.
Well depth, $D_e(O_4)$	-1130 ± 80	-94.367	Aquilanti *et al.*, 2001
$E_{vib}^0(O_2)$ (internal)	18637.15	1556.4	Long *et al.*, 1973
$E_{vib}^0(O_4)$ (1. internal O_2)	18600.0	1553.3	Long *et al.*, 1973
$E_{vib}^0(O_4)$ (2. internal O_2)	18655.1	1557.9	Long *et al.*, 1973
$E_{vib}^0(O_4)$ (4 O_2–O_2 modes)	529.2	44.194	Aquilanti *et al.*, 2001
Zero point energy, ZPE(O_4)	255	21.29	calculated from above
$B_{rot}^0(O_4)$ for $[O_2(^1\Delta_g)_{\nu=0}]_2$	0.754	0.063	Biennier *et al.*, 2000
$B_{rot}(O_2)$ for $O_2(^3\Sigma_g^-)$	17.31	1.445	Herzberg, 1989
Change in rot. energy, ΔE_{rot}^0	-66.22	-5.228	calculated from above
Calculated $\Delta H_0^T = RT +$	-941 ± 80	-78.30	Σ from above
Measured $\Delta H_0^T = RT +$	-1207 ± 83	-100.79	Pfeilsticker *et al.*, 2001

going into ΔH_0^T i.e.

$$\Delta H_0^T = -3\,RT + D_e(O_4) + \Delta E_{vib}^0 + \Delta E_{rot}^0,$$

where the term $-3RT$ is due to the release in translational energy involved in the association of two O_2's. $D_e(O_4) = -(1130 \pm 80)\,\text{J/Mol})$ is from [Aquilanti *et al.*, 1999a, b], ΔE_{vib}^0 $(= E_{vib}^0(O_2\text{–}O_2) - 2E_{vib}^0(O_2))$ is due to the change in the energies, i.e. a sum given by the net change in thermal vibrational energy (4RT) and the zero point energies (255 J/Mol). Finally ΔE_{rot} is the change in the O_2 rotational energy (-66.2 J/Mol) (for details see Table 1). Bringing together all contributions, we obtain $\Delta H_0^T = RT - (941 \pm 100)\,\text{J/Mol}$. As our method to infer ΔH is insensitive to the first term (RT) in ΔH_0^T, we then arrive at $\Delta H = -(941 \pm 100)\,\text{J/Mol}$, a value being close to our inferred $<\Delta H> = -(1207 \pm 83)\,\text{J/Mol}$ (Fig. 4).

The relatively small bond enthalpy could be interpreted in terms of the low relative abundance of truly bond O_4 dimers at atmospheric Ts. Their rovibrational states are almost equally occupied but the population, hence the contribution to absorption, of metastable and non-bonded interacting pairs dominate in atmospheric conditions, in contrast to supersonic jet or very low T experiments. This may lead to a T-dependent but smaller averaged dissociation energy $<D_0(T)>$ and hence an apparent smaller $K_{eq}(T)$ [e.g. Vigasin, 2001].

Conversely, the additional T-dependence introduced by $<D_0(T)>$ into $K_{eq}(T)$ may then only allow to infer an effective ΔH when using equation 2, which explains the notation $<\Delta H>$ used above. The value of ΔH (almost constant within the error bars in the T and spectral range covered) should then be understood as representative of the atmospheric conditions rather than as a characteristic of the truly bond dimer. Due to the short lifetime of the formed metastable dimers, the rovibrational transitions are also expected to be largely lifetime broadended. The overlap of the multitude of nearly coinciding rovibrational transitions, then leads to the observed absorption band profiles. Accordingly, the σ_i should not be structured, nor T dependent [e. g. Johnston et al., 1984]. At low T (<150 K), however, truly bonded O_2–O_2 van der Waals pairs start to be dominant [Vigasin, 2001] thus allowing for structured rovibrational transitions of the dimer, a conclusion being in agreement with the spectroscopic observations by [Long and Ewing, 1973] (at 87.3 K), and/or recently by [Campargue et al., 1998] (at 77 K).

Thus our own observation and analysis reveals that at atmospheric temperatures the UV/vis O_4 absorption features stem from collision-induced (forbidden in the isolated molecule) dipole transitions of O_2. It is only at very low T's (<150 K) that rovibrational transitions of the O_2–O_2 van der Waals molecule dominate [e. g. see Campargue et al., 1998]. It has to be acknowledged that, already in 1973 Long and Ewing somewhat predicted this spectroscopic behavior of O_4, although by then much less information on the nature of O_4 was available than today.

Our study has several implications for O_4 applications in atmospheric studies. The rather weak (but sizeable) T-dependence of the O_4 peak absorption cross sections ($\Delta\alpha = 22\%$ for a $\Delta T = 100$ K) may introduce additional uncertainties into the calculation of O_4 instantaneous solar heating rates, inferred photon path lengths, and cloud heights when not accordingly considered. For example, assuming a $\Delta T = 50$ K for the variance of atmospheric T's then the instantaneous O_4 heating rate is modified by 11% (out of total of ~ 2 W/m^2 for clear sky and overhead Sun [Solomon et al., 1988]). Or, assuming as before an atmospheric ΔT of 50 K, then the apparent photon path lengths may change by 11% as well, and inferred cloud top heights by about 440 m assuming an atmospheric O_4 scale height of 4 km.

In effect, when using O_4 spectroscopy in atmospheric applications, the O_4 absorption T-dependence add some complications to the somewhat larger effects arising from the variable atmospheric p's (typically 10%) and the O_2 partial p-square dependence of the O_4 absorptions (leading to a $\sim 21\%$ variability).

Acknowledgments

Support of the project by the EU (contracts ENV4-CT-95-0178; ENV4-CT-97-0524; ENV4-CT-97-0521; ENV4-CT-99-00047) is acknowledged. The authors are grateful for the discussion with V. Aquilanti, and E. Carmona-Novillo, (both with the University Perugia/Italy), V. Vaida and J. E. Headrick (both at CU Boulder/USA), A. A. Vigasin (Russian Academy of Sciences, Moscow/Russia) and S. Solomon (NOAA, Boulder/USA). We also thank P. Jeseck, T. Hawat, and the team of CNES/France for the assistance given to perform successfully the balloon flights. Most of the investigation was performed while the first author held a National Research Council—NOAA Research Associateship.

References

[1] Aquilanti, V., Bartolomei, M., Cappelletti, D., Carmona-Novillo, E., and Pirani, F. (2001) Dimers of the major components of the atmosphere: Realistic potential energy surfaces and quantuum mechanical prediction of spectral features, Phys. Chem. Chem. Phys., 3, 3891–3894.

[2] Aquilanti, V., Ascenzi, D., Bartolomei, M., Cappelletti, D., and Cavalli, S. (1999) Quantum interference scattering of aligned molecules: bonding in O_4 and the role of spin coupling, Phys. Rev. Lett., 82, 69–73.

[3] Aquilanti, V., Ascenzi, D., Bartolomei, M., Cappelletti, D., and Cavalli, S. Vitores, M. de C., and Pirani, F. (1999) Molecular beam scattering of aligned oxygen molecules. The nature of the bond in O_2–O_2 dimer, J. Am. Chem. Soc., 121, 10794–10802.

[4] Biennier, L., Romanini, D., Kachanov, A., Campargue, A. Bussery-Honvault, B., and Bacis, R. (2000) Structure and rovibrational analysis of the $[O_2(^1\Delta_g)]_2 \leftarrow [O_2(^3\Sigma_g^-)]_2$ transition of the O_2 dimer, J. Chem. Phys., 112, 14, 6309–6321.

[5] Blake, A. J. and McCoy, D. G. (1987) The pressure dependence of the Herzberg photoabsorption continuum of oxygen, J. Quant. Spectrosc. and Radiat. Transfer, 38, 113–120.

[6] Blickensderfer, R. P. and Ewing, G. E. (1969) Collision-induced absorption spectrum of gaseous oxygen at low temperatures and pressure. 1. The $^1\Delta_g \leftarrow {}^3\Sigma_g^-$ system, J. Chem. Phys., 51, 873–883.

[7] Blickensderfer, R. P. and Ewing, G. E. (1969) Collision-induced absorption spectrum of gaseous oxygen at low temperatures and pressure. 2. The simultaneous transition $^1\Delta_g + {}^1\Delta_g \leftarrow {}^3\Sigma_g^- + {}^3\Sigma_g^-$ and $^1\Delta_g + {}^1\Sigma_g^+ \leftarrow {}^3\Sigma_g^- + {}^3\Sigma_g^-$, J. Chem. Phys., 51, 5284–5289.

[8] Campargue, A., Biennier, L., Kachonov, A., Jost, R., Bussery-Honvault, B., Veyret, V., Churassy, S., and Bacis, R. (1998) Rotationally resolved absorption spectrum of the O_2 dimer in the visible range, Chem. Phys. Lett., 288, 734–742.

[9] Cappelletti, D., Vecchiocattivi, F., Pirani, F., Heck, E. L., and Dickinson, A. S. (1998) An intermolecular potential for nitrogen from a multi-property analysis, Mol. Phys. 93, 485–499.

[10] Dianov-Klokov, V. I. (1964) Absorption spectrum of oxygen at pressures from 2 to 35 atm in the region from 12,600 to 3600 Å, Optics and Spectroscopy, 16, 224–227.

[11] Ellis, J. W. and Kneser, H. O. (1933) Kombinationsbeziehungen im Absorptionsspektrum des flüssigen Sauerstoffs, Zeitsch. der Physik, 86, 583–591.

[12] Erle, F., Pfeilsticker, K., and Platt, U. (1995) On the influence of tropospheric clouds on zenith-scattered-light measurements of stratospheric species, Geophys. Res. Lett., 20, 2725–2728.

[13] Ferlemann, F. *et al.* (2000) A new DOAS-instrument for stratospheric balloon-borne trace gas studies, J. Applied Optics, 39, 2377–2386.

[14] Greenblatt, G. D., Orlando, J. J., Burkholder, J. B., and Ravishankara, A. R. (1990) Absorption measurements of oxygen between 330 and 1140 nm, J. Geophys. Res., 95, 18577–18582.

[15] Gorelli, F. A., Ullivi, L., Santoro, M., and Bini, R. (1999) The ϵ phase of solid oxygen: evidence of an O$_4$ molecule lattice, Phys. Rev. Lett., 83, 4093–4096.

[16] Herman, L. (1939) Spectre d'absorption de l'oxygène, Ann. Phys., 11, 548–611.

[17] Herzberg, G. (1989) *Molecular Spectra and Molecular Structure*, Robert Krieger Publsihing Company, Malabar, FL, Vol. 1, p. 585.

[18] Horowitz, A., Schneider, W., and Moortgat, G. K. (1989) The role of oxygen dimer in oxygen photolysis in the Herzberg continuum. A temperature dependence study, J. Phys. Chem., 93, 7859–7863.

[19] Janssen, J. (1885) Analyse spectrale des éléments de l'atmosphère terrestre, Compt. Rendus de l'Academie des Sciences, 101, 649–651.

[20] Johnston, H. S., Paige, M., and Yao, F. (1984) Oxygen absorption cross sections in the Herzberg continuum and between 206 and 327 K, J. Geophys. Res., 89, 11661–11665.

[21] Long, C. A. and Ewing, G. (1973) Spectroscopic investigation of van der Waals molecules. 1. Infrared and visible spectra of (O$_2$)$_2$, J. Chem. Phys., 58, 11, 4824–4834.

[22] McKellar, A. R. W., Rich, N. H., and Welsh, H. L. (1972) Collsion-induced vibrational and electronic spectra of gaseous oxygen at low temperatures, Can. J. Phys., 50, 1–9.

[23] Mlawer, E. J., (1998) Clough, S. A. Brown, P. D., Stephens, T. M, Landry, J. C., Goldman, A., and Murcray, F. J. (1998) Observed atmospheric collision induced absorption in near infrared oxygen bands, J. Geophys. Res, 103, 3859–3863.

[24] Newnham, D. A. and Ballard, J. (1998) Visible absorption cross section and integrated absorption intensities of molecular oxygen O$_2$, and O$_4$, J. Geophys. Res., 103, 28801–28816.

[25] Orlando, J. J., Tyndall, G. S., Nickerson, K. E., and Calvert, J. G. (1991) The T-dependence of collision-induced absorption by oxygen near 6 μm, J. Geophys. Res., 96, 20755–20760.

[26] Pfeilsticker, K., Erle, F., and Platt, U. (1997) Absorption of solar radiation by atmospheric O$_4$, J. Atmos. Sci., 54, 7, 939–445.

[27] Perner, D. and Platt, U. (1980) Absorption of light in the atmosphere by collision pairs of oxygen (O$_2$)$_2$, Geophys. Res. Lett., 7, 1053–1056.

[28] Platt, U. (1994) Differential optical absorption spectroscopy (DOAS), *Air Monit. by Spectr. Techniques.* Ed. by M. W. Sigrist, Chemical Analysis Series, John Wiley & Sons Inc., V. 127, pp. 27–84.

[29] Rinsland, C. P. *et al.* (1982) Stratospheric measurements of collision-induced absorption by moleculer oxygen, J. Geophys. Res., 87, 3119–3122.

[30] Rodgers, C. W. (1976) Retrieval of atmospheric temperature and composition from remote measurements of thermal radiation, Rev. of Geophys. and Space Phys., 14, 4, 609–623.

[31] Salow, H. and Steiner, W. (1936) Durch die Wechselwirkungkräfte bedingten Absorptionsspektren des Sauerstoffs, 1. Absorptionsbanden des O_2–O_2 Moleküls, Zeitsch. der Physik, 99, 137–158.

[32] Shardanand (1969) Absorption cross section of O_2, and O_4 between 2000–2800 Å, Phys. Rev., 186, 5–9.

[33] Solomon, S., Portmann, R. W., Sanders, R. W., and Daniel, J. S. (1998) Absorption of solar radiation by water vapor, oxygen and related collision pairs in the Earth atmosphere, J. Geophys. Res., 103, 3847–3858.

[34] Tabisz, G. C., Allin, E. J., and Welsh, H. L. (1969) Interpretation of the visible and near-infrared spectra of compressed oxygen as collision-induced electronic transitions, Can. J. Physics, 47, 2859–2871.

[35] Vigasin, A. A. (2001) Thermally averaged spectroscopic parameters of weakly bound dimers, J. Molec. Spectr., 205, 9–15.

[36] Wagner, T. and Platt, U. (1998) Observation of tropospheric BrO from GOME satellite, Nature, 395, 486–490.

Index